ADVANCES IN IMMUNOLOGY AND IMMUNOPATHOLOGY OF THE EYE

Edited by

G. RICHARD O'CONNOR

Francis I. Proctor Foundation for Research in Ophthalmology
University of California, San Francisco
San Francisco, California

JOHN W. CHANDLER

Corneal Disease Research Laboratory
The Swedish Hospital Medical Center
and
Department of Ophthalmology
University of Washington School of Medicine
Seattle, Washington

Distributed by
YEAR BOOK MEDICAL PUBLISHERS · INC.
35 EAST WACKER DRIVE, CHICAGO

Janeiro

Proceedings of the
Third International Symposium
on the Immunology and
Immunopathology of the Eye

Seattle, Washington
October 25–27, 1982

Library of Congress Cataloging in Publication Data
Main entry under title:

Advances in immunology and immunopathology of the eye.

  Based on the proceedings of the Third International
Symposium on the Immunology and Immunopathology of the
Eye, held at the Battelle Institute, Seattle, Wash.,
Oct. 25-27, 1982.
  Includes bibliographies and index.
  1. Eye—Diseases and defects—Immunological aspects—
Congresses. 2. Immunopathology—Congresses.
I. O'Connor, G. Richard (George Richard), 1928-
II. Chandler, John W. III. International Symposium on
the Immunology and Immunopathology of the Eye (3rd:
1982: Battelle Institute) [DNLM: 1. Eye—immunology—
congresses. 2. Eye Diseases—immunology—congresses.
W3 IN924DM 3rd 1982a/WW 140 A244 1982]
RE68.A35   1985   617.7'079   84-26081
ISBN 0-89352-224-4

ISBN 0-89352-224-4

Library of Congress Catalog Card Number 84-26081

Printed in the United States of America

Preface

The interest that has recently developed worldwide in the subject of ocular immunology knows no peer. The eye is often considered to be a special target of immunologic disease processes, but the immunopathology of the eye is much less clearly understood than that of the thyroid gland, the kidney, or the testis. Progress in research on the immunopathology of the eye has, until recently, been impeded by the lack of readily available biopsy specimens. Our inability to analyze cellular and humoral events occurring in the ocular fluids has kept us from recognizing the essential nature of certain immunologic processes within the eye. Fortunately, the development of microassays, the use of monoclonal antibodies, and the availability of new instruments such as cell sorters have radically changed this situation to the benefit of researchers and clinicians alike. The reader of this book will profit from these innovations.

Advances in Immunology and Immunopathology of the Eye summarizes the Proceedings of the Third International Symposium on the Immunology and Immunopathology of the Eye. The Symposium was held at the Battelle Institute, Seattle, Washington, October 25–27, 1982, immediately before the opening of the XXIV International Congress of Ophthalmology in San Francisco. One of the most successful elements of this meeting was the bringing together of young investigators, new in the field, with ophthalmologists and with established researchers in general immunology.

The papers presented at this meeting were organized under five headings corresponding to five major topics that were selected in advance as areas of focal interest. These were (1) lymphocyte subpopulations and functions in ocular disease, (2) immune complex-mediated diseases, (3) general ocular immunology, (4) immunology of receptors, and (5) immunology of mucosal disorders. Each of these major sessions was opened by a keynote speaker, selected because of his established expertise in that particular field. The value of this arrangement to ophthalmologists will be obvious. In each session a number of papers corresponding to the broader topics of general ocular immunology was also inserted in order to achieve balance.

The success of the meeting was due in large part to the efforts of certain individuals. Among these, we wish to thank the International Organizing Committee consisting of Wilhelm Böke, Kiel; Étienne Bloch-Michel, Paris; Shigeaki Ohno, Sapporo; and Antonio Secchi, Padua. We also wish to thank the Scientific Program Committee whose task was especially difficult due to the large number of papers submitted from all over the world. In addition to ourselves, this Committee consisted of Mathea R. Allansmith, Boston; Arthur M. Silverstein, Baltimore; Robert A. Prendergast, Baltimore; and Mitchell H. Friedlaender, San Francisco. Finally, we wish to thank the Local Arrangements Committee which deserves the major credit for a meeting that was not only scientifically but also socially successful. This Committee consisted of John W. Chandler, Julia R. Swor, George Chin, Walter R. Rotkis, and Thomas E. Gillette, all of Seattle.

We acknowledge the thoughtful, thorough, and kind assistance of Ms. Jolene Kitzerow, Ms. Sally Hawley, Ms. Lorna Slominski, and Ms. Julia Swor of the Conference Service Staff of the Battelle Seattle Conference Center. The success of the meeting and the functioning of the organizational committees were facilitated by them.

G. Richard O'Connor, *San Francisco*
John W. Chandler, *Seattle*

Senior Contributors

Mark B. Abelson, M.D., Harvard Medical School, Eye Research Institute of Retina Foundation, Boston, Massachusetts 02114

Mathea R. Allansmith, M.D., Harvard Medical School, Eye Research Institute of Retina Foundation, Boston, Massachusetts 02114

Mario R. Angi, M.D., Clinica Oculistica, Universita di Padova, Policlinico, 35100 Padova, Italy

Rubens Belfort, Jr., M.D., Escola Paulista Medicina—Brasil, C.P. 4086 São Paulo, Brazil

David BenEzra, M.D., Hadassah Hebrew University Hospital, Department of Ophthalmology, The Pediatric and Immuno-ophthalmology Unit, Jerusalem, 91120 Israel

Richard S. Berk, M.D., Department of Immunology/Microbiology, Wayne State University School of Medicine, Detroit, MI 48201

Étienne Bloch-Michel, Ophtalmologie, Institut Gustave-Roussy, 94805 Villejuif, France

Ysolina M. Centifanto-Fitzgerald, Ph.D., Louisiana State University Eye Center, New Orleans, Louisiana 70112

John W. Chandler, M.D., Swedish Hospital Medical Center, Seattle, Washington 98104

Susan M. Chant, Ph.D., 11728 Kiowa Avenue, Los Angeles, California 90049

Devron H. Char, M.D., Francis I. Proctor Foundation, University of California, San Francisco, California 94143

Joseph Colin, M.D., Centre Hospitalier, 29200 Brest, France

Noveen D. Das, Ph.D., Department of Ophthalmology, University of Florida College of Medicine, Gainesville, Florida 32610

Yvonne de Kozak, Ph.D., Laboratoire d'Immunopathologie de l'Oeil, Hôtel Dieu, 75181 Paris 04, France

Jean-Paul Dernouchamps, M.D., Université Catholique de Louvain, B-1150 Brussels, Belgium

Norman T. Felberg, Ph.D., Research Division and Neuro-Ophthalmology Unit, Wills Eye Hospital, Philadelphia, Pennsylvania 19103

Stephen C. Foster, M.D., Massachusetts Eye and Ear Infirmary, Boston, Massachusetts 02114

Rudolph M. Franklin, M.D., Louisiana State University Eye Center, New Orleans, Louisiana 70112

Leslie S. Fujikawa, M.D., Eye Institute, National Institutes of Health, Clinical Ophthalmic Immunology Section, Bethesda, Maryland 20205

Alec Garner, M.D., Institute of Ophthalmology, Department of Pathology, London EC1 9AT, England

Bryan M. Gebhardt, Ph.D., Louisiana State University Eye Center, New Orleans, Louisiana 70112

Igal Gery, Ph.D., National Eye Institute, Laboratory of Vision Research, National Institutes of Health, Bethesda, Maryland 20205

Gunther Grabner, M.D., University Eye Clinic, A 1090 Vienna, Austria

Jack V. Greiner, D.O., Ph.D., Department of Pathology, Chicago College of Osteopathic Medicine, Chicago, Illinois 60615

Olafur G. Gudmundsson, M.D., Harvard Medical School, Eye Research Institute of Retina Foundation; and Massachusetts General Hospital, Boston, Massachusetts 02114

Joan M. Hall, Ph.D., Francis I. Proctor Foundation, University of California, San Francisco, California 94143

Linda D. Hazlett, Ph.D., Wayne State University School of Medicine, Detroit, Michigan 48201

Paul A. Hunter, M.D., Moorfields Eye Hospital, London EC1V 2PD, England

Dr. H.A. Hylkema, M.D., Department of Ophthalmo-Immunology, Netherlands Ophthalmic Research Institute, 1100 AC Amsterdam, The Netherlands

Henry J. Kaplan, M.D., Department of Ophthalmology, Emory University, Atlanta, Georgia 30322

Aize Kijlstra, Ph.D., The Netherlands Ophthalmic Research Institute, 1005 EK Amsterdam, The Netherlands

George L. King, M.D., E. P. Joslin Research Laboratory, Boston, Massachusetts 02215

Dr. Ellen Kraus-Mackiw, Department of Ophthalmology, University Hospital, D-6900 Heidelberg-1, West Germany

Dr. A. Linssen, Department of Ophthalmo-Immunology, The Netherlands Ophthalmic Research Institute, 1005 EK Amsterdam, The Netherlands

Sammy Hintung Liu, M.D., The Johns Hopkins Hospital, Baltimore, Maryland 21205

Mart Mannik, M.D., Division of Rheumatology, Department of Medicine, University of Washington, Seattle, Washington 98195

George E. Marak, M.D., 2059 Huntington Avenue, Alexandria, Virginia 22303

Anne-Catherine Martenet, M.D., University Eye Clinic, University Hospital, 8091 Zürich, Switzerland

Kaoru Mizuno, M.D., Department of Ophthalmology, Osaka University Medical School, Fukushima-ku, Osaka 553, Japan

Bartly J. Mondino, M.D., University of California, L.A., Jules Stein Eye Institute, UCLA Center for Health Science, Los Angeles, California 90024

H. Konrad Müller-Hermelink, M.D., Institute of Pathology, University of Kiel, D-23 Kiel, West Germany

Jerry Y. Niederkorn, Ph.D., University of Texas Health Science Center at Dallas, Department of Ophthalmology, Dallas, Texas 75235

Robert B. Nussenblatt, M.D., Eye Institute, National Institutes of Health, Clinical Ophthalmic Immunology Section, Bethesda, Maryland 20205

G. Richard O'Connor, M.D., Francis I. Proctor Foundation, Department of Ophthalmology, University of California, San Francisco Medical Center, San Francisco, California 94143

Jang O. Oh, M.D., Francis I. Proctor Foundation, University of California, San Francisco, California 94143

Yuichi Ohashi, M.D., Laboratory of Research for Transparent Ocular Tissues, Department of Ophthalmology, Osaka University Medical School, Fukushima-ku, Osaka 553, Japan

Shigeaki Ohno, M.D., Department of Ophthalmology, Hokkaido University School of Medicine, Sapporo, Hokkaido 060, Japan

Amjad H.S. Rahi, M.D., Institute of Ophthalmology, Department of Pathology, London EC1V 9AT, England

Narsing A. Rao, M.D., Estelle Doheny Eye Foundation, Los Angeles, California 900033

John H. Rockey, M.D., Ph.D., Scheie Eye Institute, Philadelphia, Pennsylvania 19104

Noel R. Rose, M.D., Ph.D., Department of Immunology and Infectious Diseases, Johns Hopkins University School of Hygiene and Public Health, Baltimore, Maryland 21205

Patrick E. Rubsamen, M.D., Department of Ophthalmology, University of Miami School of Medicine, Miami, Florida 33101

K. Matti Saari, M.D., Institute of Clinical Sciences, University of Tampere, SF-33520 Tampere 52, Finland

Dr. Jean Sainte-Laudy, Laboratoire d'Immunologie, 75014 Paris, France

Junichi Sakai, M.D., Tokyo Medical College, Shinjuku-ku, Tokyo 160, Japan

Hans Otto Sandberg, M.D., Fr. Nansens vei 2, 1600 Fredrikstad, Norway

Mario-Cesar Salinas-Carmona, M.D., Eye Institute, National Institutes of Health, Clinical Ophthalmic Immunology Section, Bethesda, Maryland 20205

Antonio G. Secchi, M.D., Clinica Oculistica, Universita' di Padova Policlinico, 35100 Padova, Italy

Arthur M. Silverstein, Ph.D., The Johns Hopkins School of Medicine, The Johns Hopkins Hospital, Baltimore, Maryland 21205

Paul C. Stein, Ph.D., National Institutes of Health, Bethesda, Maryland 20205

E. Lee Stock, M.D., Northwestern University Medical School, Department of Ophthalmology, Chicago, Illinois 60611

J. Wayne Streilein, Ph.D., Department of Microbiology and Immunology, University of Miami School of Medicine, Miami, Florida 33101

Yoshitsugu Tagawa, M.D., Department of Ophthalmology, Hokkaido University School of Medicine, Sapporo 060, Japan

S. Takano, M.D., Tokyo Medical College, Shinjuku-ku, Tokyo 160, Japan

Alfred Tanoé, M.D., Laboratoire d'Immunopathologie de l'Oeil, Hôtel-Dieu, 75181 Paris, France

Hugh R. Taylor, M.D., The Wilmer Institute, The Johns Hopkins Hospital, Baltimore, Maryland 21205

Charles E. Thirkill, Ph.D., Ophthalmology Research, University of California, Davis, Medical Center, Sacramento, California 95817

Masahiko Usui, M.D., Tokyo Medical College Hospital, Department of Ophthalmology, Shinjuku-ku, Tokyo 160, Japan

Ruth van der Gaag, M.D., Netherlands Ophthalmic Research Institute, 1100 AC Amsterdam, The Netherlands

Waldon B. Wacker, Ph.D., University of Louisville, Kentucky Lions Eye Research Institute, Louisville, Kentucky 40202

Denis Wakefield, M.D., Department of Immunology, St. Vincent's Hospital, Darlinghurst 2010 N.S.W., Australia

John Clifford Waldrep, Ph.D., Emory University, Department of Ophthalmology, Atlanta, Georgia 30322

Richard P. Wetzig, M.D., Harvard Medical School, Department of Ophthalmology and Pathology; and Eye Research Institute of Retina Foundation, Boston, Massachusetts 02114

Judith A. Whittum, Ph.D., Immunology Laboratories, The Wilmer Institute, Johns Hopkins School of Medicine, Baltimore, Maryland 21205

Jung Koo Youn, M.D., Laboratoire d'Immunologie, Institut Gustave-Roussy, 94805 Villejuif, France

Elaine Young, Ph.D., The Wilmer Institute, Johns Hopkins University School of Medicine, Baltimore, Maryland 21205

Takenosuke Yuasa, M.D., Department of Ophthalmology, Osaka University Medical School, Osaka 553, Japan

Contents

Part II. IMMUNE COMPLEX-MEDIATED DISEASES AND OTHER IMMUNOLOGICAL DISORDERS

Part III. GENERAL OCULAR IMMUNOLOGY

Part IV. IMMUNOLOGY OF RECEPTORS

chapter 1

Cell Surface Analysis of Intraocular Lymphocytes in Idiopathic Uveitis

Henry J. Kaplan,[a] J. Clifford Waldrep,[a] Jan Nicholson,[b] and David Gordon[c]

The most frustrating clinical aspect of chronic uveitis is the recurrent nature of the disease. Repeated episodes of inflammation lead to ocular damage and eventually to loss of vision.

Since various forms of chronic uveal inflammation are often subject to an acute exacerbation of self-limited duration, it is tempting to postulate that a local abnormality in immunoregulation is responsible for such acute episodes. Clinically it appears that the immune response transiently escapes from control resulting in an acute exacerbation, but that within a relatively short period of time immunoregulation is re-established, and the acute episode spontaneously subsides. One approach to investigate if chronic uveitis represents an abnormality in immunoregulation is to determine whether the appropriate immunoregulatory lymphoid cell types are present in an eye with uveitis.

Characterization of intraocular mononuclear cell populations in uveitis is difficult, because uveoretinal tissue is not available for analysis, and the numbers of cells available for study in the aqueous humor and vitreous humor are very small. We have previously reported the absence of E-rosette forming cells (i.e., T-cells) in the vitreous cavity of eyes with chronic idiopathic uveitis. In order to determine the number of T-cells, we used a sheep RBC (SRBC)-immunobead microassay.[1] Recently, we have extended these studies to the aqueous humor as well as the vitreous humor and have used a fluorescence-activated cell sorter (the Beckton-Dickinson FACS IV) and antibodies specific for cell surface antigens for this analysis.

Material and Methods

Sixteen patients with uveitis of various etiologies were studied. Anterior chamber paracentesis was performed to obtain aqueous humor, while pars plana vitrectomy was used to obtain vitreous humor. The following antibodies were used to identify the cell surface markers on the mononuclear cells present within the aqeous humor and/or vitreous humor: monoclonal antibody OKT-3, a pan T-cell reagent; monoclonal antibody B-1, a pan B-cell reagent; and heteroantisera U937, a human macrophage reagent. After incubation with the appropriate antibody, the cells were then marked with a fluorescent-labeled anti-immunoglobulin (IG) antibody.[2]

The samples were then counted, using a microsampler and the histograms that were generated by the FACS IV.[3] Preliminary studies established that as few as 5000 peripheral blood mononuclear cells could be accurately marked and counted by means of these techniques.

In addition to FACS analysis, vitreous

[a]Department of Ophthalmology, Emory University Clinic, Atlanta, Georgia. [b]Centers for Disease Control, Atlanta, Georgia. [c]Department of Medicine, Emory Clinic, Atlanta, Georgia.

TABLE I
CSM on Mononuclear Cells in Aqueous Humor

Diagnosis	Percentage of Cells			
	T	B	M	Null
IAU[a]	67	0	4	29
IAU	51	1	9	39
IAU	33	0	20	47
IAU	30	9	18	43
HSV	75	4	11	10
PP	11	0	73	16
AI	15	0	25	60

[a]IAU = idiopathic anterior uveitis; HSV = herpes simplex uveitis; PP = pars planitis; AI = autoimmune disease.

TABLE II
CSM on Mononuclear Cells in Vitreous Humor

Diagnosis	Percentage of Cells				
	T	B	M	Plasma Cell	Null
IAU	5	0	38	28	29
TbU[a]	72	4	5	0	19
ARN	76	0	0	ND	24
TU	71	2	8	0	19

[a]TbU = tuberculous uveitis; ARN = acute retinal necrosis; TU = traumatic uveitis.

humor samples were examined by immuno-fluorescent microscopy for cytoplasmic staining with a polyvalent anti-IG antiserum. Cells that stained positively in this reaction were considered plasma cells. Mononuclear cells not stained with any of the above reagents were classified as null cells.

Results

The aqueous humor of seven patients with acute uveitis of various etiologies was studied (Table I). In each patient, regardless of etiology, the percentage of mononuclear cells marked as T-cells was greater than those indentified as B-cells. In the four patients with idiopathic anterior uveitis the largest number of labeled cells were T-cells.

In one patient with herpes simplex uveitis, 90% of the cells were labeled, and there was a marked predominance of T-cells. By contrast, the patient with pars planitis showed marking of 84% of the aqueous cells, and there was an abundance of macrophages. The only patient in whom the majority of cells were not labeled had multiple autoimmune diseases, including systemic lupus erythematosus, rheumatoid arthritis, and Sjögren's syndrome.

The vitreous humor of four patients with chronic uveitis, of various etiologies, was also studied (Table II). There was a paucity of identifiable T-cells in the one patient with idiopathic anterior uveitis, although numerous macrophages and plasma cells were observed. By contrast, the vitreous humor of each of the other three patients showed a definite majority of T-cells. Two of these patients had presumptive infectious uveitis, while the third had a traumatic uveitis.

Most importantly, five patients underwent simultaneous study of their aqueous humor and vitreous humor. They were in a quiescent phase of their chronic disease when the samples were obtained (Table III). The aqueous humor of the two patients with idiopathic anterior uveitis displayed a larger population of T-cells than B-cells. However, the reverse was observed in the vitreous humor. In both patients there were no identifiable T-cells; on the contrary, there was a predominance of cells of B-lineage (i.e., B, and/or plasma cells). A similar pattern was observed in one patient with sympathetic ophthalmia. However, in that patient the majority of vitreous mononuclear cells did not mark, and he was studied at a period before we routinely stained for plasma cells. In contrast to the patients with idiopathic anterior uveitis, the two patients with toxoplasmosis and sarcoidosis exhibited a much larger T- than B- cell population in both their aqueous and vitreous humors.

Discussion

Our results with the FACS confirm our previously reported observation.[1] Namely, vitreous mononuclear cells in chronic idiopathic uveitis are primarily of B-cell lineage (i.e., B- and/or plasma cells) with a marked paucity of identifiable T-cells. Simultaneous analysis of aqueous mononuclear cells demonstrated that T-cells are abundantly present in the anterior chamber while absent

TABLE III
CSM on Mononuclear Cells in Aqueous Humor and Vitreous Humor

Diagnosis		Percentage of Cells				
		T	B	M	Plasma Cell	Null
IAU	A	29	1	59	ND	11
	V	0	10	8	36	46
IAU	A	19	0	25	ND	56
	V	0	90	0	ND	10
Sympathetic ophthalmia	A	43	16	1	ND	40
	V	0	15	7	ND	78
Toxoplasmosis	A	45	0	63	ND	0
	V	58	0	33	11	0
Sarcoidosis	A	68	0	19	ND	13
	V	48	2	18	8	26

in the vitreous cavity. Furthermore, the pattern of T:B cell reversal seen in idiopathic uveitis was also present in sympathetic ophthalmia but not in patients with either infectious uveitis (toxoplasmosis, tuberculosis, acute retinal necrosis) or sarcoidosis.

Analysis of aqueous humor confirmed the previous report of Belfort, et al.[4] Regardless of the etiology of the uveitis there is a larger proportion of T-cells than B-cells in the anterior chamber.

There is an apparent difference between the intraocular mononuclear cell populations in eyes with traumatic uveitis and sympathetic uveitis. The majority of the cells within the vitreous humor of the eye that we examined with traumatic uveitis were T-cells. This parallels the report by Marboe et al.[5] and our own unpublished observations, in which monoclonal antibodies and immunofluorescent microscopy were used to study the mononuclear cell populations in the uvea of an enucleated traumatized eye. We found a majority of the mononuclear cells were OKT-3 positive, with a predominance of the OKT-4 subset. By contrast, our study of an eye with sympathetic ophthalmia (i.e., the eye not traumatized) demonstrated a paucity of T-cells within the vitreous cavity. It is not known how closely the mononuclear cell population in the aqueous humor and vitreous reflect the cell populations in the iris and choroid, respectively. However, the similarity of our results in the vitreous and uvea in a traumatized eye suggests that there may be a dependable correlation.

Finally, what is the significance of the T:B cell reversal observed in the aqueous humor and vitreous humor in chronic idiopathic uveitis? It is well established that the immune response is a complex dynamic event with multiple regulatory controls. One pathway of regulation involves close cell to cell interaction and is primarily effected by suppressor T-cells.[6] It seems reasonable to postulate that the paucity of T-cells within the vitreous cavity may allow unregulated B-cell activity. Such undesirable B-cell function might precipitate an acute phase of inflammation. Upon the entrance of multiple types of new inflammatory cells (including T-cells) into the posterior segment, T-cell immunoregulation of B-cell function may once again be established, signaling an end to the clinical phase of inflammation.

Acknowledgments

This work was supported by NIH Grant EY-03723 and by an unrestricted departmental grant from Research to Prevent Blindness, Inc., New York City.

We thank Ms. Martha Johnson for excellent technical assistance.

References

1. Kaplan HJ et al.: *Arch Ophtalmol* **100**:585-587, 1982.
2. Kung PC et al.: *Science* **206**:347-349, 1979.
3. Reinherz EL et al.: *Proc Natl Acad Sci USA* **76**:4061-4065,1979.
4. Belfort R Jr et al.: *Arch Ophthalmol* **100**:465-467, 1982.
5. Marboe C et al.: *Invest Ophthalmol* **22(3)**:171, 1982.
6. Cantor H and Gershon RK: *Fed Proc* **38**:2058-2063, 1979.

chapter 2

Uveitis in Sarcoidosis

Immunological Studies at the Site of Disease Activity

Mario R. Angi,[a] Antonio Pezzutto,[b] Antonio G. Secchi,[a] Giovanni Pizzolo,[c] and Gianpietro Semenzato[b]

Ocular involvement has been reported to occur in 10–29% of systemic sarcoidosis patients and includes the uvea, the retina, the conjunctiva, and the lacrimal gland.[1] Its evaluation represents a useful clinical tool in establishing the diagnosis and in defining the different phases of the disease.

The purpose of the present study is to evaluate whether ocular involvement in sarcoidosis patients is characterized by a T-cell redistribution similar to that reported in lung[1a] or lymph nodes.[1b] This analysis includes a comparative study of T-cell subsets in the blood, in conjunctival follicles, and in the aqueous humor of patients with acute anterior uveitis.

Materials and Methods

Forty-four patients with histologically proven systemic sarcoidosis were studied, including 20 males ranging between the ages of 18 and 55 years (mean 33) and 24 females ranging between the ages of 21 and 56 years (mean 39). The activity of sarcoidosis was determined through the evaluation of clinical features, bronchoalveolar lavage findings, angiotensin converting enzyme serum levels, [67]gallium scans, and chest x-rays.

Uveal manifestations were studied by

ophthalmoscopy, both direct and indirect, and by slit-lamp examination using a contact-lens equipped with a scleral depressor (Haag-Streit, Bern). The activity of the inflammatory lesions, as well as the involvement of the retinal pigmented epithelium, was evaluated by fluoroangiography and fundus photography.

The immunological methods used to determine T-cell subpopulations in the blood have already been described in detail.[2,3,4,5,6] Before any local or systemic therapy had been prescribed, aqueous humor cytology was studied in two patients that presented an acute anterior uveitis preceded by erythema nodosum. Detection of lymphocyte subpopulations, defined by Fc-receptors, has already been reported.[2]

Conjunctival biopsies were only performed in patients with clinical evidence of follicles. The specimens were cut in half. One of the halves was fixed in 10% formalin and then processed for paraffin embedding; the other half was embedded in Ames OCT Compound (Miles Lab., Inc., Elkhart, IN) and snap frozen in liquid N_2 within 30 minutes after excision. Five micron sections were cut on a cryostat, transferred to glass slides, air dried, and fixed in 95% ethanol (for 5 minutes at 4°C) before staining. Staining procedures for immunohistological studies with anti-T-cell subset monoclonal antibodies were described previously.[3] In brief, sections were covered with monoclonal antibodies (0.1–0.2 µg in 5–10 µl of

Department of Ophthalmology[a] and Department of Clinical Medicine, First Medical Clinic and Clinical Immunology Branch,[b] Department of Haematology, Verona University School of Medicine,[c] Verona, Italy.

TABLE I
Distribution of Ocular Signs in 44 Sarcoidosis Patients[a]

	Major (Symptomatic)	Minor (Asymptomatic)
Ocular Adnexa		
Conjunctivitis	2	6
Conjunctival follicles	2	9
Anterior Segment		
Acute uveitis	2	—
Chronic uveitis	1	1
Nodular trabecular infiltration	—	1
Anterior synechiae	—	1
Cataract	2	—
Oral Zone and Vitreous		
Perivasculitis/ vasculitis	—	8
Pigmentary changes	—	3
Creamy nodules	—	2
Vitreous opacities	1	5
Posterior Segment		
Perivasculitis/ vasculitis	1	4
Candle wax drippings	1	7
RPE changes	—	4
Macular edema	1	—

[a]In some patients different signs may be associated.

fluid), incubated for 30 minutes, washed three times, and stained with goat antimouse immunoglobulin antibody. A second layer was coupled with fluorescein isothiocyanate (G-anti-MIg-FITC) for 30 minutes. After further washing, the sections were mounted with a drop of glycerol and a coverslip and then examined under a fluorescence microscope.

Immunological studies were always performed before treatment. Twelve healthy volunteers, averaging 37 years of age, were used as controls. Statistical analysis compared the mean and standard error of the mean using Student's t test.

Results and Discussion

The frequency and the distribution of the ocular signs detected are shown in Table I.

The age and sex distribution of our patients is similar to that reported elsewhere.[1,7] Conjunctival follicles were found in 11 cases (25%); 9 of these patients did not have any symptoms, while the remaining 2 complained of moderate itching or burning sensation. Nine biopsies were performed; in 5 of them (55%) a typical sarcoid noncaseating granuloma was found on histological examination, whereas in 4 others, a lymphocytic infiltrate with rare macrophages, monocytes, and neutrophils, was noticed.

Anterior uveitis was present in only 4 of the cases; in three of these it appeared along with erythema nodosum. This surprisingly low incidence was also reported by other Italian authors.[8]

Immunological studies of the blood revealed that these patients had T-lymphopenia associated with an imbalance of T-cell subpopulations, as demonstrated by Fc-receptors and monoclonal antibodies (Table II). In particular, a net increase of T_G and a marked decrease of T_M cells was found. On the other hand, increased values of OKT8 and a reduced number of OKT4 positive cells was observed, resulting in a reduction of the OKT4/OKT8 ratio in comparison to controls. However, we have to emphasize that values obtained in the blood with monoclonal antibodies did not reveal statistically significant differences, whereas results obtained using Fc-receptors did. Furthermore, this observation strongly argues for the heterogeneity of cell populations defined by Fc-receptors and available monoclonal antibodies. T_G/OKT8 and T_M/OKT4 positive cells have been reported to define, respectively, suppressor-cytotoxic and helper-inducer populations.[4,9] However, the above assays probably include, but do not identify, suppressor and helper cells. In clinical immunology, functional studies must always be performed before one assumes that certain cells have immunoregulatory activity. Thus, in a previous paper[2] we demonstrated that in sarcoidosis, despite the T_G cell heterogeneity, the increase of lymphocytes bearing the Fc–IgG receptors includes cells that effectively provide suppressor activity *in vitro*.

In our opinion, studies of the blood are

TABLE II
Peripheral T-Lymphocyte Subsets in 44 Sarcoidosis Patients

	n	T_G	T_M	T_8	T_4
Sarcoidosis	44	346 ± 41	509 ± 52	327 ± 34	568 ± 73
Controls	12	196 ± 17	896 ± 64	342 ± 30	792 ± 61
p		<0.005	<0.001	NS	<0.05

Results are expressed as number of cells/mm³ ± standard error; NS = not significant; p = according to Student t test.

insufficient for the documentation of immunocompetence in various disease states. The few cells recovered from the aqueous humor of two patients affected by anterior uveitis were almost completely characterized by the OKT4 phenotype. The ratio OKT4/OKT8 of the cells present in the aqueous humor is remarkably different from the values obtained in the blood of the same two patients.

Further evidence for the hypothesis of a T-cell redistribution at the eye level is provided by an immunohistological study of conjunctival follicles. As shown in Figure 1, cells infiltrating the conjunctiva, and sometimes the sarcoid granulomas, were almost completely represented by lymphocytes bearing the helper-related phenotype (Leu-3, OKT4 equivalent). Only few cells bearing the suppressor/cytotoxic related phenotype have been observed.

Further studies are needed in order to clarify both of the mechanisms leading to this T-cell redistribution and their specificity. In particular, other diseases characterized by lymphocytic infiltration of the eye must be evaluated so as to determine whether the pattern we discovered is spe-

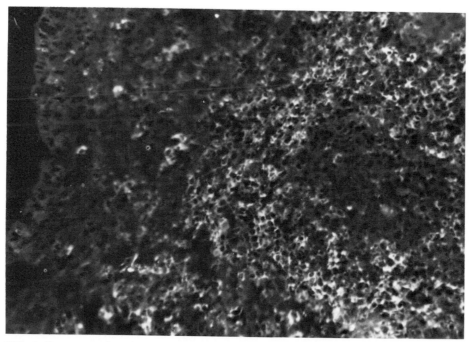

FIG. 1. Cryostat section of a conjunctival nodule from a sarcoidosis patient. On the left side, the epithelium can be observed; on the right, a granuloma. Most lymphocytes surrounding the sarcoid granuloma express a helper-related phenotype (Leu-3 monoclonal antibody positivity).

cific for sarcoidosis or not. Nevertheless, the T-cell redistribution that we observed at the ocular level in sarcoidosis patients represents additional evidence of the systemic nature of the disease and further emphasizes the necessity for an evaluation of different organs as possible sites of involvement. This is not only significant in the search for new insights into the pathogenesis of sarcoidosis, but, more importantly, it may represent a useful tool in the staging of disease activity, which is crucial in the choice of therapy.

Acknowledgments

The authors thank Dr. A. Cipriani and all the physicians of the Pulmonary Department of Padua Hospital who permitted the study of most of these sarcoidosis patients, Mrs. C. Vincenzi for expert technical assistance, and Miss I. Caridakis for typing the manuscript.

References

1. Siltzbach LE et al.: *Am J Med* **57**:847, 1974.
1a. Hunninghake GW and Crystal RG: *N Engl J Med* **305**:429, 1981.
1b. Semenzato G et al.: *N Engl J Med* **306**:48, 1982.
2. Semenzato G et al.: *Clin Immunol Immunopathol* **19**:416, 1981.
3. Pizzolo G et al.: *Cancer* **46**:2640, 1980.
4. Reinherz EL and Schlossman SF: *Cell* **19**:821, 1980.
5. Evans RL et al.: *Proc Natl Acad Sci* **78**:544, 1981.
6. Ledbetter JA et al.: *J Exp Med* **153**:310, 1981.
7. James DG et al.: *Trans Ophthalmol Soc UK* **96**:133, 1976.
8. Maccolini E, Iannetti D, Falcone F, and Minore GC: In *Sarcoidosis,* International Siena Sarcoidosis Symposium, Lenzini L, and Rottoli P (Eds.), 1979, p. 271.
9. Moretta L et al.: *J Exp Med* **146**:184, 1977.

chapter 3

Immunoregulation in Uveitis

**A.G. Secchi,[a] I. Fregona,[a] A. Corsano,[b] N. Malacarne,[a]
R. Corradini,[a] and U. Fagiolo[b]**

Previous studies by us have demonstrated that active anterior nongranulomatous uveitis is related to a decrease in "total" T-lymphocytes (i.e., those forming E-rosettes) and a diminution in suppressor cell activity. In posterior focal chorioretinitis, on the other hand, no such changes were observed. In acute active uveitis a slight increase in B-lymphocytes, expressed as EA-rosette-forming cells, was also seen.[1] In the present study further investigations are reported concerning the behavior of lymphocyte subsets in anterior and intermediate uveitis and in Behçet's disease.

Materials and Methods

Patient selection

Twenty-five patients with acute anterior nongranulomatous uveitis were evaluated; in seven, the inflammation was probably related to Herpes Simplex Virus (HSV) infection; in six, high titers of antibodies to Streptococcus were detected. Four cases were related to an arthropathic status, and in eight "idiopathic" cases all other clinical and laboratory tests were negative. In all cases the disease was at an acute phase with at least 3^+ of flare and cells in the anterior chamber. Sixty percent of the patients were having their first episode of inflammation, while 40% were recurrences. None of the patients was receiving corticosteroids, nor had they received such medication for at least one month prior to participation in the study.

Of the 25 patients, 11 have been tested two or more times in a longitudinal study set up to compare, within the same patient, the immunological backgrounds during inflammation and quiescence. In these cases, the second evaluation was performed after either topical, periocular or systemic steroid administration, at the minimal doses capable of controlling the inflammation. No immunosuppressive drugs were ever used.

Twelve patients with intermediate uveitis were evaluated. The status of the inflammation in almost all patients was chronic, as is usually the case in this disease. In most cases the tests were performed prior to initiating therapy. In only two cases has a longitudinal study been possible.

Sixteen patients with Behçet's disease were studied; 11 were of the "complete" type, while five were considered incomplete on the basis of currently applied criteria. Of all cases, 68% were HLA-B5 positive. In all, uveal inflammation was active; therapy ranged from none to maximal steroid and/or immunosuppressive.

Healthy adult donors provided blood samples for the control group.

Mononuclear cell separation

Mononuclear cells were separated from heparinized peripheral blood samples by Lymphoprep density sedimentation (Nyegaard & Co. As, Oslo, Norway), washed with phosphate-buffered saline (PBS) and incubated for 30 minutes at 37°C in a plastic

Departments of Ophthalmology[a] and Clinical Medicine[b], Research Unit for Allergy and Ocular Immunology, University of Padova School of Medicine, Padova, Italy.

Petri dish. After collection they were washed three times with PBS and adjusted to a final concentration of 3×10^6 cells/ml and 5×10^6 cells/ml.

Surface membrane immunoglobulin-positive cells (SmIg⁺)

SmIg⁺ cells were prepared and counted as described in detail by Gupta et al.[2]

T-cells reacting with monoclonal antibodies OKT3, OKT4, and OKT8

The percentage of T-cells reacting with these antibodies was determined in lymphoid cell suspensions by indirect immunofluorescence, according to the technique described by Reinherz et al.[3]

Spontaneous E-rosette forming cells (E-RFC)

A lymphoid cell suspension of 100 μl (3×10^6 cells/ml) was mixed with an equal volume of SRBC 1% v/v in a mixture of 80% Tc 199 (Difco Lab., Detroit, Michigan) and 20% fetal calf serum (Eurobio, Paris), incubated for 10 minutes at 37°C, centrifuged for 5 minutes at 200 g, and incubated overnight at 4°C. The sediment was gently resuspended and 200 cells per sample were counted. Each determination was performed in duplicate. Lymphocytes with three or more adherent SRBC were considered as rosettes. Lymphocytes that were aggregated were not counted.

E-rosette forming cells with neuraminidase-treated SRBC (Eₙ-RFC) and purification of T-lymphocytes.

The preparation of E-rosette forming cells with neuraminidase treated SRBC was performed following the techniques described by Weiner et al.[4] and Semenzato et al.[5] T-lymphocytes were separated from non-rosette forming cells (NRC) by Ficoll-Urovison SG 1094 gravity sedimentation in order to reduce the contamination of the rosette forming cells (RFC) (Stocker et al.[6]).

T-cells with Fc receptors for IgG and IgM

The preparation of Ox-RBC antibody complexes as well as the counting procedures Tγ and Tμ were performed as described elsewhere.[2]

T-cells bearing Ia antigens

Ia⁺ T-cells were identified in purified suspensions of T-lymphocytes by indirect immunofluorescence using a mouse monoclonal antihuman Ia-common structure antibody (gift of Dr. S. Ferrone). The features of this antibody (n 513) are reported elsewhere.[7]

Results and Discussion

"Total" T-lymphocytes were measured by two different techniques: (1) rosette formation with SRBC, which takes into account a membrane receptor, and (2) reactivity to OKT3 monoclonal antibody, which is related to a membrane antigen that is present in 85–95% of peripheral T-lymphocytes. It has been demonstrated that in patients with acute anterior uveitis (AAU), intermediate uveitis (IU), and Behçet's disease the level of E-rosette forming cells (E-RFC) is significantly lower than in healthy controls. However, the numerical differences are small, and the values obtained from the population of patients, although lower than those from controls, are well within the normal range.

In our study no difference between patients and controls was found with regard to the level of OKT3 positive cells. In our opinion, this finding suggests that the diminution in E-RFC detected during the acute phases of uveitis may not be a real diminution in T-lymphocytes, but may only reflect a moderate change in the receptor characteristics of T-cells.

The normal numbers of helper/inducer (OKT4) lymphocytes may be reduced in the course of viral infections,[8] but we did not find this even in our small series of patients with probable HSV uveitis. On the basis of our data, it may therefore be concluded that at least as far as surface markers are concerned, T-lymphocyte subsets, which are supposed to include helper/inducer function, do not change in the peripheral blood of uveitis patients.

TABLE I
T-Lymphocyte Subpopulations in Various Forms of Uveitis

	Controls %	AAU %	IU %	Behçet %
E_{St}-RFC	61.1 ± 2.8	54.7 ± 6.5 **P<0.01**	56.1 ± 2.6 **P<0.001**	54.5 ± 5 **P<0.001**
E_N-RFC	72.6 ± 3.0	67.0 ± 6.2 **P<0.01**	67.0 ± 3.1 **P<0.001**	65.0 ± 7.1 **P<0.001**
OKT3	62.0 ± 5.2	62.8 ± 4.8 NS	58.2 ± 4.9 NS	62.1 ± 4.7 NS
OKT4	42.2 ± 3.0	42.5 ± 2.8 NS	39.9 ± 3.9 NS	44.6 ± 4.9 NS
Tμ	43.7 ± 8.1	46.0 ± 6.8 NS	38.4 ± 11.0 NS	39.3 ± 4.8 NS
OKT8	23.8 ± 2.6	24.5 ± 2.4 NS	25.3 ± 2.3 NS	25.1 ± 2.4 NS
Tγ	18.1 ± 2.2	15.8 ± 4.0 NS	18.2 ± 2.6 NS	16.0 ± 3.55 0.05>P>0.01
"Suppressor" activity index	2.7 ± 1.1	1.2 ± 0.7 **P<0.01**		
OKT4/OKT8	1.81 ± 0.30	1.65 ± 0.39 NS	1.61 ± 0.35 NS	1.99 ± 0.48 NS
TIa$^+$	4.17 ± 1.52	7.00 ± 3.19 **P<0.01**	6.33 ± 4.01 NS	7.10 ± 2.57 **P<0.001**
SmIg$^+$	12.7 ± 3.4	16.5 ± 4.3 0.05>P> 0.01	14.1 ± 6.2 NS	15.3 ± 4.0 NS

E_{St}-RFC: spontaneous E-rosette forming cells.
E_N-RFC: neuraminidase treated E-rosette forming cells.
OKT3: lymphocytes reacting with OKT3 monoclonal antibody.

In both acute anterior uveitis and Behçet's disease populations the Tγ subset was lower than in controls, although the difference was of low statistical significance. OKT8 positive cells were, on the contrary, essentially normal in all of the three populations of uveitis patients examined by us. It may, therefore, be concluded that in acute anterior uveitis and in Behçet's disease there is probably a slight trend toward a reduction in T-cell subsets possessing suppressor activity.

It is believed that Ia positive T-lymphocytes are cell populations that are activated by specific antigenic stimulation; it can, therefore, be reasonably assumed that their level should increase in any situation in which immune insults play a role. Our data, in fact, show that Ia$^+$ T-cells increase significantly during acute anterior uveitis and

Behçet's disease; they also increase (but not to a statistically significant level) in the course of intermediate uveitis. All OKT4/OKT8 ratios and B-lymphocyte populations were within the normal range (Table I).

The results of our longitudinal investigation in the course of anterior uveitis showed that the level of E-RFC, reduced during inflammation, increases slightly during remission. In each of our populations, however, the E-RFC level was always well within the normal range. "Total" T-lymphocytes (OKT3 positive cells) did not change in the different stages of the disease. Furthermore, the OKT4 and OKT8 subsets did not show any change during the different stages of uveitis.

Ia positive T-cells were seen to increase significantly in the course of uveitis, with no further change during remission. This

TABLE II
T-Lymphocyte Subpopulations Analyzed According to the Activity of Uveitis: "Total" T-Cells, Helper T-Cells

	"Total" T-Lymphocytes		"Helpers"	
	E_{St}-RFC %	OKT3 %	OKT4 %	Tμ %
Controls	61. 1 ± 2.8 P<0.01	62.0 ± 5.2 NS	42.2 ± 3.0 NS	43.7 ± 8.1 NS
Acute anterior uveitis	54.7 ± 6.5 NS	62.8 ± 4.8 NS	42.5 ± 2.8 NS	46.0 ± 6.8 P<0.01
Remissions	57.4 ± 2.7	58.4 ± 5.6	45.8 ± 4.2	38.3 ± 2.2

Suppressor T-Cells, Ia-Positive Cells, and B-Lymphocytes

"Suppressors"					B-Lymphocytes	
OKT$_8$	Tγ	Activity Index[a]	Ia$^+$ T-Cells	SmIg$^+$ Cells	EA Rosettes[a]
23.8 ± 2.6 NS	18.2 ± 2.2 NS	2.7 ± 1.08 P<0.01	4.17 ± 1.52 P<0.01	12.7 ± 3.4 0.05>P>0.01	6.1 ± 3.3 P<0.01
24.5 ± 2.4 NS	15.8 ± 4.0 NS	1.2 ± 0.7 P<0.01	7.0 ± 3.19 NS	16.5 ± 4.3 NS	13.0 ± 7.6 P<0.01
29.8 ± 6.1	14.7 ± 3.8	2.5 ± 1.4	7.50 ± 3.02	13.2 ± 4.4	6.4 ± 4.5

[a]From previous investigation (Secchi et al.[1]).

finding would suggest that in the course of anterior uveitis, immunopathological events occur that are substantially unaffected by medical treatment (see Table II).

In summary, 25 patients suffering from acute anterior uveitis (AAU), 12 cases of intermediate uveitis (IU), and 16 cases of Behçet's disease have been evaluated with regard to their peripheral lymphocyte subpopulations. The results showed a decrease in E-RFC's in the acute phase of the disease, a slight reduction in "suppressor" activity during active inflammation, and an increase in Ia$^+$ T-cells. Other parameters studied did not differ from healthy controls. The changes detected by us were always numerically small, even though in some instances they were statistically significant.

References

1. Secchi AG et al.: *Boll Ocul* **59**:559-566, 1980.
2. Gupta S et al.: *Clin Exp Immunol* **38**:342-347, 1979.
3. Reinherz EL and Schlossmann SF: *Cell* **19**:821-831, 1980.
4. Weiner MS et al.: *Blood* **42**:939-948, 1973.
5. Semenzato G et al.: *Lab J Res Lab Med* **2**:129-131, 1978.
6. Stocker J W et al.: *Tissue Antigens* **13**:212-222, 1979.
7. Molinaro GA et al.: *Transplant Proc* **11**:1748-1759, 1979.
8. Bendinelli M: *MV Flash* **5**:31-36, 1980.

chapter 4

Quantitation of the Soluble Receptor of Human T-Lymphocytes for Sheep Erythrocytes in Aqueous Humor and Serum of Patients with Ocular Diseases

Rubens Belfort Jr., Ieda M. Longo, Nayla C. Moura, and Nelson F. Mendes

Human T-lymphocytes carry a membrane receptor for sheep erythrocytes[1–5] which can be detected in a soluble form in the serum.[6,7] Abnormally high serum levels of the soluble receptor (R_s) are found in patients with tumors, uremia and lepromatous leprosy.[8] The role played by R_s *in vivo* seems to be related to immunoregulation, since high levels of R_s inhibit the proliferation of lymphocytes in the presence of mitogens or allogenic cells.[9,10] We have previously shown that a much higher number of T-lymphocytes than B-lymphocytes is present in the aqueous humor of patients with different types of uveitis.[11] In this study, we have quantitated R_s by electroimmunodiffusion in the aqueous humor and serum from patients with uveitis and other ocular diseases.

Materials and Methods

Ocular anterior chamber paracentesis was performed on 23 patients whose diagnoses are indicated in Table I. The keratocenteses were performed after topical application and subconjuntival injection of lidocaine hydrochloride (xylocaine) as described elsewhere.[11]

Serum samples were obtained from the patients indicated above and from 43 adult normal individuals.

Electroimmunodiffusion (*Rocket* immunoelectrophoresis)[12]: An anti-R_s serum was obtained by immunizing an adult sheep with autologous erythrocytes sensitized with R_s, as described previously in this laboratory.[7] For electroimmunodiffusion, rectangular glass plates (75 X 50 mm) were covered with 7 ml of the following mixture at 56°C: 0.2 ml of anti-R_s, 1.8 ml of saline, and 5.0 ml of 1.5% agarose diluted in 3 parts of veronal buffer and 2 parts of distilled water. Seven wells of 3 mm diameter were made at a distance of 1cm from one of the edges of the plate. Each well received 10µl of the serum or aqueous humor to be tested, and one well on each plate was filled with 10µl of control serum. The plates were subjected to 250 volts for 3 hours in an electrophoresis chamber (Gelman Instrument Company, Ann Arbor, MI) containing 1 liter of veronal buffer (pH 8.6). The migration of R_s was from the cathode to the anode. After migration, the plates were washed in saline for 24 hours at ambient temperature, and were then dried at 37°C and stained with amido black. The resulting *rockets* were measured in millimeters. The concentration of R_s in the samples tested was proportional to the height of the *rockets* obtained.

Escola Paulista de Medicina, Disciplina de Oftalmologia e Disciplina de Imunologia, São Paulo, Brasil.

TABLE I

Quantitation of R_s by Electroimmunodiffusion in the Aqueous Humor and Serum from Patients with Ocular Diseases

Patient Number	Clinical Diagnosis	Cell Grading in AC	R_s in Aqueous Humor	R_s in Serum[b]
1	Cataract	0	0[a]	7[a]
2	Cataract	0	0	15
3	Cataract	0	0	0
4	Cataract	0	0	6
5	Vitreous hemorrhage	2+	0	5
6	Bacterial endophthalmitis	3+	0	9
7	Posterior uveitis	3+	0	3
8	Toxoplasmosis	1+	0	7
9	Toxoplasmosis	1+	0	3
10	Behçet	3+	0	5
11	Behçet	4+	0	6
12	V.K. Harada	3+	0	9
13	Chronic cyclitis	0	0	10
14	Diffuse granulomatous uveitis	1+	0	9
15	Eales	3+	21	15
16	NGC I–C	1+	3	10
17	NGC I–C	4+	13	10
18	NGC I–C	3+	8	10
19	NGC I–C	2+	0	0
20	NGC I–C	2+	0	ND
21	NGC I–C	1+	8	ND
22	GC I–C	4+	4	ND
23	CG I–C	4+	0	ND

[a] Numbers express the height in mm of the *rockets* obtained.
[b] Serum R_s in 43 normal controls: mean = 5.0 mm; range = 0–8 mm.
AC = Anterior Chamber.
NGC I–C = non granulomatous chronic iridocyclitis.
GC I–C = granulomatous chronic iridocyclitis.
ND = not done.

Results

The results of quantitation of R_s in the aqueous humor and serum by electroimmunodiffusion are shown in Table I. By this method, R_s could be detected and quantitated in the aqueous humor of six patients with uveitis (Eales' disease, one patient; nongranulomatous chronic iridocyclitis, four patients; and granulomatous chronic iridocyclitis, one patient). In two of these patients the level of R_s in the aqueous humor was higher than in the simultaneously sampled serum. The serum levels of R_s were within the limits observed in normal individuals in 10 cases, but in the remaining nine patients the levels were above 8 millimeters.

Discussion

In the work reported here, R_s was detected in the serum and aqueous humor of some patients with ocular diseases. In most patients, R_s could be detected only in the serum, which demonstrates that in the aqueous humor, R_s is usually absent or present in a concentration below the sensitivity limit of electroimmunodiffusion. High concentrations of R_s in the aqueous humor, observed in a few patients, may reflect increased local synthesis by activated T-lymphocytes or local lymphocyte destruction. The levels of R_s in the aqueous humor proved to be independent of the serum levels in the same patient. In certain disease states, a significant increase of the serum levels of R_s has been associated with depressed cell-mediated immunity.[8,10] The clinical significance of high concentrations of R_s in the aqueous humor of some patients is still unknown, but it opens a new line of research.

Acknowledgments

This work was supported by FINEP, CNPq, CAPES, and FAPESP.

References

1. Lay WH et al.: *Nature* **230**:531, 1971.
2. Silveira NPA, et al.: *J Immunol* **108**:1456, 1972.
3. Mendes NF et al.: *J Immunol* **111**:860,1973.
4. Mendes NF et al.: *J Immunol* **113**:531, 1974.
5. Mendes NF: *Lymphology* **10**:85, 1977.
6. Mendes NF et al.: *Cell Immunol* **17**:560, 1975.
7. Mendes NF et al.: *Cell Immunol* **72**:143, 1982.
8. Moura NC et al.: *Experientia* **39**:306, 1983.
9. Mussatti CC et al.: *Clin Immunol Immunopathol* **17**:323, 1980.
10. Mussatti CC, et al.: *Clin Immunol Immunopathol* **14**:403, 1979.
11. Belfort Jr R et al.: *Arch Ophthalmol* **100**:465, 1982.
12. Laurell CB: *Anal Biochem* **15**:45, 1966.

chapter 5

Lymphocyte Subpopulations in Graves' Ophthalmopathy

**Norman T. Felberg, Robert C. Sergott, Peter J. Savino,
John J. Blizzard, and Norman J. Schatz**

Graves' ophthalmopathy has many immunologic findings distinguishing it from the form of the disease characterized only by thyrotoxicosis. We have previously detected differences in T-lymphocyte populations between patients with severe ophthalmopathy and those with only minimal signs and symptoms.[1] We have also observed changes in T-lymphocyte populations during successful corticosteroid therapy suggesting that T-lymphocyte enumeration may be a useful parameter to evaluate potential therapy for Graves' ophthalmopathy.[2]

To further delineate the T-lymphocyte dysfunction(s), we examined T-lymphocyte subpopulations determined by monoclonal antibodies in patients with various forms of Graves' disease.

Methods

Patients with the diagnosis of Graves' disease were classified according to the American Thyroid Association system.[3] Monoclonal antibodies specific for T-lymphocyte antigens included: LYT3 PAN, a pan-T-lymphocyte monoclonal antibody that recognizes the sheep E-receptor; OKT3 PAN, a pan-T-lymphocyte monoclonal antibody that recognizes an antigen believed to be present on all T-lymphocytes, but not

Research Division and the Neuro-Ophthalmology Unit of the Wills Eye Hospital, Philadelphia; and the Department of Medicine, Lankenau Hospital, Philadelphia, Pennsylvania.

the E-receptor; OKT4 IND, an antibody that recognizes inducer-helper cells; and OKT8 SUP, an antibody that recognizes suppressor-cytotoxic T-lymphocytes. T-lymphocytes were detected manually according to the manufacturer's instructions. Control values were subtracted to correct for nonspecific binding.

Results

T-lymphocyte subpopulations for the normal volunteers and the patients with Graves' disease are compared in Table I. Statistically significant pretherapy elevations were seen in the number of OKT4 IND + cells in all patients with class 4–5 ophthalmopathy ($p<0.005$). Those patients with class 4–5 ophthalmopathy who responded to corticosteroid therapy were responsible for the elevation with 36.22 ± 16.54 OKT4 IND + cells ($p<0.05$). In general, patients with class 4–5 ophthalmopathy who responded to corticosteroids had decreased numbers of pretherapy LYT3 PAN + cells and showed an increase during therapy. There were normal numbers of OKT3 PAN + cells in the pretherapy corticosteroid responsive patients and no trend was seen during therapy. These patients did have statistically elevated pretherapy level of OKT4 IND + cells that decreased to normal during therapy (Table I). The number of pretherapy OKT8 SUP + cells was elevated for these patients, though not statistically, and decreased to normal during

TABLE I
T-lymphocyte Subpopulations[a] in Normal Volunteers and Patients with Graves' Disease

Subpopulation		Normal Volunteers	Graves' Disease			
			Class 2	Class 3	Class 4–5	Class 6
Subjects	(No.)	24	8	4	31	12
LYT3 PAN	(AVE)	48.67	44.14	55.25	43.50	52.67
	(SD)	15.12	21.97	15.44	16.55	14.64
OKT3 PAN	(AVE)	47.83	41.75	59.00	51.39	53.83
	(SD)	12.56	12.67	9.59	13.24	10.99
OKT4 IND	(AVE)	23.63	24.50	40.75	31.97[b]	27.08
	(SD)	8.67	13.88	20.06	12.57	9.81
OKT8 SUP	(AVE)	16.38	12.50	16.00	19.06	17.50
	(SD)	9.06	8.12	4.08	9.98	9.16
T4/T8	(AVE)	1.83	2.39	2.59	1.92	2.07
	(SD)	1.02	1.38	1.30	0.99	1.21

[a] Percentage of positive cells.
[b] $p < 0.005$ compared to normal volunteers.

therapy. Because of the elevated pretherapy level of OKT8 SUP + cells, the T4/T8 ratio was initially depressed, rising to normal levels during therapy.

Patients with class 4–5 ophthalmopathy who did not respond to corticosteroid therapy showed an increase in LYT3 PAN + cells and a decrease in OKT3 PAN + cells during the first week. When the corticosteroid dosage was tapered, these subpopulations returned to pretherapy values. There were slightly more OKT4 IND + cells in the pretherapy nonresponders, but there were no trends during therapy. The numbers of OKT8 SUP + cells were normal in pretherapy patients and showed only a slight downward trend, if any, during therapy. In spite of the generally flat curves for the numbers of OKT4 IND and OKT8 SUP + cells, the T4/T8 ratio showed a slightly elevated pretherapy value that increased at the beginning of therapy and decreased when corticosteroids were tapered.

Patients with class 6 ophthalmopathy who responded to corticosteroid therapy showed an initial decrease in the number of LYT3 PAN and OKT3 PAN + cells. After the first week of therapy, both subpopulations increased to stable levels. Pretherapy levels of OKT4 IND and OKT8 SUP + cells were slightly elevated and showed no trends during therapy. Pretherapy ratios of T4/T8 + cells were only slightly elevated and showed only a moderate downward trend during therapy.

Discussion

The present study used monoclonal antibodies to enumerate T-lymphocytes and confirms our previous observations of abnormal numbers of T-lymphocytes in patients with severe Graves' ophthalmopathy. Decreased numbers of T-lymphocytes were detected with the LYT3 PAN monoclonal antibody in patients with class 4–5 ophthalmopathy responding to corticosteroid therapy. During therapy, this subpopulation increased, similar to the E-rosette formation results.

The difference in the number of LYT3 PAN and OKT3 PAN + cells in the corticosteroid responsive patients with class 4–5 ophthalmopathy suggests that the T-lymphocytes are present in the peripheral blood, but some do not have the E-receptor nor its antigen. This implies that the E-receptor may be nonfunctional either by an intrinsic cell membrane defect or by an extrinsic factor, blocking access of the antibody and sheep erythrocytes to the cell. Significant elevations in numbers of OKT4 IND + cells were found in these patients. The roles of the subpopulation lacking the LYT3 PAN (E−) receptor or the excessive numbers of OKT4 IND + cells is unknown. The reappearance of the E-receptor during therapy may be due to the removal of the blocking substance or the completion of synthesis and/or maturation of the E-receptor.

This preliminary investigation reaffirms that there is a defect in the T-lymphocyte E-receptor function in corticosteroid responsive patients with class 4–5 ophthalmopathy. These abnormalities are corrected with oral corticosteroids, whereas the same anti-inflammatory therapy has little effect on the T-lymphocyte subpopulations in the corticosteroid resistant patients. Defining further these imbalances may allow the development of precise theraputic regimens based on the treatment of specific T-lymphocyte subpopulation abnormalities.

In summary, previous studies have demonstrated that patients with severe ophthalmopathy had different levels of T-lymphocytes measured by spontaneous active and total rosette forming cells (A-RFC and T-RFC) compared to normal individuals and to patients with Graves' thyrotoxicosis. Patients with severe ophthalmopathy responding to corticosteroids show increases in A-RFC and T-RFC.

To further define these T-lymphocyte abnormalities, we have examined T-lymphocyte subpopulations using monoclonal antibodies. Our preliminary studies show statistically elevated levels of OKT4 IND + cells in all patients with Werner's class 4–5 ophthalmopathy who respond to corticosteroid therapy.

Serial changes were observed during therapy in other classes. LYT3 PAN + cells initially were decreased in patients with class 4–5 ophthalmopathy who responded to corticosteroids. OKT8 SUP + cells were slightly elevated in patients with class 4–5 ophthalmopathy responding to corticosteroids resulting in a low T4/T8 ratio. Those patients who did not respond to corticosteroid therapy showed decreases in the number of OKT3 PAN + cells.

Acknowledgments

Supported in part by a grant from the Public Health Service (National Eye Institute 2 RO1 EYO2631) and the Pennsylvania Lions Sight Conservation and Eye Research Foundation, Inc.

References

1. Sergott RC et al.: *Invest Ophthalmol Vis Sci* **18**:1245-1251, 1979.
2. Sergott RC et al.: *Invest Ophthalmol Vis Sci* **20**:173-182, 1981.
3. Werner SC: *Am J Ophthalmol* **83**:725-727, 1977.
4. Kung PC et al.: *Science* **206**:347-349, 1979.
5. Reinherz EL et al.: *Proc Natl Acad Sci USA* **77**:1588-1592, 1980.
6. Reinhertz EL et al.: *N Engl J Med* **301**:1018-1022, 1979.
7. Strelkauskas AJ et al.: *Proc Natl Acad Sci USA* **75**:5150, 1978.

chapter 6

Lymphocyte Subpopulations in Uveitis Patients

R. Van der Gaag,[a] B.J. Christiaans,[b] A.Rothova,[b] G.S. Baarsma,[c] M.R. Dandrieu,[d] and A.Kijlstra[a]

Both clinical and experimental evidence suggest that immunological mechanisms are involved in the pathogenesis of endogenous uveitis.[1,2] In many diseases with underlying abnormalities in the immune system the number of T-lymphocytes in peripheral blood is decreased.[3] In uveitis both normal and decreased values have been reported for T-lymphocytes in the peripheral blood.[4,5,6,7] Recently, it has become possible to characterize T-lymphocytes and T-cell subsets using monoclonal antibodies directed to distinctive cell surface antigens present at various stages of differentiation.[8] Therefore, we decided to reinvestigate uveitis patients using OKT monoclonal antibodies to detect T-cells, inducer T-cells, and suppressor T-cells.

Increased levels of B-cells have been reported in patients with Vogt–Koyanagi–Harada syndrome using antibody-coated erythrocytes for rosette formation.[7] But T-cells with Fc receptors and monocytes may also form rosettes in this assay; therefore F(ab')2 fragments of polyvalent anti-human immunoglobulin which do not bind to Fc receptors are a more accurate way to assess B-cell levels.[9] We used this method to de-

termine the B-cell levels in our uveitis patients.

Materials and Methods

Heparinized blood samples were obtained from 67 uveitis patients during an active episode of their disease. Blood samples were also taken from 22 healthy controls and handled in the same way. The mononuclear cell fraction was isolated by means of density gradient centrifugation on Ficoll-Paque, washed twice, and resuspended in Roswell Park Memorial Institute (RPMI) 1640 medium containing hepes, glutamine (300 mg/liter) penicillin (1 U/ml), streptomycin (100 μg/liter) and 20% fetal calf serum. The cells were frozen in the vapor phase of liquid nitrogen in the presence of 10% dimethylsulfoxide[10] and when needed thawed, washed, incubated for 3 hours in medium, washed again, and used for membrane marker assays and HLA-B27 typing. After this procedure viability always exceeded 85% using the trypan blue exclusion test. Freezing and thawing does not qualitatively affect the membrane characteristics and functional properties of mononuclear cells provided that, after thawing, the cells are incubated in culture medium for at least 3 hours.[11,12] The mononuclear cells used for cell membrane marker studies were fixed with 1% paraformaldehyde and washed twice prior to use. T-cells were detected using the indirect immunofluorescence test with monoclonal antibodies against periph-

[a]Department of Ophthalmo-Immunology of the Netherlands Ophthalmic Research Institute, Amsterdam; [b]Department of Ophthalmology, Wilhelmina Gasthuis, Amsterdam; [c]Eye Hospital, Erasmus University, Rotterdam; and [d]Department of Ophthalmology of the Free University of Amsterdam, the Netherlands.

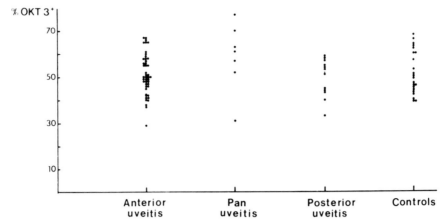

FIG. 1. Lymphocyte subpopulations in uveitis patients and healthy controls. T-cell distribution in peripheral blood.

eral T-cells, inducer T-cells, and suppressor T-cells respectively known as OKT 3, OKT 4, and OKT 8.[8,13] B cells were detected using the direct immunofluorescence test with fluorescein isothiocynate-labeled F(ab′)2 fragments of polyvalent sheep antihuman immunoglobulin serum. All the cells were counterstained with eriochrome black. Each lymphocyte subpopulation determination was performed in duplicate and at least 200 cells were counted for each assay.

The normal values for cryopreserved cells from 22 healthy donors were as follows: % T-cells (OKT 3^+): 51.5 ± 9.4, % inducer T-cells (OKT 4^+):36.8 ± 8.7, % suppressor T-cells (OKT 8^+): 17.1 ± 6.4 and % B-cells: 6.2 ± 2.6.

The determination of HLA-B27 antigen on mononuclear cells was performed using the two step cytotoxicity test according to a procedure described at the National Institutes of Health.[14] Statistical analysis of the data was performed using the Student t test.

Results

The patients admitted to this study were classified as follows: anterior uveitis (n = 48), posterior uveitis (n = 12) and pan uveitis (n = 7) according to the localization of their ocular inflammation. The distribution of T-lymphocytes, T-cell-subpopula-

tions, and B-cells in these patients and in healthy controls is shown in Figure 1. When monoclonal antibodies were used to determine peripheral blood T-cells (OKT3 positive cells), inducer T-cells (OKT4 positive cells) and suppressor/cytotoxic T-cells (OKT8 positive cells), no significant (p>0.1) differences were found between any of the uveitis patients and the healthy controls.

The number of surface immunoglobulin bearing B-cells, demonstrated by the use of FITC labeled F(ab′)2 fragments of a polyvalent sheep antihuman immunoglobulin serum, was normal in acute anterior uveitis patients. However it increased in both posterior uveitis and pan uveitis patients (Fig. 2). The number of patients in these two groups is small, but the difference from healthy controls is statistically significant (p<0.01).

A strong association has been reported between acute anterior uveitis and the HLA-B27 antigen.[15] In these proceedings Linssen and others report that 90% of the AAU patients positive for HLA-B27 have ankylosing spondylitis or one or more lesser symptoms of this disease. Therefore, we studied the lymphocyte subpopulations in AAU patients on the basis of their HLA-B27 genotype. Table I shows that there is no difference in lymphocyte subpopulation distribution between HLA-B27 negative and HLA-B27 positive AAU patients.

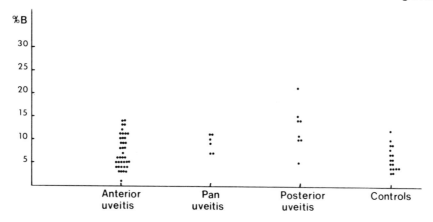

FIG. 2. B-cell distribution in peripheral blood of uveitis patients and healthy controls.

Discussion

The results of this investigation show that all uveitis patients have normal circulating levels of T-cells and T-cell subsets during the acute phase of their disease. Byrom et al[6] have reported longlasting T-lymphocytopenia in uveitis patients, detectable as a decrease of spontaneous rosette formation with sheep red blood cells.

Using the same assay Char[5] reported decreased active and total T-cell levels in Vogt–Koyanagi–Harada patients, but Boone[4] and Yokoyama[7] could not confirm these results. Rosette formation is a poorly understood, complex phenomenon, and the number of rosette-forming cells found depends on many critical conditions during the

assay.[16] In our laboratory we also found sheep red blood cell rosette formation quite variable (unpublished observations), whereas the indirect immunofluorescence test using OKT monoclonal antibodies was very reproducible, even on blood samples stored for many months in liquid nitrogen. Thus far, we have no indication that there is a systemic dysregulation at the T-cell level in uveitis patients. Nor could we confirm the investigations reported by Byrom[4] and Nilsson[4] that patients with ankylosing spondylitis with or without AAU have abnormal T-cell levels, either decreased or increased.

The levels of B-cells detected with our assay are somewhat lower than those reported by others because we do not erro-

TABLE I
Lymphocyte Subpopulation Distribution in HLA-B27 Positive and Negative Acute Anterior Uveitis Patients

	T-Cell (OKT3[+])	Inducer T-Cell (OKT4[+])	Suppressor T-Cell (OKT8[+])	B-Cell
HLA-B27 positive	51.9%	33.4%	17.9%	6.8%
	±	±	±	±
AAU (n = 20)	8.4	9.7	7.3	3.6
HLA-B27 negative	52.3%	35.6%	18.1%	7.5%
	±	±	±	±
AAU (n = 22)	10.1	9.1	6.9	3.5
	NS	NS	NS	NS

neously count as B-cells, cells that have receptors for the Fc fragment of immunoglobulin on their membrane (monocytes and certain T-cells). Acute anterior uveitis patients have B-cell levels within the limits of normal, whereas elevated levels are found in a small group of patients where the posterior pole of the eye is involved in the inflammatory process. With different assays, elevated levels have also been observed in acute anterior uveitis whether or not it is associated with ankylosing spondylitis[6] and in the Vogt–Koyanagi–Harada syndrome.[7] The significance of increased B-cells is unknown, but may indicate a response to an antigenic stimulus. Based on these studies, it seems there are no gross nonspecific alteration in the general immune status of uveitis patients, but a specific or nonspecific dysregulation at the local level cannot be ruled out. Kaplan[17] and Belfort[18] have already typed lymphocytes isolated from the vitreous in a small number of uveitis patients and have found a local imbalance of lymphocyte subpopulation. Studies with monoclonal antibodies on histological material will help to elucidate further the extent to which dysregulation at a local level may be responsible for the initiation and chronicity of uveitis.

In summary, during episodes of uveitis total T-lymphocytes, T-lymphocyte subsets, and B-lymphocytes were determined in 67 patients. All the blood samples were collected during the acute phase and the isolated cells were cryopreserved until used. Monoclonal antibodies against distinct cell surface antigens were used to determine T-cells, inducer T-cells, and suppressor T-cells. All the uveitis patients had normal levels of T-cells and of T-cell subsets. The determination of the B-cell population showed that, in the small group of patients in whom the choroid was involved in the inflammatory process, the level of B-cells was significantly increased as compared to normal controls and to acute anterior uveitis patients.

Acknowledgment

This study was supported by a research grant from the Netherlands Association against Rheumatism.

References

1. Char DH: In *Current Ophthalmology Monographs,* Grune & Stratton, 1978, p.54-64.
2. Faure JP: *Curr Top Eye Res* 2:215-302, 1980.
3. Wybran J and Fudenberg HH: *J Clin Invest* 52:1026-1032, 1973.
4. Boone WB et al.: *Invest Ophthalmol* 15:957-960, 1976.
5. Char DH et al.: *Invest Ophthalmol* 16:179-181, 1977.
6. Byrom NA et al: *Lancet* 2:601-603, 1979.
7. Yokoyama MM et al.: *Invest Ophthalmol Vis Sci* 20:364-370, 1981.
8. Reinherz EL and Schlossman SF: *Cell* 19:821, 1980.
9. Decary F et al.: Thesis. Amsterdam, 1977.
10. Wood N et al.: *Tissue Antigens* 2:27-31, 1972.
11. Slease RB et al.: Cryobiology 17:523-529, 1980.
12. Bom-van Noorloos AA et al.: *Cancer Res* 40:2890-2894, 1980.
13. van Wauwe J and Goossens J: *Immunology* 42:157-164, 1981.
14. N.I.H., Bethesda, MD *DHEW Publication No. (NIH) 75-545:* 67-74, 1974.
15. Brewerton DA et al.: *Lancet* 2:994, 1973.
16. Bernard A et al.: *J Exp Med* 155: 1317-1333, 1982.
17. Kaplan HJ et al.: *Arch Ophthalmol* 100:585-587, 1982.
18. Belfort R. et al.: *Arch Ophthalmol* 100: 465-467, 1982.

chapter 7

Cyclosporin A-Induced Clinical and Immune Alterations in Experimental Autoimmune Uveitis

**Robert B. Nussenblatt, Merlyn M. Rodrigues,
Mario C. Salinas-Carmona, William C. Leake, Peter E. Franklin,
Byron H. Waksman, Waldon B. Wacker, and Igal Gery**

The concept of ocular autoimmunity has become increasingly accepted as a mechanistic explanation to the development of intraocular inflammatory disease. One major piece of evidence to support this notion has been the purification of retinal antigens, one of which has marked uveitogenic properties.[1] The retinal S-antigen (S-Ag), a glycoprotein with a molecular weight of approximately 55,000 has been previously shown to be able to induce an immune-mediated uveitis in both lower mammals and nonhuman primates when injected into a site far from the globe.[2,4] Because there is generalized activation of the immune system with S-Ag immunization,[3,5,6] several hypotheses could be forwarded as to which element of this complicated network plays a mandatory role in the induction of this disease. Cyclosporin A (CsA), a fungal product, has been demonstrated to affect only the T-cell component of the immune system.[7] This drug has been used, therefore, as an immune probe in order to define better and to understand more thoroughly the mechanisms involved in the development of experimental autoimmune uveitis (EAU) in rats.

Clinical Branch and Laboratory of Vision Research, National Eye Institute, NIH, Bethesda, Maryland; Multiple Sclerosis Society, New York; and the University of Louisville, Louisville, Kentucky.

Materials and Methods

Female Lewis rats, 6 weeks of age and weighing approximately 200 grams, were utilized for this series of experiments. Experimental animals were immunized once with 20 μg of bovine S-Ag (prepared as described elsewhere[1]), emulsified in complete Freund's adjuvant (CFA, GIBCO, Grand Island, NY) and augmented with *Mycobacterium tuberculosis* H37 RA (Eli Lilly, Indianapolis, IN) so that the Mycobacterium concentration was 25 mg per 10 ml of adjuvant. The injections were made into the hind footpads. Control animals received CFA alone in the same manner. CsA (a generous gift of SANDOZ, Ltd., Basel, Switzerland) was dissolved in olive oil, and the drug was administered by subcutaneous injection into the thigh. Animals received 2 mg of CsA per day in 0.1 ml volume. Control animals received injections of olive oil in the same manner. The animals were treated daily, or on alternate days, with CsA or plain olive oil for 14 days, starting on the day of S-Ag immunization. The rats were followed for signs of disease and were sacrificed on day 14 after S-Ag immunization. EAU was passively transferred to naive rats by injecting $1.5–3.0 \times 10^7$ uveitis donor lymph node cells that had been preincubated with S-Ag for 3 days as described elsewhere.[8] CsA treatment of the recipients included 7

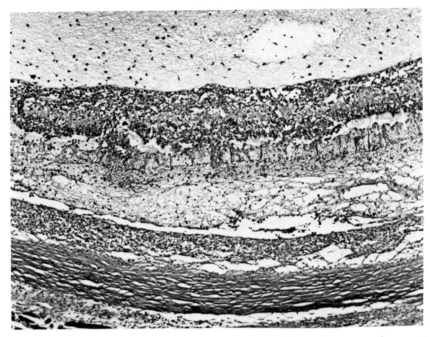

FIG. 1. Photomicrograph showing extensive destruction of posterior segment from a rat immunized with 20 μg S-antigen 14 days previously. (H & E, × 80.)

daily injections of 3 mg/rat, starting 1 day before the cell transfer. Skin testing was performed on specially designated animals. Animals received intracutaneous injections of S-Ag (20 μg) and PPD (20 μg) and the results were recorded at 4 and 24 hours later.

Eyes were taken at various intervals and fixed in 10% formalin or glutaraldehyde. After fixation, the eyes were prepared for light as well as electron- and scanning electron-microscopy, as described elsewhere.[9] Lymph nodes from CsA treated and non-treated animals were also taken at the time of sacrifice. Lymph nodes designated for histologic examination were first carefully cleaned, their size determined, and then fixed in 10% formalin. Paraffin sections stained with hematoxylin and eosin were examined by light microscopy. Cell cultures from lymph nodes were prepared as described elsewhere.[3,10] Briefly, cells were cultured in microtiter wells at a concentration of 3×10^5 cells per well in 10% heat-inactivated AB positive human serum. Cultures were terminated after 4 days with tritiated thymidine, added 14 hours prior to the

termination of culture. Serum from control rats and from those receiving daily CsA therapy were substituted for human serum in some experiments. In others, CsA (1 μg/ml) was added directly to the *in vitro* cultures at their initiation.

Results

Histology

All S-Ag immunized Lewis rats not receiving CsA therapy developed ocular inflammatory disease by day 14 after immunization. The histologic appearance was characterized by severe bilateral anterior and posterior inflammatory infiltrates (Fig. 1). Posteriorly, the retina showed a consistent loss of the photoreceptor layer, and often a perivasculitis was evident. Transmission electron microscopy revealed acute and chronic inflammatory cells, predominantly polymorphunuclear cells with lesser numbers of lymphocytes and macrophages. Scanning electron microscopy disclosed loss of outer retinal layers with marked disor-

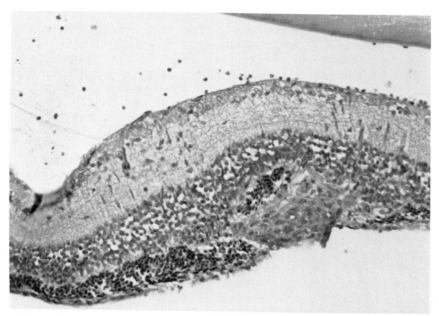

FIG. 2. Photomicrograph of retina from a rat that received 1 mg of CsA per day. Note the attenuated form of the disease and particularly the granuloma in the retina. (H & E, 220K.)

ganization and edema of retina and choroid. This was in sharp contrast to animals receiving 2 mg of CsA per day or every other day, or 8 mg per day of CsA starting 7–10 days after S-antigen immunization. These animals demonstrated no histologic evidence of inflammation. However, some animals treated daily with suboptimal doses (O.5–1 mg) of CsA developed an attenuated form of the disease, including localized granulomatous retinitis in occasional animals (Fig.2).

Skin Testing

S-Ag immunized animals demonstrate good delayed hypersensitivity skin responses to both the S-Ag and PPD 24 hours after the intradermal injection. Additionally, an Arthus response to the S-Ag was observed 4 to 6 hours after injection. Delayed hypersensitivity skin responses to both the S-Ag and PPD were fully abrogated in the animals treated with 2 mg of CsA daily. However, some animals did manifest positive Arthus responses to the S-Ag even with CsA therapy.

Lymph Node Size and Histology

The popliteal lymph nodes draining the site of S-Ag immunization from CsA treated and nontreated rats demonstrated significant differences in size when compared with each other 14 days after immunization. The average size of the control lymph nodes was 25 mm, while that of the CsA group was 17 mm. In some experiments utilizing 8 mg of CsA daily, starting from the day of S-Ag immunization, the draining lymph node was sometimes not visible. Histologically, however, the normal general architecture of an activated lymph node was maintained in the animals receiving CsA. A marked hypocellularity of the T-cell dependent areas was evident, however. In addition, there was absence of a granulomatous response surrounding the oil droplets of the adjuvant, a characteristic finding in normally activated lymph nodes.

Draining Lymph Node Lymphocyte Culturing

Draining lymph nodes from S-Ag immunized animals demonstrated alterations in the kinetics of their responses to the immunizing antigen when compared to controls. Though higher doses (8 mg per day) markedly decreased the proliferative responses,[10] more moderate dosages appeared to shift the peak day when the lymph node proliferative responses were seen, from day 10 to day 14. Furthermore, the *in vitro* addition of CsA at levels similar to those found in the serum of treated animals markedly and consistently depressed the proliferative responses to both antigens and mitogens *in vitro*.

Inhibition of EAU Transfer

Cyclosporin A (CsA) treatment of recipients of lymph node cells from donors with EAU prevented completely the development of the passive transfer of EAU, while all rats of the untreated control group showed severe disease. In addition, the CsA-treated recipients did not exhibit the delayed type skin responses to S-Ag, observed in the control recipients. On the other hand, the treated recipients resembled the controls in exhibiting both high levels of serum antibodies and Arthus skin reaction to S-Ag.

Discussion

Our studies with CsA in the EAU model would strongly support the necessary participation of T-cells in this model for human disease. CsA was effective in preventing the onset of this disease, even when utilized on an every other day schedule, or begun long after immunization. Of note was the alteration of the histologic appearance of some animals' lymph nodes treated with doses of the drug that were not completely protective. In addition to a general diminution of the ocular inflammatory response, a granulomatous pattern of disease was also seen. This might suggest a change in the recruitment pattern of effector cells to the eye caused by the administration of CsA. Further evidence of a diminution of recruitment of immunocompetent cells could be seen in the decrease in lymph node size and in the altered kinetic cell-mediated response in CsA treated animals. In association with the histologic changes seen in the lymph node of CsA treated animals, it would appear that the drug is impeding the central cell in the induction of this disease. The finding that CsA also inhibits the transfer of EAU to naive rats indicates that the drug may affect processes such as clone expansion and the recruitment of lymphocytes and inflammatory cells. These lymphokine-mediated processes are presumably needed for the development of the passive disease.

References

1. Wacker WB et al.: *J Immunol* **119**:1949-1958, 1977.
2. Nussenblatt RB et al.: *Am J Ophthalmol* **89**:173-179, 1980.
3. Nussenblatt RB et al.: *Invest Ophthalmol* **19**:686-690, 1980.
4. Nussenblatt RB et al.: In *Immunology of the Eye, Workshop II* (Special suppl to *Immunology Abstracts*), Helmsen RJ et al. (Eds.). Bethesda, MD, 1981, pp. 49-65.
5. Faure JP and de Kozaky: In *Immunology of the Eye, Workshop II* (Special suppl to *Immunology Abstracts*), Helmsen RJ et al. (Eds.). Bethesda, MD, 1981, pp. 33-48.
6. de Kozak Y et al.: *Eur J Immunol* **11**:612-617, 1981.
7. Brent L: *Transplant Proc* **12**:234-238, 1980.
8. Richert JR et al.: *J Neuroimmunol* **1**:195, 1981.
9. Nussenblatt RB et al.: *Arch Ophthalmol* **100**:1146-1149, 1982.
10. Nussenblatt RB et al.: *J Clin Invest* **67**:1228-1231, 1981.
11. Nussenblatt RB et al.: *Am J Ophthalmol* **94**:147-158, 1982.
12. Nussenblatt RB et al.: *Arch Ophthalmol* **99**:1090-1092, 1981.

chapter 8

Adoptive Transfer of Experimental Allergic Uveitis with Antigen-Stimulated Guinea Pig Peritoneal Exudate Cells

Waldon B. Wacker,[a] Carolyn M. Kalsow,[b] and Gregory T. Stelzer[c]

Experimental allergic uveitis (EAU) can be produced in several mammalian species by immunization with purified retinal S antigen (S-Ag) in complete Freund's adjuvant (CFA).[1,2] The disease is associated with both humoral and cellular immune responses,[2,3] but their respective roles in pathogenesis remain poorly understood. Successful adoptive transfer of EAU was first obtained in strain 13 guinea pigs using lymph node cells from donors sensitized with uveal extract[4] or rod outer segments.[5] More recently, similar experiments carried out with Lewis rats using lymph node cells from donors immunized with purified S-Ag have also been successful.[2] The disease in both species was generally less intense than in actively sensitized animals, affected only a portion of the recipients, and required a large number of cells of the order of 10^8 to 10^9 to effect transfer.

Recently, it was reported[6] that transfer of the prototype autoimmune disease, experimental allergic encephalomyelitis (EAE), was greatly augmented in guinea pigs if peritoneal exudate cells (PEC) from sensitized donors were cultured in vitro with specific antigen. Since then, cultured spleen or lymph node cells have been used to transfer EAE in rats,[7] autoimmune thyroiditis in

guinea pigs[8] and, while this work was in progress, the transfer of EAU in rats.[9] In the present study, we used this approach to facilitate the transfer of EAU in guinea pigs so that the nature of the effector cell(s) may ultimately be defined.

Materials and Methods

Strain 13 guinea pigs weighing 300–700 g (Crest Caviary, Raymond, CA; Altick Associates, River Falls, WI) were used in all experiments. Donors were immunized with 50 µg bovine S-Ag or saline in 0.1 ml of complete Freund's adjuvant (CFA) by injection of one hind footpad. Ten days later, the animals were injected intraperitoneally with 25 ml sterile mineral oil. After 3 days the animals were exsanguinated under ether anesthesia, and their peritoneal cavities were washed with 200 ml of Hanks balanced salt solution (HBSS).

The cells were washed, incubated, and cultured according to the procedures of Driscoll et al.[6] After two washings with HBSS, the cells were suspended in RPMI 1640 medium containing 5% fetal calf serum, 300µg/ml of glutamine, 100U/ml of penicillin, and 100µg/ml of streptomycin. The cells were plated at a concentration of 2×10^7/ml in 100 mm plastic tissue culture dishes and incubated for 2 hours in a humidified air mixture containing 5% CO_2. After incubation, the nonadherent cells were harvested, pooled, centrifuged, and resus-

[a]Department of Ophthalmology, University of Louisville, Louisville, Kentucky; [b]Department of Biology, Hope College, Holland, Michigan; [c]Tissue Typing Laboratory, Jewish Hospital, Louisville, Kentucky.

pended in culture medium. Viable cells were counted in the hemocytometer using trypan blue, and the number of phagocytic cells was determined by latex bead ingestion. The cells were then incubated for 72 hours in flat-bottomed plastic trays (Costar Tissue Culture Cluster,[24] Cambridge MA) at a concentration of 1×10^6/ml, with or without the addition of 10µg/ml bovine S-Ag.

After the incubation period, the cells were harvested, counted, and the number of phagocytic cells determined. The final suspensions used for transfer contained on average about 60% viable nonphagocytic cells. Cultured cells were then injected intraperitoneally into strain 13 guinea pigs, which were examined daily with the ophthalmoscope for clinical signs of EAU. Tissues were collected 12 days after transfer for histologic examinations.

Estimation of the lymphocyte stimulation index and the fractionation of PEC on columns of nylon wool were done as previously described.[10] The stimulation index (2 or greater) of cells from S-Ag sensitized guinea pigs, incubated in culture with S-Ag, indicated proliferation of at least some S-Ag specific cells. In contrast, cells from control animals incubated in culture with S-Ag failed to show significant stimulation.

Results

The cell transfer experiments are summarized in Table I. S-Ag sensitized cells cultured in the presence of antigen transferred EAU in each animal when 4×10^7 cells were used, and in about 50% of animals when fewer cells were used. Onset of disease occurred between 4 to 8 days after transfer in most animals, and developed much earlier than that previously observed in actively sensitized guinea pigs. In contrast, cells from S-Ag sensitized donors cultured without antigen, and cells from CFA-sensitized donors cultured with antigen, did not transfer the disease. The latter result rules out the possibility of active sensitization of recipients by S-Ag nonspecifically bound to cells.

The pineal gland, an extra-ocular site of S-Ag, was also inflamed. Pineal involvement

TABLE I

Transfer of EAU in Strain 13 Guinea Pigs with Cultured PEC

Donor Sensitization	S Antigen in Culture 10 µg/ml	No. Cells Transferred 10^7	Histo-pathology EAU	Histo-pathology Pineal
S antigen/CFA	+	4.0	3/3	3/3
	+	2.0	2/4	2/4
	+	1.0	5/10	6/10
	+	0.5	1/4	1/4
	+	0.1	0/4	0/4
	−	4.0	0/4	2/4
	−	2.0	0/2	0/2
CFA	+	4.0	0/3	0/3
	+	2.0	0/2	0/2
	−	4.0	0/3	0/3
	−	2.0	0/2	0/2

was found in all animals with EAU, as well as in a few lacking ocular disease. The latter finding was limited to three recipients of S-Ag sensitized cells cultured with or without S-Ag. This finding suggests that in this model the pineal is more susceptible to immunologic attack than the eye.

In order to investigate whether the effector cells resided in the thymus-derived population of lymphocytes (T-cells), cultured stimulated cells were fractioned on a column of nylon wool. The nonadherent fraction, enriched in T-cells and markedly depleted of B-cells and macrophages, was injected into four guinea pigs at a concentration of 2×10^7/recipient. This population of cells termed peritoneal exudate lymphocytes (PEL) was relatively more effective in transfer than was PEC.

When challenged intradermally with S-Ag at 8 days following transfer, recipients of PEL demonstrated marked reactions of delayed hypersensitivity. Humoral antibodies were not detected.

The ocular lesions varied from mild to severe. In animals with mild disease, there was a small area of inflammation in the posterior choroid adjacent to the optic nerve and in the region of pars plana and anterior choroid, with cells extending into the vitreous. In moderate disease the infiltration extended from the anterior choroid into the ciliary

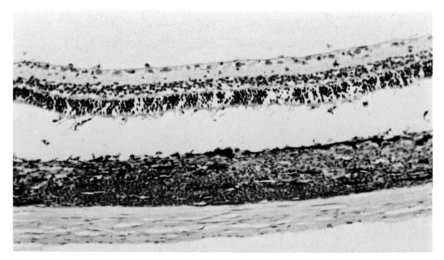

FIG. 1. Marked diffuse infiltration of the choroid with loss of the overlying rod outer segments.

body. Focal peripheral choroiditis was also observed. More severe disease affected both anterior and posterior uveal segments. There was diffuse, heavy infiltration of the choroid, with cellular invasion of the outer retina and loss of photoreceptor outer segments (Fig. 1). The character of the infiltrates in all degrees of severity was predominantly mononuclear. There was scant evidence of the acute inflammatory elements previously seen in the eyes of highly sensitized guinea pigs after active immunization.

Discussion

This study has demonstrated the successful transfer of EAU in inbred guinea pigs using cultured PEC stimulated with S-Ag. The frequency of transfer and severity of disease are similar to that previously reported[4,5] for transfer in strain 13 guinea pigs using sensitized lymph node cells, but far fewer cells were needed to effect transfer with antigen stimulated PEC. The ease of transfer was further enhanced by the use of nonadherent cells obtained after fractionation of PEC on nylon wool columns. In this case, the incidence and severity of disease developing in all animals was similar to that

occurring after active sensitization with S-Ag.[11]

The role of the T-cell as an effector mechanism in EAU has long been suspected. This notion was recently supported by prevention of EAU by treatment with Cyclosporin-A[12], an inhibitor of T-cell mediated diseases. In addition, transfer in the rat was recently effected using antigen-stimulated and cultured spleen and lymph node cells,[9] in which the recipients developed delayed hypersensitivity in the absence of detectable circulating antibody. Similar findings were noted in the present study after transfer of PEL, further implicating the T-cell in pathogenesis. On the other hand, hyperimmune serum from enucleated donors has been reported to transfer EAU in guinea pigs[13] and rats[2] in some instances. Immune complexes have also been reported[14] to contribute to the severity of EAU in highly sensitized guinea pigs. Thus, it is possible that EAU may be initiated by effector T-cells, by antibody, or by a combination of these, depending upon the experimental situation.

Previously, investigations into the mechanisms of EAU were greatly hampered by the large number of cells required to effect transfer. This study has shown that as few as 2×10^7 cultured cells can induce disease comparable to that following active sensitization. With this method it should now be possible to define more clearly the cell populations and immunologic mechanisms which contribute to the pathogenesis of EAU.

In summary, incubation with S-antigen greatly enhanced the ability of peritoneal exudate cells and lymphocytes from sensitized donors to transfer experimental allergic uveitis in guinea pigs. The disease occurred within 4–8 days after transfer and was associated with the development of delayed hypersensitivity in the recipients. The relatively small number of cells required increases the feasibility of *in vitro* studies to determine mechanisms of the disease process.

Acknowledgments

This work was supported by United States Public Health Service Research Grants EY-00254 and EY-01516 from the National Eye Institute.

The authors thank Lois Cecil and Betty Short for their skilled technical assistance and Henrietta Philpot for typing the manuscript.

References

1. Wacker WB: In *Immunology of the Eye, Workshop II* (Special suppl to *Immunology Abstracts*), Helmsen RJ et al. (Eds.). Bethesda, MD, 1981, pp. 11-32.
2. Faure JP and de Kozak Y: In *Immunology of the Eye, Workshop II* (Special suppl to *Immunology Abstracts*), Helmsen RJ et al. (Eds.). Bethesda, MD, 1981, pp. 33-48.
3. Wacker WB and Lipton MM: *J Immunol* **101**:157-165, 1968.
4. Aronson SB and McMaster PRB: *Arch Ophthal* **86**:557-563, 1971.
5. Meyers RL: *Mod Prob Ophthalmal* **16**:41-50, 1976.
6. Driscol RF and Kies MW: *Science* **203**:547-548, 1979.
7. Richert JR et al.: *J Immunol* **122**:494-496, 1979.
8. Braley-Mullen H et al.: *J Immunol* **127**:1767-1771, 1981.
9. Gery I, Shichi H, Robison WG Jr, El-Saied M, and Nussenblatt RB: Transfer of experimental autoimmune uveitis by cultured lymphocytes. *Invest Ophthalmol Vis Sci (suppl)* **22**:213, 1982.
10. Nussenblatt RB et al.: *Invest Ophthalmol Vis Sci* **19**:686-690, 1980.
11. Wacker WB et al.: *J Immunol* **119**:1949-1958, 1977.
12. Nussenblatt RB et al.: *J Clin Invest* **67**:1228-1231, 1981.
13. de Kozak Y et al.: *Mod Probl Ophthal* **16**:51-58, 1976.
14. Marak, GE Jr et al.: *Ophthalmic Res* **11**:97-107, 1979.

chapter 9

An ELISA Test to Detect Antibody Production by Rabbit Ocular Cells

Joan M. Hall and James F. Pribnow

In previous experiments[1] we showed that corneal and uveal tract cells from rabbits immunized intravitreally with protein antigens produced specific antibody. We also found[2] that corneal and conjunctival cells from rabbits immunized topically with ovalbumin produced specific antiovalbumin antibody. The direct plaque assay we used in those studies is said to detect cells producing IgM antibody. The fact that we were able to detect plaque-forming cells for several weeks after immunization raises the possibility that some of the cells were actually producing IgG antibody. Previous immunodiffusion studies had shown that the aqueous humor and vitreous humor from rabbits immunized with bovine gamma globulin (BGG) contained IgG antibody. The indirect plaque assay, in which anti-IgG antibody facilitates lysis of the target erythrocytes, was not uniformly reproducible in our system. For these reasons, we have adapted the ELISA method described by Kelley et al[3] and later used by Kilburn et al[4] to document specific IgG antibody production by the ocular cells of rabbits immunized intravitreally with ovalbumin.

Materials and Methods

Immunization. Rabbits immunized by injection of 1.5 mg sterile ovalbumin into the

vitreous cavity of both eyes were killed at various times after immunization. Cell suspensions were prepared from the lymph nodes, spleens, uveal tracts, and corneas (limbal tissue) by methods described previously.[2] The medium used for preparation of the cells was RPMI 1640 supplemented with 5% gamma globulin-free fetal bovine serum. Gentamycin was also added to the medium to control possible bacterial contamination.

ELISA procedure. Preliminary experiments were done to determine the optimal amounts of the various reagents to be used in the ELISA test to detect anti-OA antibody. We found that alkaline phosphatase conjugated goat antirabbit IgG (Miles) used in dilutions of 1:2000 to 1:3000 would detect at least 0.1 microgram of rabbit IgG. The ELISA procedure was essentially that described by Voller et al.[5]

Plaque assay. The direct plaque assay was carried out as described previously. It was done on the tissues of many rabbits killed from days 8 through 17. The results are expressed as the member of plaques per million cells in each suspension. In some cases both direct and indirect plaque assay were done.

Antibody determinations. We used the passive hemagglutination test[2] to determine the antibody titers in the sera, and in the aqueous and vitreous humors from each rabbit. Titers are expressed as the log 2 of the highest dilution of the serum or ocular fluid that showed agglutination. In some

Francis I. Proctor Foundation for Research in Ophthalmology, University of California, San Francisco, California.

TABLE I

Hemolytic Antibody Titers in Serum and Ocular Fluids of Rabbits Immunized Intravitreally with Ovalbumin

Day	Number of Rabbits	Average Titer ($\log_2$)		
		Serum	Aqueous	Vitreous
7	2	1.5[a]	1[b]	1
8	4	3.5	1	1
10	5	4.2	8.2	1
11	2	3.5	4.0	1
12	5	7.0	8.2	3.2
13	3	2.2	2.0	1
14	1	8.0	8.0	8.0
15	4	2.0	1.0	1.0
17	2	1.3	0.5	7.5
19	2	3.5	3.5	9.0
21	2	4.5	3.5	8.0
29	2	3.5	6.0	8.0

[a] Numbers indicate average $\log_2$ titer.
[b] No agglutination in sample diluted 1:2.

TABLE II

ELISA Readings and PFC in Lymph Nodes of Rabbits Immunized Intravitreally with Ovalbumin

Day	No. of Rabbits	No. of Cells Cultured	ELISA Reading	PFC per Million Cells
5	1	3.0×10^6	Positive	ND[a]
6	2	1.0×10^6	Negative	ND
7	2	1.0×10^6	0.51	210
8	4	1.0×10^6	Negative	493 (2)[b]
10	5	1.0×10^6	0.38 (3/5)[c]	500 (1)
11	3	1.0×10^6	Negative	315
12	5	1.0×10^6	0.31 (3/4)	132
13	5	1.1×10^6	Negative	0 (3)
14	3	1.0×10^6	0.78 (2/2)	219 (1)
15	3	1.0×10^6	0.21 (1/3)	32
17	2	1.0×10^6	Negative	67 IgM, 142 IgG (1)
19	2	1.0×10^6	0.55 (1/2)	74 (1)
21	2	1.3×10^6	Negative	ND
29	2	1.0×10^6	0.72	142
44	1	1.0×10^6	0.54	ND
64	1	1.0×10^6	0.36	ND

[a] Not done.
[b] Numbers in parentheses refer to number of rabbits on which assay was done.
[c] Numbers in parenthesis refer to number of positive assays/number of tissues assayed. Where no number is given all tissues were assayed.

cases the ELISA method was also used to determine antibody titers.

Results

Antibody determinations. The results of the hemagglutination titers performed on the aqueous humor, the vitreous humor and the serum are given in Table I. Antibody was detected first in the serum and aqueous humor and later in the vitreous humor. Vitreous titers were higher than those in serum and aqueous humor.

Plaque assays and ELISA assays. The results of the plaque assay and the ELISA assays on the lymph node cells of the rabbits are given in Table II. IgM antibody producing cells were usually found in the lymph nodes by day 7. Lymph node plaque numbers, as in other experiments, were rather low. The ELISA test showed that small amounts of IgG antibody were produced by the lymph nodes of some of the rabbits by day 7. Spleen cell suspensions also contained some IgG and IgM antibody producing cells, but the plaque numbers and ELISA readings were low.

Table III shows the results of the assays on the uveal tract tissues of the rabbits. Small numbers of plaque-forming cells

(PFC) were detected as early as day 7, and by day 10 many specific PFC were found in the uveal tract tissue. The ELISA tests were positive by day 10. In general, the color development with the ocular cells was greater than with the lymph node cells, and the spectrophotometer readings were higher, even though many fewer cells were cultured. IgG and IgM (PFC) producing cells seemed to be present in the same tissues. This can be seen on day 17, when the indirect plaque assay was carried out successfully. We detected IgG production by cultured uveal tract cells for at least 64 days after injection of the ovalbumin.

Table IV shows the results of the PFC and ELISA tests on the corneal tissues of the rabbits. IgM producing cells were first found by day 7, and high numbers of such antibody forming cells were seen by day 12. IgG production by relatively few corneal cells was noted by day 10. As with the uveal tract, IgG producing and IgM producing cells were present in the same tissues. Early in the response similar numbers of cultured cells did not produce antibody.

Discussion

These experiments showed that the ELISA method could be used to document IgG antibody production by cultured rabbit lymphoid and ocular cells. The assay is capable of detecting small amounts of antibody. A reading of 1.0 or higher was achieved when amounts of rabbit IgG from 1 mg to 1 microgram were coated onto the wells. The high readings obtained with uveal and corneal cells, therefore, probably indicate the minimum amount of antibody produced. The fact that high readings were obtained by culturing relatively few ocular cells, and that negative readings were obtained with many times more spleen and lymph node cells, might provide further

evidence for the sensitivity of the ELISA test. We were able to dilute lymph node cells and obtain successively lower readings, and in some cases reach an endpoint. We were also able to dilute a few of the uveal tract cell suspensions and obtain lower readings.

One objection might be that we are detecting immune globulins on cells that might have adhered to the wells. This does not seem likely, since the wells were washed several times using a mechanical washer-aspirator that seemed to be quite effective. Readings on the control plates were always less than 0.1. If cells remained on the plates we might have expected to obtain positive reading at later dates when at least 1×10^6 spleen or lymph node cells were cultured. It must be noted that with all tissues the readings on the control were always less than 0.1.

In cases where sufficient ocular cells are available (simultaneous plaque assays are not done) it should be possible to dilute the ocular cells and express the results as the

TABLE III

ELISA Readings and PFC in Uveal Tissue of Rabbits Immunized Intravitreally with Ovalbumin

Day	No. of Rabbits	No. of Cells Cultured	ELISA Reading	PFC per Million Cells
5	1	1.8×10^4	Negative	ND
6	2	5.6×10^4	Negative	ND
7	2	1.4×10^5	Negative	259
8	4	3.1×10^5	Negative	1107 (2)[a]
10	5	5.0×10^5	0.64 (4/5)[b]	4500
11	3	2.8×10^5	0.77	6200
12	5	5.0×10^5	0.98	6900
13	5	3.8×10^5	0.47	8000 (3)
14	3	2.3×10^5	0.75	2239 (1)
15	3	2.9×10^5	0.79	480 (1)
17	2	7.2×10^5	Positive[c]	6000 IgM, 11,000 IgG
19	2	1.1×10^5	1.13	2945
21	2	2.5×10^5	0.77	ND
29	2	1.0×10^5	0.88	1930
44	1	4.8×10^4	0.29	ND
64	1	5.8×10^4	0.61	ND

[a] Numbers in parentheses refer to number of rabbits on which assay was done.
[b] Numbers in parentheses refer to numbers of positive assays or number of tissues assayed. Where there is no number, all tissues were assayed.
[c] Reading not obtained, but color development noted.

TABLE IV

ELISA Readings and PFC in Corneal Tissue of Rabbits Immunized Intravitreally with Ovalbumin

Day	No. of Rabbits	No. of Cells Cultured	ELISA Reading	PFC per Million Cells
5	1	2.0×10^4	Negative	ND
6	2	2.4×10^4	Negative	ND
7	2	6.0×10^4	Negative	83
8	4	7.8×10^4	Negative	948 (2)[a]
10	5	1.0×10^5	0.36 (3/5)[b]	500 (1)
11	3	7.8×10^4	0.46 (1/3)	10,300 (3)
12	5	1.5×10^5	0.89	9970
13	5	1.2×10^5	0.41	3500 (3)
14	3	8.0×10^4	0.67	9625 (1)
15	3	7.3×10^4	0.47 (2/3)	1200 (1)
17	2	3.4×10^4	Positive[c]	20,000
19	2	7.1×10^4	1.02	7026 (2)
21	2	1.8×10^5	0.64	ND
29	2	4.1×10^4	0.60	4040
44	1	1.7×10^4	Negative	ND
64	1	2.4×10^4	0.61	ND

[a] Numbers in parentheses refer to number of rabbits on which assay was done.
[b] Numbers in parentheses refer to numbers of positive assays/number of tissues assayed. Where there is no number, all tissues were assayed.
[c] Reading not obtained, but color development noted.

lowest numbers of cultured cells that give a positive reading.

The results also show that, as predicted by previous plaque assays, uveal tract and corneal cells do produce IgG antibody. Cultured cells produced IgG antibody at about the same time as large numbers of direct PFC were detected. It is still not certain whether some PFC are not actually producing IgG antibody. The indirect plaque assay was successful at times, and the results were consistent with those of the ELISA assay.

In summary, the ELISA test has been used to document the production of IgG antiovalbumin antibody by ocular cells of rabbits immunized intravitreally with ovalbumin. The method will permit determination of the smallest number of cells that are necessary to produce antibody. With appropriate double sandwich methods, or the ultimate use of conjugated antibody specific for other rabbit immune globulin classes, it will be possible to determine the relative contribution of other classes of antibody to the ocular immune response. In preliminary experiments, we have also detected IgG production by the conjunctival cells of rabbits immunized topically with ovalbumin.

Acknowledgment

This research was supported in part by Grant EY-01182 from the National Institutes of Health, Bethesda, Maryland.

References

1. Hall, JM et al.: *Invest Ophthalmol* **10**:775, 1971.
2. Hall JM and Pribnow JF: *Invest Ophthalmol Vis Sci* **21**:753-775, 1981.
3. Kelley BS et al.: *Immunology* **37**:45-52, 1979.
4. Kilburn DT et al.: *J Immunol Methods* **44**:301-310, 1981.
5. Hall JM and O'Connor GR: *J Immunol* **104**:440, 1970.
6. Voller A et al.: In *Manual of Clinical Immunology*, Rose NR, and Friedman H (Eds.). American Society for Microbiology, Washington D.C., 1976, pp. 506-512.

chapter 10

Antiretinoblastoma Monoclonal Antibodies

**Devron H. Char, Neelam Rand, Edward L. Howes, Jr.,
Howard Lyon, and Craig T. Morita**

We and others have previously demonstrated retinoblastoma patient reactivity towards antigens expressed on both retinoblastoma tissue and cell lines; however, the antigens responsible for this reactivity are poorly characterized.[1,2]

Hybridoma techniques have facilitated the production of monoclonal antibodies that can be specifically directed against a single antigenic determinant.[3] Monoclonal antibodies may help to characterize the nature of tumor-associated antigens, and radiolabeled monoclonal antibodies can be used both diagnostically and therapeutically.

We have performed mouse–mouse somatic cell hybridizations and produced monoclonal antibodies directed against retinoblastoma-associated antigens.

Methods

The methods used in hybridization, screening, and specificity testing have previously been described.[4] Briefly, after IP and IV immunization of mice with 10^7 retinoblastoma derived tissue culture cells, the spleen cells were fused with 50% polyethylene glycol solution plus 5% DMSO at room temperature with a cloned BALB/c myeloma cell line, Sp2/0. Sensitivity and specificity tests were performed using both an ELISA and a cytotoxicity assay for reactivity with both test and control extracts and cell lines.

Ocular Oncology Unit, Department of Ophthalmology, and Francis I. Proctor Foundation, University of California, San Francisco.

Monoclonal antibody binding was also studied with direct immunofluorescence to measure binding to fresh or snap-frozen retinoblastoma and control tissues using standard techniques.

Isotyping was performed in a standard manner. Antigen–antibody precipitates in agar gels were prepared by Ouchterlony's method using immunodiffusion plates with 0.5 g agar in 50 ml borate buffer at pH 8.0.

Monoclonal antibodies reactive with retinoblastoma cells were tested for anti-HLA activity using a modification of standard HLA typing techniques.[5]

Results

Of 162 wells that were positive for growth, 25 were reactive against retinoblastoma-derived tissue culture cells, four were reactive against only allogeneic retionoblastoma cells, and one was reactive only towards the immunizing retinoblastoma cell line.

Sensitivity and specificity testing were performed using both a Cr^{51} release and an ELISA assay. The results in both assays were almost always concordant: The chromium release assay was technically unsuitable for specificity testing with some adherent tissue culture cell lines (high background due to spontaneous release), and with those lines testing was limited to the ELISA.

During initial screening for hybridoma antibody production, a positive clone was defined as one which had 4–5 times background activity against a retinoblastoma derived tissue culture cell line. Most positive

TABLE I

Patterns of Antibody Binding of Specific and Nonspecific Monoclonal Antibodies Toward Human Tissue Culture Retinoblastoma and Control Lines and Extracts in an ELISA Assay[a]

| | | RETINOBLASTOMA CELL LINES | | | HUMAN TUMOR CELL LINES | | | | | | |
| | | | | | Neuroblastoma | | Glioblastoma | Melanoma | T-Cell Molt. | Normal Fibro. | Retinal Extract |
		Y79	NgRb	McARb	IMR-5	B.Jones	209	HT 144			
ANTI-RETINOBLASTOMA MONOCLONAL Abs	p16	+ +	+	+	−	−	−	−	−	−	−
	p21	+ +	+	+	−	−	−	−	−	−	−
	p22	+ −	−	−	−	−	−	−	−	−	−
	p38	+ +	+ −	+	−	−	−	−	−	−	−
	p50	+ −	+ −	+ −	−	−	−	−	−	−	−
ANTI-TUMOR CROSS-REACTING MONOCLONAL Abs	p13	+	+	+	−	+ −	−	+	+ −	−	+
	p25	+ +	+	+	−	−	−	+	−	−	−
	p34	+ +	+	+	+ −	+ −	−	+	+ −	−	−
	p42	+ +	+ +	+ +	−	−	−	+	−	−	−
	p56	+ +	+	+	−	−	−	+	−	−	−
	p63	+	+	+	−	−	−	+ −	−	−	−
	f.21	+ +	+	+	−	−	−	+ −	−	−	−
NON-SPECFIC MONOCLONAL Abs	p 5	+ +	+ +	+ +	+ −	+	+	+	+	+	+
	p36	+ +	+	+	+ −	+	+	+	+	+	nd

[a] + − is less than 2 × control.

wells had ELISA or Cr[51] activity between 7–10 times background. A negative clone was defined as one that had less than two times background activity. Most negative wells were at or below medium control. Table I demonstrates the reactivity patterns of monoclonal antibodies from selected hybridomas, after limited dilution cloning, against various retinoblastoma cell lines, other tumor cell lines, and allogeneic pooled retinal extracts. The Fab and Fab$_2$ monoclonal antibodies specific for retinoblastoma cells had the same specificity pattern as the complete antibody.

The titers of the antiretinoblastoma monoclonal antibodies produced in tissue culture wells varied. P38 retained activity greater than 4 times background at a 1:256 dilution. Antibody titration curves for antigens expressed on three different allogeneic retinoblastoma cell lines are shown in Figures 1 and 2. Higher titers were obtained when hybridomas were grown in ascites fluid (data not shown).

Direct immunoflourescence studies demonstrated binding of P38 antibody to a number of fresh and cryopreserved retionoblas-

toma tissue and single cell suspensions, but not to melanoma tissue, pooled allogeneic retina, or different lymphocytes of known HLA A, B, C, and D loci.

The isotypes of the monoclonal antibodies which reacted only with retinoblasoma cells were IgG$_1$ in three cases and an IgM in one case. HLA and Ia-like antigen cytotoxicity typing studies did not demonstrate any anti-HLA or anti-Ia activity in any monoclonal antibodies that had specific antiretinoblastoma reactivity (courtesy of H. Perkins, UCSF).

Discussion

We used murine somatic cell hybridization techniques to produce four monoclonal antibodies which reacted specifically with retinolastoma-associated antigens. The exact nature of the antigens towards which these antibodies were reactive remains unclear. None of the antibodies was positive in the chromium release cytotoxicity assay, immunofluorescence (data not shown). ELISA assays with control extracts, fresh,

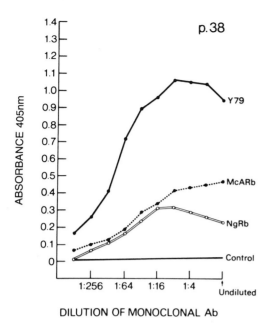

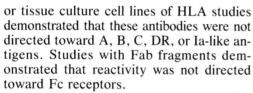

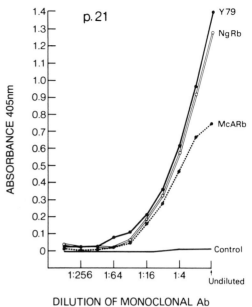

Fig. 1. Cross-reactivity pattern and titers of P-38 monoclonal antibody reactivity with three retinoblastoma cell lines (Ng, McA, Y-79).

Fig. 2. Cross-reactivity pattern and titers of P-21 monoclonal antibody reactivity with three retinoblastoma cell lines (Ng, McA, Y-79).

or tissue culture cell lines of HLA studies demonstrated that these antibodies were not directed toward A, B, C, DR, or Ia-like antigens. Studies with Fab fragments demonstrated that reactivity was not directed toward Fc receptors.

Monoclonal antibodies can react with small antigenic subunits. Untoward cross reactivity can occur secondary to different tissue, tumor, viral, or fetal antigens sharing antigenic substructures.[6–8]

Two groups have reported that monoclonal antibodies produced as a result of immunizations with human neuroblastoma or melanoma had cross reactivity with retinoblastoma, neuroblastoma, and glioblastoma cell lines. [9, 10] We did not observe this cross reactivity in our studies. This difference was probably due to a random selection of hybridoma clones which produced monoclonal antibodies not directed towards shared antigenic determinants. Other groups obtained results similar to ours in other tumor systems. Monoclonal antibodies have been described that are reactive with allogeneic human tumors.[11–13]

Acknowledgments

Supported in part by an unrestricted grant from That Man May See, and from NIH grants EY01441, EY01759, EY03675, EY02162, and GMO 7618-02. Dr. Char is a recipient of NIH Research Career Development Award K04-EY00117.

References

1. Char DH *Immunology of Uveitis and Ocular Tumors*, Grune & Stratton, New York 1978.
2. Stein PC et al.: *Invest Ophthalmol Vis Sci* **21**:550–553, 1981.
3. Milstein C et al.: *Cell Biol Int Rep* **3**:1–16, 1979.
4. Char DH et al.: *Invest Ophthalmol Vis Sci* (in press), 1983.
5. Mittal KK et al.: *J. Transplantation* **6**:913–927, 1968.
6. Crawford LV et al.: *Proc Natl Acad Sci USA* **78**:41–45, 1981.
7. Brown JP et al.: *Proc Natl Acad Sci USA* **78**:539–543, 1981.
8. Dippold WG et al.: *Br J Cancer* **43**:561, 1981.
9. Liao SK et al.: *Eur J Immunol* **11**:450–454, 1981.
10. Kenneth RH and Gilbert F: *Science* **203**:1120–1121, 1979.
11. Carrel S et al.: *Cancer Res* **40**:2523–2528, 1980.
12. Schnegg JF et al.: *Cancer Res* **41**:1209–1213, 1981.
13. Imai K et al.: *Transpl Proc* **12**:380–383, 1980.

chapter 11

The Role of Antibody and Mononuclear Cells in the Pathogenesis of Recurrent Herpes Simplex Uveitis

In vitro Production of Chemotactic Factor

Jang O. Oh, Petros Minasi, and Marcela Kopal

Our previous study indicated that the interaction of herpes simplex virus (HSV) antigen, sensitized lymphocytes, and anti-HSV antibody is essential in the induction of recurrent herpes simplex uveitis in rabbits.[1] The data to be presented in this paper show that the interaction of these three immune components are also required in the *in vitro* inductions of chemotactic factors for polymorphonuclear leukocytes (PMNs) and macrophages.

Materials and Methods

We used partially purified, type 1 HSV, Shealey strain, with an infectivity of 10^8 50% tissue culture infectious dose $(TCID_{50})$.[2] HSV antigen was made by inactivating the virus by ultraviolet light as described in a previous paper.[1] To obtain sensitized lymph node (LN) cells, inbred rabbits (III/J strain) were inoculated intravitreally in both eyes with 0.05 ml of HSV containing $10^3 TCID_{50}$, and on postinfection day 12, superficial and

Francis I. Proctor Foundation, University of California, San Francisco.

deep cervical LNs were aseptically dissected out. LN cells were separated according to a previous method [1] and 10^6 LN cells in 1 ml of RPMI medium were then mixed with various combinations of HSV antigen, anti-HSV rabbit serum, or normal rabbit serum as outlined in Tables I and II. The sera were inactivated at 56°C for 1 hour, and the concentration of serum in the medium was 2%. They were incubated at 36°C for 48 hours, and supernatant fluids from these cultures were tested for chemotactic activity for rabbit peritoneal PMNs and macrophages in a modified Boyden's chemotaxis chamber.[3] To quantitate the chemotactic activity for PMNs, the distance of migration from the top of the filter was measured in micrometers by determining the leading front of the cells in the filter as outlined by Zigmond and Hirsch.[4] For macrophages, the cells that migrated completely through the filter onto its distant surface were magnified × 600, visually identified, and counted. Five random fields were used to quantitate the chemotactic activity, and an average value per field was calculated from duplicate tests of each specimen.

TABLE I
Chemotactic Activity of Various Reaction Products

Reaction Product	Chemotaxis for PMN[a]	
	Exp. 1	Exp. 2
LN + HSV AG + Antiserum	52.6	51.4
LN + HSV AG + Normal Serum	30.8	31.0
LN + Normal Serum	30.2	29.2
HSV AG + Antiserum	ND	45.0
MEDIUM + Normal Serum	35.6	22.8

[a] Leading front in *um*.

TABLE II
Chemotactic Activity of Various Reaction Products

Reaction Product	Chemotaxis for MØ*	
	Exp. 1	Exp. 2
LN + HSV AG + Antiserum	193.0	30.0
LN + HSV AG + Normal Serum	2.4	9.2
LN + Normal Serum	0.4	0.6
HSV AG + Antiserum	ND	4.5
Medium + Normal Serum	3.4	1.2

[a] No. of macrophages per field at × 600

Results and Discussion

As shown in Tables I and II, strong chemotactic activity for PMNs and macrophages was noted in the culture medium of sensitized LN cells mixed with HSV antigen and anti-HSV serum. Sensitized LN cells with only HSV antigen or sensitized LN cells alone failed to produce chemotactic factors. The mixture of HSV antigen and anti-HSV serum without sensitized LN cells also failed to exert any significant chemotactic activity for PMNs and macrophages. Thus, we have shown that the *in vitro* production of chemotactic factors by sensitized lymphocytes with HSV is dependent on the presence of HSV antigen as well as anti-HSV antibody. Further work will be needed to establish which cell or group of cells are involved in the production of chemotactic factor in this rabbit system.

Viability of LN cells after a 2-day incubation with HSV antigen, anti-HSV serum (or without them) ranged between 80 and 92% as indicated by dye-exclusion method.

Recurrent herpes simplex (HS) uveitis is an important eye disease in man, yet its exact pathogenic mechanisms are poorly understood. In our previous study[6], we showed that the rabbits that had had primary HS uveitis aquired both humoral and cellular immunity against HSV. The affected eyes showed high concentrations of antibody against HSV in both aqueous humor and vitreous humor and a heavy infiltration of lymphocytes in the uveal tissue. The interaction of these lymphocytes and antibody with HSV antigen appears to induce uveal inflammation of this animal model.[1]

The events that lead to the uveal inflammation following the interaction of HSV antigen, antibody, and sensitized lymphocytes are still unknown. However, our present experiment clearly shows that chemotactic activities for both PMNs and macrophages were detected in the culture medium of sensitized LN cells only when these cells were cultured with HSV antigen and anti-HSV serum. Therefore, chemotaxis of these inflammatory cells to the reaction site (uvea) following interaction of these three immune components appears to be a contributing factor in the pathogenesis of recurrent herpetic uveitis.

In summary, the *in vitro* production of chemotactic factors for PMNs and macrophages by sensitized lymphocytes was studied using inbred rabbits (III/J strain). A significant chemotactic activity for both PMNs and macrophages was detected in the culture medium of sensitized LN cells only when LN cells were incubated with HSV antigen and anti-HSV serum.

Acknowledgment

This work is supported in part by USPHS research grants, EY-00964 and EY-01578 from the National Eye Institute.

References

1. Oh JO et al.: *Invest Ophthalmol Vis Sci Suppl* **22**:99, 1982.
2. Oh JO: *Invest Ophthalmol Vis Sci* **17**:769, 1978.
3. Oh JO and Kopal M: *Invest Ophthalmol Vis Sci* **18**:206, 1979.
4. Zigmond SH and Hirsch JG: *J Exp Med* **137**:389, 1973.
5. Wilson BS et al.: *J Immunol* **116**:1306, 1976.
6. Oh JO: *Surv Ophthalmol* **21**:178, 1976.

chapter 12

T-Lymphocyte Subsets in Patients with Herpes Simplex Ocular Infection

Joseph Colin, Pierre Youinou, Pierre Miossec, Yvon Pennec, Francine Colin, and Claude Chastel

Cell mediated immunity has been implicated as one of the most important factors in the pathogenesis of herpes simplex virus (HSV) keratitis.[1] As HSV can affect the cornea in many ways, the aim of this work was to study the T-lymphocyte subsets in patients with superficial or stromal keratitis, either primary or recurrent, compared to healthy individuals.

Methods

Studies were performed on 60 patients (37 males, 23 females, mean age: 51, range: 7–73 years). During the active phase of an ocular HSV infection, there were 37 patients with superficial dendritic ulcers and 23 with active stromal keratitis or kerato-uveitis, 20 with a primary attack and 40 with recurrences of the corneal disease. The control group consisted of 101 healthy individuals (mean age: 47, range: 21–78 years) without evidence or history of HSV infection.

Lymphocytes were isolated by Ficoll-Hypaque density gradient centrifugation. Adherent cells were removed following a 45 minute incubation on plastic petri dishes.

Total E-rosette forming cells (E-RFC) were enumerated using JF Bach's method[2] with sheep red blood cells (SRBC). The lymphocyte ratio was 1:30. Active-E-rosette-forming cells (E-act-RFC) were

enumerated according to the Wybran and Fudenberg method[3] with an SRBC per lymphocyte ratio of 1:8, and immediate determination. The results were expressed in mean percentages $\pm$ 1 SD and the comparison was made using Student t test.

Results

Peripheral blood lymphocyte counts were done, and the mean lymphocyte count in patients with ocular HSV infections was shown to be similar to that of the control group (Tables I and II).

The mean percentage of total E-RFC of all patients was lower than that of the control group ($p < 0.01$, Table I). The levels are lower in patients with stromal involvement ($n = 23$, $p < 0.02$), than in patients with superficial keratitis only ($n = 32$, $p < 0.05$), when compared to the control group. When considering their previous history, patients with primary infection had normal results, and those with recurrent infection had significantly decreased levels ($n = 40$, $p < 0.001$, Table II).

The mean percentage of E-act-RFC of the patients was lower than that of the control group ($p < 0.01$, Table I). The decrease depended on the lower results for stromal keratitis ($p < 0.001$ vs. controls), considering that the epithelial keratitis results were not different from those of the controls. Moreover, patients with recurrent infection had lower levels than the controls ($p < 0.01$), yet patients with primary infection were al-

Departments of Ophthalmology, Immunology, and Virology, Brest University Hospital, Brest, France.

TABLE I
Total Lymphocyte Count, E-RFC, and E-act-RFC in Percentage of Patients with HSV Keratitis vs. Controls

	Lymphocyte Count	E-RFC	E-act-RFC
Controls n = 101	2384 ± 827	63.7 ± 5.8	24.1 ± 4.7
HSV Keratitis n = 60	2965 ± 1608[NS]	59.6 ± 10.6[a]	20.1 ± 9.4[a]
Superficial Keratitis n = 37	2720 ± 1539[NS]	60.2 ± 9.8[a]	23.5 ± 7.9[NS]
Stromal Keratitis n = 23	3575 ± 1740[NS]	58.7 ± 8.9[b]	16.5 ± 8.5[c]

NS = not significant; [a] = $p < 0.05$; [b] = $p < 0.02$; [c] = $p < 0.001$.

TABLE II
Total Lymphocyte Count, E-RFC and E-act-RFC in Percentage of Patients with Primary or Recurrent HSV Ocular Infections vs. Controls

	Lymphocyte Count	E-RFC	E-act-RFC
Controls n = 101	2384 ± 827	63.7 ± 5.8	24.1 ± 4.7
Primary n = 20	3284 ± 1869[NS]	62.1 ± 8.7[NS]	21.1 ± 7.4[a]
Recurrent n = 40	2846 ± 1529[NS]	53.6 ± 6.4[c]	19.7 ± 8.2[b]

NS = not significant; [a] = $p < 0.05$; [b] = $p < 0.01$; [c] = $p < 0.001$.

most equal to the controls ($p < 0.05$, Table II).

When considering the clinical features of HSV keratitis and T-lymphocyte subsets, we found that (1) stromal keratitis cases showed decreased E-RFC and particularly E-act-RFC levels, and (2) epithelial keratitis cases were not clearly distinguishable from the controls.

The patients with primary infection were found to be almost the same as the controls. On the other hand, patients with recurrent infections had decreased E-act-RFC and were above all the E-RFC levels. Thus, the most severe cases of keratitis, including both recurrent and stromal keratitis, seem to show abnormalities of cellular immunity.

Discussion

The decrease of E-FRC and E-act-RFC levels in stromal keratitis may represent one of several systemic cell-mediated immuno-

deficiencies encountered in other viral or malignant diseases. The reason for the recurrence of HSV infections is still unclear, as abnormalities are found in the same patients even in the quiescent stage.[4] Longitudinal studies may explain whether the impairment of the host defenses plays a role in the recurrence or whether it represents the immunological consequence of this impairment, due to persistent and high HSV virulence.[5] Some abnormalities that we found with easily used methods may represent indicators with prognostic value and may be of help in the choice of the treatment.

References

1. Easty DL et al: *Br J Ophthalmol* **65**:82–88, 1981.
2. Bach, JF: *Transplant Rev* **16**:196, 1973.
3. Wybran J and Fudenger HH: *J Clin Invest* **52**:1026, 1973.
4. Centifanto YM et al.: *Invest Ophthalmol Vis Sci* **17**:863–868, 1978.
5. Kaufman, HE: *Am J Ophthalmol* **94**:119–121, 1982.

chapter 13

T-Cell Subsets and Langerhans' Cells in Conjunctival Lesions

Leslie S. Fujikawa, Atul K. Bhan, and C. Stephen Foster

Characterization of lymphocyte subsets with distinct immunologic functions at the site of inflammatory reactions may provide valuable information regarding the pathogenesis of the disease process. The development of monoclonal antibodies against human T-cells and T-cell subsets has made it possible to differentiate the functional subsets of human lymphocytes in tissue sections. In this study, we utilized a series of monoclonal antibodies to analyze the infiltrating lymphocytes in frozen tissue sections of inflammatory conjunctival lesions.

Materials and Methods

Conjunctival specimens. Specimens were obtained from the inferior bulbar conjuctiva of 14 patients and were frozen after embedding in OCT (ornithine carbamyl transferase) compound. We studied conjunctiva from four cases of active vernal conjunctivitis, four cases of active ocular pemphigoid, two cases of active Mooren's ulcer, one case of conjunctival inflammation associated with chronic graft-vs.-host disease after bone marrow transplantation for aplastic anemia, and three normal conjunctiva.

Monoclonal antibodies. We used a series of monoclonal antibodies that react with T-

Department of Ophthalmology, Pacific Medical Center, San Francisco, California; Department of Pathology, Massachusetts General Hospital, The Eye Research Institute of Retina Foundation, Harvard Medical School; and the Massachusetts Eye and Ear Infirmary, Boston, Massachusetts.

lymphocyte surface antigens.[1, 2] In brief, the monoclonal antibodies anti-T1 and anti-T3 react with all peripheral T-cells and intensely with 10% of thymocytes.

Anti-T6 antibody reacts with 70% of thymocytes but not with peripheral T-cells. Anti-T6 antibody also stains Langerhans' cells in the skin and lymph nodes.[3]

Anti-T4 antibody reacts with about 60% of circulating T-cells and defines inducer/helper T-cells. Anti-T8 antibody reacts with about 30% of circulating T-cells and defines suppressor/cytotoxic T-cells. We also used another monoclonal antibody, anti-T4B (clone No. SFC1-12T4D11), which stains the same cell population as anti-T4, but more intensely.[4]

Anti-T10 antibody reacts with most thymocytes, a significant proportion of activated T-cells, and only a minority of peripheral T-cells.[5] Anti-T10 antibody also stains a subpopulation of B-cells and the cytoplasm of plasma cells in tissue sections,[2, 6] and is therefore not specific for T-cells.

Anti-Ia monoclonal antibodies (anti-I1 and anti-I2) react with a nonpolymorphic region of human Ia-like antigens (HLA-DR)[8] and stain a wide variety of cells, including activated T-cells, marcrophages, Langerhans' cells, and endothelial cells. Anti-IgM monoclonal antibody was used to identify B-cells with surface or cytoplasmic IgM.[6]

Immunoperoxidase technique. A four-step immunoperoxidase technique employing peroxidase-antiperoxidase complexes was used as described previously.[8] In brief, 4-μ

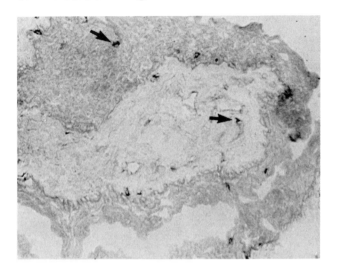

FIG. 1. Suppressor/cytotoxic T (T8 +) cells in normal human conjunctiva. There are T8 + cells in the epithelium and substantia propria (arrows) × 150. (From Bhan et al., published with permission from *The American Journal of Ophthalmology* **94**: 205–212, 1982. Copyright by The Ophthalmic Publishing Company.)

thick frozen sections were fixed in acetone, incubated with ascitic fluid containing the monoclonal antibodies, and then incubated with rabbit antimouse immunoglobulins, and peroxidase–rabbit–anti-peroxidase reagent. Sections were stained with 3-amino-9-ethyl-carbazole, [6, 8] which gives a red-brown product.

Results

Monoclonal antibodies anti-T1, anti-T3, anti-T4, anti-T4B, and anti-T8 stained the periphery of specific cells in a manner compatible with staining of the lymphocyte membrane. Together, these antibodies stained most of the lymphocytes in the specimens, indicating that most of the lymphocytes were T-cells. Anti-IgM antibody produced peripheral staining of cells, in a pattern similar to that obtained with small B-lymphocytes in primary follicles and the mantle-zone of secondary follicles in lymph nodes.[6]

Normal conjuctiva. Most of the stained cells in the substantia propria were T10 + plasma cells. Few cells stained with anti-T4 antibodies (define inducer/helper T-cells), and anti-T8 antibody (which defines suppressor/cytotoxic T-cells) stained only slightly more cells in the substantia propria than did anti-T4 antibodies. T8 + cells were also found in and adjacent to the epithelium

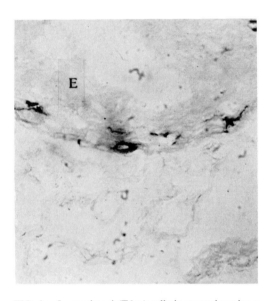

FIG. 2. Langerhans' (T6 +) cells in normal conjunctiva. T6 + cells are present in the epithelium of normal conjunctiva (E) × 400. (From Bhan et al., published with permission from *The American Journal of Ophthalmology* **94**: 205–212, 1982. Copyright by The Ophthalamic Publishing Company.)

(Fig. 1). There was little or no staining with anti-IgM antibody, indicating that few IgM + B-cells were present in these normal conjunctival specimens.

The cytoplasm of large cells with dendritic processes in the epithelium stained with anti-T6 antibody (Fig. 2), and these were considered to be Langerhans' cells. There were no lymphocytes or other conjuctival cells that stained with this antibody.

Anti-Ia antibodies stained a few large cells (Langerhans' cells) in normal conjunctival epithlium. Staining in the substantia propria was restricted primarily to endothelial cells of blood vessels and a few cells in the stroma considered to be "fibrohistiocytic cells."

Vernal conjunctivitis. Infiltrates composed of a large number of mononuclear cells and a few eosinophils were present in the lesions of vernal conjunctivitis. T4+ (inducer/helper) cells predominated in these infiltrates (Fig. 3). A large number of T8+ (suppressor/cytotoxic) cells were also present in the infiltrate but these were less numerous than the T4+ cells.

B-cells and plasma cells were both increased above normal levels in the substantia propria. Langerhans' cells, identified by staining with anti-T6 antibody, appeared to be slightly increased in the substantia propria in one case. In adddition to endothelial cells, a variable number of cells in the cellular infiltrate and "fibrohistiocytic cells" were Ia+.

Ocular pemphigoid. Large mononuclear infiltrates were present in the substantia propria in which T4+ cells predominated. The T4+ cells were found in the infiltrate, in groups around blood vessels, and scattered in the substantia propria near the epithelium. T8+ cells were more common than in the vernal conjunctivitis lesions, but there were more T4+ cells than T8+ cells. The two cell populations made up almost all of the mononuclear infiltrate.

A few B-cells stained with anti-IgM antibody were scattered in all specimens and were present in aggregates in only one specimen. Plasma cells, shown by anti-T10 antibody, were increased above normal levels. Langerhans' cells, defined by anti-T6 antibody, were increased both in the epithelium and in the substantia propria in two of the three cases (Fig. 3). Ia+ cells were mostly located in the substantia propria and included lymphocytes, large cells (fibrohistiocytic cells), and endothelial cells.

Mooren's ulcer. In these lesions, there was a mild to moderate mononuclear infiltrate in which T4+ and T8+ cells were both increased in the substantia propria (slightly more T4+ than T8+ cells). B-cells were present at normal levels. Plasma cells were increased in one specimen. A few Langerhans cells, defined by anti-T6 antibody, were present in the epithelium and in the substantia propria; and in one case, a large number of cells in the basal region of the surface epithelium also stained with anti-Ia antibody.

Chronic graft-vs.-host disease. In this tissue, the amount of fibrous tissue was greatly increased. A mild to moderate infiltrate of mononuclear cells, slightly increased over normal, was present. In this infiltrate, T8+ cells predominated over T4+ cells in the substantia propria (Fig. 4). B-cells and plasma cells were not increased in number. A few T6+ dendritic cells (Langerhans' cells) were found in the epithelium and substantia propria. Ia+ large cells similar to "fibrohistiocytic" cells were present in the substantia propria in increased numbers.

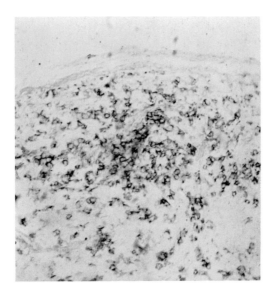

FIG. 3. Inducer/helper T (T4+) cells in vernal conjunctivitis. Large numbers of T4+ cells are present in the substantia propria ×256. (From Bhan et al.; published with permission from *The American Journal of Ophthalmology* **94**:205–212, 1982. Copyright by The Ophthalmic Publishing Company.

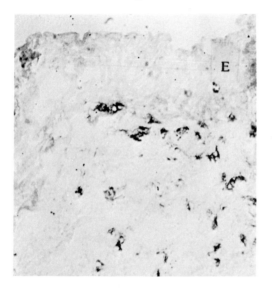

FIG. 4. Suppressor/cytotoxic T (T8+) cells in chronic graft-vs.-host disease. There are increased numbers of T8+ cells in the epithelium (E) and substantia propria ×400. (From Bhan et al.; published with permission from *The American Journal of Ophthalmology* **94**:205–212, 1982. Copyright by the Ophthalmic Publishing Company.

Discussion

In normal conjunctiva, most of the cells in the substantia propria were T10+ plasma cells. A few T8+ suppressor/cytotoxic cells were found under the epithelium. T8+ cells are also present in the mucosal epithelium of tonsils and intestines. (A.K. Bhan,[6] unpublished data), the significance of which is not clear. Dendritic cells, scattered in the epithelium, stained diffusely with anti-T6 antibody, and were considered to be Langerhans' cells.[3] A few T4+ cells and IgM+ cells were found in normal conjunctiva.

In all the specimens of diseased conjunctiva (vernal conjunctivitis, ocular pemphigoid, Mooren's ulcer, and graft-vs.-host disease), a cellular infiltrate was imposed upon the cellular profile for normal conjunctiva. In this infiltrate, T-cells were a major component, while IgM+ B cells were present in only small numbers. Plasma cells, defined by reactivity with anti-T10 antibody, were either scattered throughout the substantia propria or in aggregates.

In most diseased conjunctival lesions (vernal conjunctivitis, ocular pemphigoid, and Mooren's ulcer), inducer/helper T4+ cells formed the major component of the T-cells in the infiltrate. In contrast, T8+ (suppressor/cytotoxic) cells formed only a minor component of the infiltrate in all the above lesions. One possible explanation for the predominance of T4+ inducer/helper T-cells in the cellular infiltrates is that a cell-mediated reaction of the delayed hypersensitivity type may be involved in the induction of these lesions. This is supported by the presence of Langerhans' cells in some of these lesions since these cells function as antigen-presenting cells to T-lymphocytes.[9] Vernal conjunctivitis is considered to be allergic in origin, and is probably mediated in part by IgE antibody. The predominance of T4+ cells in the cellular infiltrate, along with the results of other studies [10] in which basophils with associated endothelial changes and fibrin deposition were found, suggests that cell-mediated reactions of the delayed hypersensitivity type may also be involved.

In ocular pemphigoid and Mooren's ulcer, T4+ cells may also be involved in the induction of the lesions. However, there is no evidence to date that suggests that delayed hypersensitivity plays an important role in the pathogenesis of these lesions. Ocular pemphigoid is considered to be an auto-immune disease because patients have circulating or bound antibodies to the epithelial basement membrane.[11] T4+ cells may play an important role in the generation of auto-antibodies and in the pathogenesis of the lesions. The cause of Mooren's ulcer is unknown and no consistent immunologic abnormality has been found.[12]

The predominance of T8+ cells in a case of graft-vs.-host disease suggests a role of cytotoxic/suppressor T-cells in the induction of conjunctival lesions in the disease. However, it is important to note that our methods do not distinguish between cells that initiate the reaction and cells that may accumulate secondarily because of the liberation of factors at the reaction site.[13] Both cell groups would be expected in an active lesion, while in the resolving stage of a lesion, the cells responsible for inducing injury may have disappeared. In the case of graft-vs.-host disease, T8+ cells may not be responsible for inducing fibrosis in the substantia propria, but may function as suppressor cells and may help control a reaction originally induced by other (T4+) cells. Further studies, including specimens from sequential biopsies obtained early in and throughout the clinical course of these diseases, may help determine the exact role of T-cell subsets in these and other ocular inflammatory diseases.

Acknowledgments

This study was supported in part by grants AM-18729, HL-18646, CA-29601, EY-03063, and RR-05527 (Drs. Foster and Bhan) from the National Institutes of Health; by National Eye Institute Fellowship F32-05405 (Dr. Fujikawa); and by grants from Fight for Sight, Inc., New York, and the Massachusetts Lions Eye Research Funds, Inc.

References

1. Reinherz EL and Schlossman SF: *N Engl J Med* **303**:370, 1980.
2. Reinherz EL and Schlossman SF: *Cell* **19**:821, 1980.
3. Murphy GF et al.: *Lab Invest* **45**:465, 1981.
4. Reinherz EL et al.: *J Immunol* **124**:463, 1982.
5. Terhorst C et al.: *Cell* **23**:771, 1981.
6. Bhan AK et al.: *J Exp Med* **154**:737, 1981.
7. Bhan AK et al.: *J Exp Med* **152**:771, 1980.
8. Reinherz EL et al.: *J Exp Med* **150**:1472, 1979.
9. Stingl G et al.: *J Invest Dermatol* **74**:315, 1980.
10. Dvorak HF et al.: *Lab Invest* **31**:111, 1974.
11. Furey N et al.: *Trans Am Acad Ophthalmol Otolaryngol* **81**:806, 1976.
12. Foster CS et al.: *Am J. Ophthalmol* **88**:149, 1979.
13. McCluskey RT and Bhan AK: In *Mechanisms of Tumor Immunity*, Green I et al.(Eds.). New York, John Wiley and Sons, 1977, p. 1.

chapter 14

Immune Complex-Mediated Diseases

Mart Mannik

Many disorders are caused by immune complexes. The affected organs include the eyes, renal glomeruli, renal tubules, blood vessels of varying sizes, thyroid gland, choroid plexus, and many other organs. In a given organ the pathogenic immune complexes are either deposited from the circulation or the complexes are locally formed. In local formation of immune complexes, the antigens are part of the target organ, or antigens unrelated to the target organ are selectively deposited in the organ. Subsequently, circulating antibodies combine with the antigens to form immune complexes that then cause inflammation and tissue damage. Diseases that result from local formation of immune complexes generally are confined to one organ or tissue. In disorders mediated by circulating immune complexes, the complexes are formed in circulation or enter the circulation from other compartments. The circulating immune complexes then deposit in tissues and cause inflammation and tissue damage. The disorders caused by this mechanism tend to involve multiple organs in contrast to the diseases caused by local immune complex formation.

The purpose of this article is to review briefly some characteristics of antigen–antibody complexes and to consider one target organ as an example of mechanisms that lead to presence of immune complexes. The renal glomeruli were chosen as the example because, in recent years, considerable progress has been made in understanding how the presence of immune complexes develops in these structures.

Since the space for this review is limited, the reader may benefit from more extensive recent reviews that consider the characteristics and detection of immune complexes [1-3] and mechanisms of autoimmune diseases.[4]

Characteristics of immune complexes

The essential properties of immune complexes in relation to pathogenic mechanisms are complement activation, interaction with Fc and complement receptors on phagocytic cells, and tissue deposition. Immune complexes also alter lymphocyte functions and the immune response, but these properties are beyond the scope of this review. The listed properties of immune complexes are influenced by the ingredients of immune complexes, namely the antigens, the antibodies and the nature of the union between these reactants.

Antigen. This term is defined as a substance that interacts specifically with available antibodies or with sensitized lymphocytes. The actual portion of an antigenic molecule or substance that reacts with the antibody combining site or with the specific receptor on a sensitized lymphocyte is defined as the *antigenic determinant*. A given antigenic molecule may have one or more specific antigenic determinants or several different antigenic determinants. The number of antigenic determinants defines the antigenic valence of a molecule in respect to one antibody specificity; a given macromolecule may be multivalent with respect to one or more antibody specificities.

Division of Rheumatology Department of Medicine, University of Washington, Seattle, Washington.

Antibodies. These are the other essential ingredients of immune complexes and may belong to IgG, IgA, IgM, IgD or IgE classes of immunoglobulins. Antibodies have a valence of two except polymeric IgA and IgM, which have higher valences. Complement activation occurs principally by IgG and IgM classes of antibodies. The interaction with the Fc receptors is confined to IgG, but different receptors may exist for other classes of antibodies. Thus, it is evident that the nature of antibodies in a given immune complex will alter the biological properties of that complex. If the antibodies in the immune complex do not activate complement, then these complexes are not able to generate chemotactants and inflammation through complement activation. If the antibodies in an immune complex do not interact effectively with the Fc receptors on phagocytic cells, they may persist longer in circulation and the phagocytosis of these complexes may be impaired after deposition in tissues.

When the antigen–antibody union occurs, a variety of immune complexes may form, ranging from an immune complex of one antigen molecule and one antibody molecule (Ag_1Ab_1) to immune complexes containing many molecules of each reactant.

Lattice of immune complexes. This is defined as the number of antigen and number of antibody molecules in each complex. The lattice of an immune complex in turn influences its biological properties. For example, with increasing lattice complement activation, interaction with Fc receptors becomes more effective. Similarly, as will be discussed later, the lattice of immune complexes influences their deposition in tissues.

The lattice of immune complexes is influenced by the valence of antigens, the valence of antibodies, the ratio of antigen and antibody, the absolute concentration of each reactant, and the avidity between the reactants. If the valence of an antigen is one, then maximally one antigen can combine with each of the two antibody combining sites, yielding Ag_2Ab_1 complexes, and larger complexes cannot form. If the antigenic molecule has a valence of two, then a linear complex may form. Alternatively, if the distance between the two antigenic determinants is appropriate, then both antibody conbining sites will interact with antigenic determinants on the same molecule,[5] yielding an Ag_1Ab_1 complex. Only multivalent antigenic molecules can form large immune complexes or immune precipitates. As already mentioned, the valence of most antibodies is two. With a multivalent antigen, a precipitin curve can be constructed, yielding maximal precipitation at the so-called equivalence point. When increasingly more antigen is added, soluble immune complexes will form. At low degrees of antigen excess, the soluble complexes are large-latticed (greater than Ag_2Ab_2). With very high degrees of antigen excess, the complexes become small-latticed (Ag_1Ab_1 and Ag_2Ab_2) as described for human serum albumin (HSA) as antigen and rabbit antibodies to HSA.[6] When the concentrations of the reactants in such a system are markedly decreased to less than μg per ml, then at small degrees of antigen excess small-latticed complexes are more common than would be formed at higher concentrations of the same reactants. Finally, when the avidity between an antigen and the specific antibody are low, then small-latticed complexes are more likely to form than large-latticed complexes. This occurs even though the concentration and antigen–antibody ratio would favor formation of large-latticed immune complexes.

In addition to the lattice of immune complexes and the characteristics of antibodies, the nature of the antigen in immune complexes may alter the biological properties of immune complexes. For example, the removal of small-latticed immune complexes from circulation may be enhanced by the presence of DNA as the antigen, since DNA alone is quickly removed from circulation.[7, 8] Similarly, some asialoproteins are quickly removed from circulation due to interaction with galactose receptors on hepatocytes. Small-latticed immune complexes made with such antigens are also quickly removed from the circulation by hepatocytes.[9] As shown in experimental animals, the fate of circulating immune complexes depends, in addition, on host factors. For example, with increasing doses of immune complexes, the removal of immune complexes by the Kupffer cells (the mononuclear phagocyte system) became saturated, leading to prolonged circulation of the large-latticed complexes [10] and enhanced deposition in glomeruli.[11] Saturation of the mononuclear phagocyte system did not occur with small-latticed immune complexes.[12]

Glomerular deposition of immune complexes

Over 80% of glomerulonephritis in humans is caused by immune complexes or by antibodies to glomerular structures.[13] The presence of immune complexes in glomeruli can arise by two mechanisms. (1) Immune complexes may deposit in glomeruli from circulation. (2) Circulating antibodies can combine with antigens in glomeruli, thus leading to the presence of immune complexes by local formation. The second mechanism can be subdivided into local immune complex formation due to antibodies combining with structural antigens in the glomerular basement membrane and into local immune complex formation due to planted antigens. Goodpasture's syndrome is an excellent example of local immune complex formation due to antibodies combining with structural antigens in the glomerulus. On immunofluorescence microscopy the glomerular basement membrane stains diffusely for the presence of immunoglobulins. In contrast, the staining of immunoglobulins in lesions caused by immune complexes deposited from circulation and in lesions resulting from local formation of immune complexes with planted antigens tend to have a beaded or lumpy-bumpy pattern by immunofluorescence microscopy.

By electron microscopy the presence of immune complexes is recognized by the finding of electron dense deposits. These deposits may be present in the subendothelial, mesangial, or subepithelial area (Fig. 1). The mesangial deposits may occur alone, but when subendothelial deposits are present then mesangial deposits are found also. Studies in experimental animals have suggested that the subendothelial and mesangial deposits arise from circulating immune complexes. The subepithelial deposits, on the other hand, develop due to local formation of immune complexes. Evidence for these conclusions will be presented in brief.

In experimental models of chronic serum sickness—induced by chronic administration of antigen—subendothelial, mesangial, and subepithelial deposits of immune complexes were observed.[14] These deposits were associated with circulating immune complexes. The injection of preformed immune complexes into nonimmunized animals, however, revealed that the circulating complexes localized transiently in the subendothelial area and were principally found in the mesangial area. In these experiments no subepithelial deposits were found. These studies were conducted in mice using rabbit IgG antibodies to bovine serum albumin,[15] rabbit antibodies to human serum albumin [11] or rabbit antibodies to DNP wtih DNP-bovine serum albumin as the antigen.[16]

Furthermore, several lines of evidence indicated that only large-latticed immune complexes (greater than Ag_2Ab_2) deposited in the subendothelial and mesangial areas. First, glomerular deposition of immune complexes progressed only while large-latticed immune complexes persisted in circulation when a mixture of small-latticed and large-latticed complexes were injected into mice.[11] Second, when only small-latticed immune complexes, prepared with 50-fold excess antigen, were injected into mice, no glomerular deposits evolved.[17] Third, when preformed, large-latticed immune complexes were allowed to deposit in glomeruli and 12 or more hours later a large excess of antigen was administered to these mice, then the deposits in glomeruli were released.[18] This occurred presumably due to conversion of the large-latticed immune complexes to small-latticed complexes. These observations collectively indicated that the lattice of immune complexes has a significant role in the development of glomerular deposits of immune complexes.

Of note is that the electron-dense deposits of immune complexes as seen by electron microscopy were much larger than the injected complexes, suggesting that the injected complexes must have undergone further rearrangement or condensation to

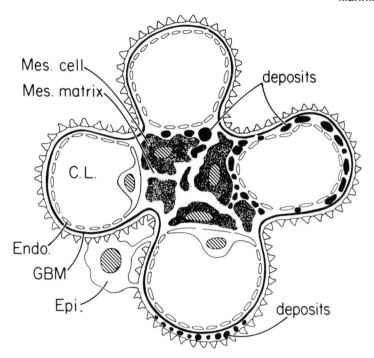

FIG. 1.   Schematic representation of glomerular capillary loops and the different locations of immune complexes. The glomerular capillary wall is a size and charge barrier for macromolecules with fixed negative charges along the glomerular basement membrane (GBM). The endothelial cells (Endo.) are fenestrated, thus permitting relatively easy access of macromolecules from the capillary lumen (C.L.) to the GBM. The foot processes of the epithelial cells (Epi.) cover the outer aspect of the GBM. The mesangial cells (Mes. cell) and mesangial matrix (Mes. matrix) occupy the space between the glomerular capillary loops. Immune complexes may occur in the mesangial matrix or in the subendothelial area along with deposits in the mesangial matrix as shown respectively in the capillary loops at 12 o'clock and 3 o'clock positions. Subepithelial deposits, as shown in the capillary loop at 6 o'clock position, develop by local formation of immune complexes.

become the electron-dense deposits. In order to provide experimental proof for this, an antigen–antibody system was utilized in which the preformed immune complexes could be covalently cross-linked in the antibody combining site to prevent rearrangement and very large complex formation.[19] Mice were injected with preparations of the noncross-linked and cross-linked complexes that had comparable degrees of lattice formation. During the first hour after injection both preparations showed comparable glomerular deposits by immunofluorescence microscopy. In mice that received the cross-linked complexes, the deposits declined after 4 hours and were absent at later times.

In contrast, in mice given the noncross-linked preparation the complexes persisted and evolved into large electron-dense deposits.[20] These observations indicated that only a precipitating antigen—antibody system leads to the formation of subendothelial and/or mesangial electron-dense deposits. Nonprecipitating immune complexes, with sufficient lattice, however, may transiently adhere to glomerular structures as recognized by immunofluorescence microscopy. But these complexes are not visualized by electron microscopy. Furthermore, each electron-dense deposit was immunospecific, as shown by simultaneous administration of ferritin and fibrinogen as antigens. The immune deposits that developed in glomeruli contained either ferritin or fibrinogen, and mixed deposits were not found.[21]

The above experiments clearly suggested that covalently cross-linked complexes attach transiently to golmerular capillary walls, even though they could not condense or rearrange into complexes that persisted. Since the glomerular capillary walls have fixed negative charges,[22] the possible role of the charge of antibodies in the complexes was examined on attachment to glomeruli. For this purpose antibodies to human serum albumin were cationized with ethylene diamine so that the isoelectric point was equal to or greater than 9.3. When soluble, large-latticed immune complexes were prepared with these positively-charged antibodies and injected into mice, complexes bound avidly to the glomerular capillary walls and persisted in glomeruli for several days. These complexes were present principally in the subendothelial area by electron microscopy and gradually translocated to the mesangial matrix.[23] When small-lattice complexes (Ag_1Ab_1 or Ag_2Ab_2) with cationized antibodies were prepared and injected into mice, these complexes persisted in glomeruli only for a short period of time, comparable to the cationized antibodies alone.[23] Thus, electrostatic interactions can contribute to the initial attachment of immune complexes to the glomerular capillary wall, but the importance of lattice and formation of large deposits remained evident.

Several kinds of experiments indicate that the immune deposits in the subepithelial area, as seen in membranous glomerulonephritis, arise from local formation of immune complexes. When antibodies to the Fx1A antigen, present in the brush border of proximal tubules and in the subepithelial area, were administered to rats in a noncirculating perfusion system, subepithelial immune deposits developed, indicating local formation of immune complexes.[25] Fixed-negative charges are present in the lamina rara externa of the glomerular basement membrane. When cationic proteins were administered to animals, these proteins remained attached to the basement membrane for a few hours,[26] but when antibodies were administered, the antigen and antibody persisted in the subepithelial area as immune complexes with a half-life of 12 days.[26] Finally, the development of subepithelial electron-dense deposits also requires that the involved antigen–antibody interaction is a precipitating system.[27] Thus, the available experimental data indicate that in human diseases the antigens that can plant in the subepithelial area of renal glomeruli most likely are cationic molecules smaller than 900,000 in molecular weight and that these antigens must be able to form precipitates with their specific antibodies.

The planting of antigenic molecules is not confined to the subepithelial area. Intravenously injected aggregated IgG or aggregated albumin accumulated in the mesangial area. If antibodies to the deposited proteins were administered later, immune complexes formed in the mesangium and acute inflammation ensued in the mesangial areas of glomeruli.[28] In addition, intravenously injected concanavalin-A bound to the glomerular capillary wall, and when antibodies to concanavalin-A were administered, lumpy-bumpy deposits of immune complexes developed.[29]

These comments summarize some features of antigen–antibody complexes and their biological properties. Renal glomeruli were used as an example of how the characteristics of immune complexes can alter their deposition in tissues. The mechanisms and variables involved in the development of immune deposits in the structures of the eye are not known. Some of the discussed approaches, however, may be useful in extending the knowledge of the mechanisms of immune complex-mediated diseases of the eye.

References

1. Haakenstad AO and Mannik M: In: *Autoimmunity* Talal N (Ed.). Academic Press, New York, 1977, pp. 277–360.
2. Zubler RH and Lambert PH: *Prog Allergy* **24**:1–48, 1978.
3. Theofilopoulos AN and Dixon FJ: *Adv Immunol* **28**:89–220, 1979.
4. Theofilopoulos AN and Dixon FJ: *Am J Pathol* **108**:319–365, 1982.

5. Crothers DM and Metzger H: *Immunochemistry* **9**:341–357, 1972.
6. Arend WP et al.: *Biochemistry* **11**:4063–4072, 1972.
7. Emlen W and Mannik M: *J Exp Med* **147**:684–699, 1978.
8. Emlen W and Mannik M: *J Exp Med* **155**:1210–1215, 1982.
9. Finbloom DS et al.: *J Clin Invest* **68**:214–224, 1981.
10. Haakenstad AO and Mannik M: *J Immunol* **112**:1939–1948, 1974.
11. Haakenstad AO et al.: *Lab Invest* **35**:293–301, 1976.
12. Jimenez RAH et al.: *Immunology* **48**:205–210, 1983.
13. Wilson CB and Dixon FJ: *Kidney Int* **5**:389–401, 1974.
14. Germuth FG Jr and Rodriguez E: *Immunopathology of the Renal Glomerulus*, Little Brown and Co, Boston, 1973.
15. Okumura K et al.: *Lab Invest* **24**:383–391, 1971.
16. Koyama A et al.: *Lab Invest* **38**:253–262, 1978.
17. Haakenstad AO et al.: *Immunology* **47**:407–414, 1982.
18. Mannik M and Striker GE: *Lab Invest* **42**:483–489, 1980.
19. Mannik M and David KA: *J Immunol* **127**:1999–2006, 1981.
20. Mannik M et al.: *Exp Med* **157**:1516–1528, 1983.
21. Kubeš L: *Virchows Arch B Cell Pathol* **24**:343–354, 1977.
22. Brenner BM et al.: *N Engl J Med* **298**:826–833, 1978.
23. Gauthier VJ et al.: *J Exp Med* **156**:766–777, 1982.
24. Couser WG et al.: *J Clin Invest* **62**:1275–1287, 1978.
25. Fleuren G et al.: *Kidney Int* **17**:631–637, 1980.
26. Oite T et al.: *J Exp Med* **155**:460–474, 1982.
27. Agodoa LYC et al.: *Exp Med* **158**:1259–1271, 1983.
28. Mauer SM et al.: *J Exp Med* **137**:553–570, 1973.
29. Golbus SM and Wilson CB: *Kidney Int* **16**:148–157, 1979.

chapter 15

Etiology of Scleritis

Narsing A. Rao, Terry M. Phillips, Vernon G. Wong, Anthony J. Sliwinski, and George E. Marak

Scleritis is a chronic, recurring inflammatory disease of unknown etiology which can lead to blindness from its complications.[1] It is one of the most severely destructive diseases of the eye.

Spontaneously occurring animal models for scleritis do not exist and the scleral inflammation is not present in the experimental arthritis produced by adjuvant or by type II collagen.[2–5] Recently we have been able to induce an anterior necrotizing scleritis in strain 2 guinea pigs that resembles necrotizing scleritis. The purpose of this report is (1) to describe the histologic features of the experimental scleritis, and (2) to show the immunologic differences between individuals with idiopathic scleritis and those with rheumatoid scleritis.

Materials and Methods

Development of an animal model. Thirty adult strain 2 guinea pigs were sensitized by intradermal injection of 1 mg of bovine soluble scleral extract in complete Freund's adjuvant. A total of three injections were given at weekly intervals. The soluble scleral antigen was prepared by homogenizing sclera with small volumes of distilled water in an ultramixer (Ivean, Sorvall, Newton, CT). The homogenate was centri-

Georgetown University Medical Center and George Washington University Medical Center, Washington, D.C.; and the Estelle Doheny Eye Foundation, Los Angeles.

fuged for 10 minutes at 10,000 g. The protein concentration of the supernatant was determined by the Lowry method.[6]

The animals were examined twice a week for the development of ocular complications. Two animals developed circumcorneal injection of the vessels, loss of hair, and loss of body weight. These two were sacrificed and the eyes, joints, and kidneys were obtained for histopathologic study. One other animal showed engorged conjunctival vessels followed by slight proptosis. This animal was sacrificed 8 weeks after the last injection of the antigen. For controls, 30 strain 2 guinea pigs were injected with complete Freund's adjuvant intradermally, followed clinically, and sacrificed along with the above experimental animals.

Immunological investigations in patients with scleritis. Three patients with rheumatoid necrotizing scleritis and two patients with idiopathic necrotizing scleritis were evaluated for the presence of circulating autoantibodies and immune complexes. The complexes were detected by polyethylene glycol precipitation, Raji cell method, and C1q precipitation. Following recovery, these complexes were dissociated in acid and separated by double electrophoresis in agarose.[7] The specific reactivity of the antibody (obtained from the complexes) to a soluble human scleral extract was determined. Peripheral leukocytes from these patients were also studied for lymphocyte proliferation in the presence of human

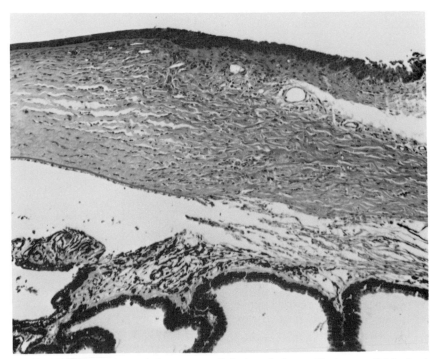

FIG. 1. Infiltration of acute inflammatory cells in the episclera and in the superficial scleral lamellae near the limbus. (H&E ×100.)

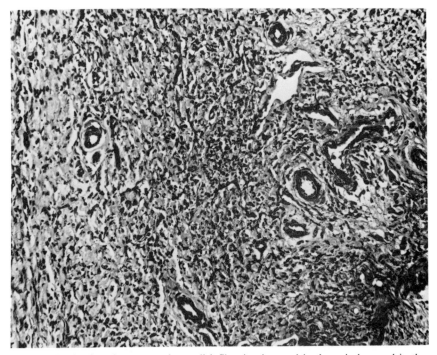

FIG. 2. Predominantly mononuclear cell infiltration is noted in the episclera and in the necrotic sclera. (H&E ×160.)

TABLE I

Immunological Investigations in Patients with Rheumatoid and Idiopathic Scleritis

Type of Investigation	Three Patients with Rheumatoid Scleritis	Two Patients with Idiopathic Scleritis
Circulating immune complexes as detected by:		
(1) P.E.G.[a]	Present	Present
(2) Raji Cell	Present	Present
(3) C1q	Present	Present
Nature of antibody in complexes	Complement binding IgM type	Complement binding IgM type
Antibody obtained from the dissociated complexes	Nonreactive to human soluble scleral extract	Reactive to human soluble scleral extract
Delayed hypersensitivity to the human soluble scleral antigen as determined by proliferation index of peripheral leukocytes in the presence of the scleral extract	Absent	Present

[a]P.E.G. = polyethylene glycol precipitation.

scleral extract in a manner similar to the bovine scleral antigen. For controls, soluble extracts of human retina and choroid were used.

Results

Histologicaly, the four eyes obtained from the two animals with clinical signs of circumcorneal injection and weight loss showed infiltration of acute inflammatory cells in the episclera and in the superficial scleral lamellae near the limbus. Inflammatory cells were also present in the adjacent corneal stroma (Fig. 1). The overlying conjunctival and corneal epithelium were intact.

The other animal with proptosis showed marked infiltration of lymphocytes and other mononuclear cells in the episcleral tissue and in the adjacent necrotic scleral lamellae (Fig. 2). There was mild lymphocytic infiltration of the iris and ciliary body, and no evidence of vasculitis. The sections of the joints and kidneys from the experimental animals were unremarkable. None of the control animals developed clinical or histopathologic scleritis.

In the patients with rheumatoid scleritis, there was no evidence of cell-mediated immune responses to scleral tissues. The patients with idiopathic scleritis did demonstrate a cell-mediated response to scleral

antigen. Only nonspecific immune complexes were detected in the sera of patients with rheumatoid scleritis, whereas the patients with idiopathic scleritis had specific antiscleral immunoglobulin as a component of the isolated complexes (Table I). None of the patients' peripheral leukocytes showed proliferation in the presence of retinal or uveal extracts.

Discussion

For understanding the etiology of scleritis, a new approach recognizing that autoimmune disease results from an altered self-tolerance should be considered. We could abrogate the tolerance to scleral antigen in strain 2 guinea pigs by repeated injections of bovine soluble scleral extract in complete Freund's adjuvant. In some of the guinea pigs the disease was characterized by the development of an acute scleritis in some animals and a chronic disease process in one animal. Histologically, the scleritis begins as an episcleral inflammation with subsequent extension into the adjacent sclera. The inflammation is associated with the peripheral keratitis and a mild anterior uveitis. Histopathologically, the induced scleritis is very similar to human idiopathic necrotizing scleritis (Table II).

The immunologic investigations in our patients with scleritis indicate distinct dif-

TABLE II
Histopathologic Similarities of Experimental Scleritis and Human Idiopathic Necrotizing Scleritis

Histopathologic Findings	Experimental Scleritis	Human Idiopathic Necrotizing Scleritis
Diffuse infiltration of chronic inflammatory cells in the episclera and the sclera	Present	Present
Necrosis of the sclera at the site of inflammation	Present	Present, and it is a prominent feature
Anterior uveitis	Present	Present
Vasculitis of the episcleral vessels	Absent	Absent

ferences between immunopathogenesis of rheumatoid scleritis and idiopathic scleritis (Table I). Rheumatoid scleritis does not appear to be a specific autoimmune response to sclera but rather a manifestation of systemic disease. The specific reactivity of the dissociated antibody (from the immune complexes) to the human soluble scleral extract and the presence of delayed hypersensitivity specific to the human sclera in patients indicate that abrogation of tolerance to a scleral antigen may play a significant role in the pathogenesis of idiopathic scleritis. Further studies on patients with scleritis and in-depth immunopathologic investigations in experimental models induced by the scleral antigens are required. It would enable us to evaluate the significance of immune complexes and the role of delayed hypersensitivity in the pathogenesis of idiopathic necrotizing scleritis.

Acknowledgment

This work was supported in part by an unrestricted research grant from Research to Prevent Blindness, Inc., New York City.

References

1. Watson PG and Hayreh SS: *Br J Ophthalmol* **60**:163–190, 1976.
2. Trentham, DE et al.: *J Exp Med* **146**:857, 1977.
3. Caulifield JP et al.: *Lab Invest* **46**:321–343, 1982.
4. Trentham DE et al.: *J Clin Invest* **66**:1109, 1980.
5. Wacksman BH and Bullington SJ: *Arch Ophthalmol* **64**:751–762, 1960.
6. Lowry OH et al.: *J Biol Chem* **193**:265, 1951.
7. Pesce AJ et al.: *J Urology* **123**:486–488, 1980.

chapter 16

Ocular Localization of Preformed Immune Complexes

H. A. Hylkema and A. Kijlstra

Uveitis is a relatively common disease of uncertain etiology. A role for circulating immune complexes in human uveitis emerges from several clinical studies.[1-4] Due to the fact that biopsies cannot be taken to confirm the presence of immune complexes in the uvea of patients, the elucidation of an immune complex pathogenesis comes mainly from animal studies. The ciliary process with its filtering function resembles the glomerulus in the kidney and could, in analogy with the latter, be the site of immune complex deposition. Experimental studies using the serum sickness model have indeed shown that complexes can be localized in the ciliary body of rats and rabbits.[5,6] In the investigation reported here, the ocular deposition of immune complexes made at four different antibody-to-antigen ratios was studied. These immune complexes were made in vitro and subsequently injected intravenously into mice.

Materials and Methods

Bovine thyroglobulin (BTG) was extracted with saline from fresh thyroids kept at 0° and purified by gel filtration on Sephadex S300.[7] BTG was labeled with fluorescein isothiocyanate (FITC) at a fluorescein to protein ratio of 1:3.[8] Anti-BTG antibodies were raised in New Zealand rabbits by hyperimmunization, and anti-BTG purification

was carried out by immunoabsorption.[9] The preparation was shown to contain only 7 S antibody and no detectable IgM. Four different kinds of immune complexes were made and these complexes were injected into the tail vein of outbred Swiss mice (20–30 g).

Immune complex solutions were made at equivalence and at 5, 10, and 20 times antigen excess by adding one volume of various concentrations of the antibody (10, 2, 1, and 0.5 mg anti-BTG per ml) to one volume of a constant amount of antigen (10 mg BTG-FITC per ml). A volume of 400 μl was injected into the tail vein immediately after mixing of the reagents. Control mice received 400 μl BTG (5 mg/ml) alone. The mice were killed at various times after injection, and their eyes were removed and snap frozen in liquid nitrogen. Cryostat sections were cut, air dried on gelatin-coated slides, and fixed in acetone for 10 minutes. The presence of rabbit IgG and mouse C3 in the eye was demonstrated making use of tetramethylrhodamine isothiocyanate (TRITC) labeled goat antirabbit IgG (Nordic Tilburg, The Netherlands) and TRITC labeled goat antimouse C3 (U.S.B., Cleveland, Ohio). The deposition was considered to be an immune complex when the green fluorescence of the BTG-FITC and the red fluorescence for the antibody localized at exactly the same site.

Blood vessels were identified in the section by staining the vascular endothelial cells with rabbit antihuman blood clotting factor VIII[10] followed by a TRITC labeled goat

Department of Ophthalmo-Immunology, The Netherlands Ophthalamic Research Institute, Amsterdam.

The Effect of Antibody-to-Antigen Ratio on Ocular Deposition of Immune Complexes 1 Hour After Inoculation

| Ab/Ag | N | Number of mice with Icx deposition in | | | |
		Ciliary body	Intrascleral capillary area	Choroid	Iris
Equivalent	2	2	—	2	1
$\frac{1}{5}$	4	3	4	3	3
$\frac{1}{10}$	4	3	4	2	2
$\frac{1}{20}$	4	—	—	—	—

antirabbit IgG. All tissue sections were examined with a Leitz orthoplan fluorescence microscope. Using a double exposure technique, both the FITC labeled BTG and the TRITC labeled endothelial cells were photographed in one frame.

Results

Because the deposition of immune complexes gave the most clear cut results at 1 hour after injection (data not shown), the effect of the antibody-to-antigen ratio on the ocular deposition of immune complexes was determined at this time (Table I). Immune complexes made at equivalence, 5 or 10 times antigen excess tended to deposit in the ciliary body, the choroid, and the episcleral area near the ciliary body (Fig. 1). No depositions were seen in the retina. On the other hand, complexes made at 20 times antigen excess showed no detectable deposition in immune complex form and the fluorescence pattern in these cases was very much like that seen in the control mice, cloudy depositions of BTG-FITC. The antibody was identified using TRITC labeled goat antirabbit IgG. In all the eyes of mice that received immune complexes made at equivalence of 5 or 10 times antigen excess the antibody could be detected at exactly the same location as the FITC labeled antigen. Antibody could not be detected in the eyes of mice that received immune complexes made at 20 times antigen excess, or in the eyes of control mice that received the fluoresceinated antigen alone.

The same staining procedure was followed to identify the presence of complement in the immune complexes. In all the eyes where antibody was present, complement could be demonstrated using goat antimouse C3.

In mice receiving immune complexes at 20 times antigen excess, and in control mice, no complement deposits were demonstrable. To determine the site of immune complex localization in the ocular vessels, the vas-

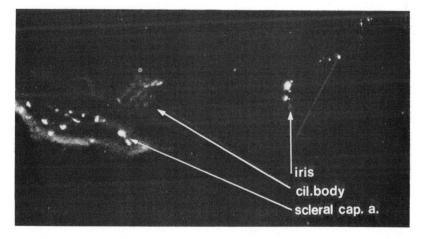

FIG. 1. Deposition of fluorescent BTG–aBTG complexes in the iris, the ciliary body, and the scleral capillary area of the mouse eye. The animals were sacrificed 1 hour after inoculation. The complexes were made at five times antigen excess. (Original magnification × 250.)

cular endothelium of eyes containing immune complexes was stained with rabbit antifactor VIII using indirect immunofluorescence. Immune complexes appeared to be deposited in the vascular walls of the iris, ciliary body, and choroid as well as in both the intravascular and intrastromal space in the intrascleral capillary area.

Discussion

Immune complexes have been found in sera and aqueous humour of uveitis patients.[1,2] Immune complex deposits have also been found at autopsy in the eyes of patients with immune complex diseases.[3] These data suggest that immune complexes may have a role in the pathogenesis of uveitis in humans.[2,4] High levels of circulating immune complexes are found in diseases such as rheumatoid arthritis (RA) and systemic lupus erythematosus (SLE). These diseases have different manifestations of ocular involvement: episcleritis in RA patients, retinal vasculitis in SLE patients.[11] Nevertheless, Aronson[3] demonstrated immune complex depositions in the uvea of SLE patients lacking any signs of uveitis. Therefore, it seems that the presence of high levels of circulating immune complexes alone is not the sole prerequisite for deposition of immune complexes and that other factors are needed for the initiation of an inflammatory reaction.

Analysis of the role of circulating immune complexes in uveitis experimental work was done by Peress[5,6] using the serum sickness model. Deposition of immune complexes could be demonstrated in the ciliary body and the choroid of rats and rabbits; cellular infiltrates were also detected in both of these locations and in the subconjunctival area. The investigations using the serum sickness model were, however, hampered by the fact that variables such as size, composition, and levels of the immune complexes formed could not be monitored. These factors strongly influence the rate of clearance and the tissue deposition of the complexes.[12] We used a passive immune complex model where the size and composition of the administered complexes could be standardized prior to inoculation.

Apart from the known deposition sites in ciliary body, iris, and choroid, the most striking observation was the strong immune complex deposition in the intrascleral capillary area. The question of whether these deposits were indeed present in the tissue and not only a reflection of the complexes circulating in the bloodstream was analyzed by perfusion experiments. In these experiments whole body perfusion was performed, but this did not result in removal of the deposited immune complexes (data not shown). On the other hand, the cloudy deposition of BTG-FITC alone seen in the eyes of the controls and the eyes of mice receiving complexes at 20 times antigen excess disappeared after perfusion.

The inability of immune complexes made at 20 times antigen to deposit in the vessel wall could be due to the fact that these complexes are small, do not fix complement, and generally cannot initiate inflammation. Further study is necessary to elucidate the deposition mechanism of immune complexes in the eye and the preferential deposition of immune complexes in the intrascleral capillary area.

References

1. Char DH et al: *Am J Ophthalmol* **87**:678–681, 1979.
2. Dernouchamps JP et al.: *Am J Ophthalmol* **84**:24–31, 1977.
3. Aronson AJ et al.: *Arch Intern Med* **139**:1312–1313, 1979.
4. Rahi AHS et al.: In *Immunology and Immunopathology of the Eye*. Silverstein AM and O'Connor GR (Eds.). Masson Publishing USA, New York, 1979.
5. Peress NS and Tompkins DC: *Arch Pathol Lab Med* **102**:104–107, 1978, chapter 6.
6. Peress NS: *Exp Eye Res* **30**:371–378, 1980.
7. Kijlstra A et al.: *J Immunol Meth* **17**:263–277, 1977.
8. Weir DM.: *Handbook of Experimental Immunology*. Blackwell Scientific Publications, Oxford.
9. Bethell et al.: *J Biol Chem* **254**(8):2572–2574, 1979.
10. Jaffe EA: *N Eng J Med* **296**(7):377–383, 1977.
11. Corwin JM and Weiter JJ: *Surv Ophthalmol* **25**(5):287–305, 1981.
12. Mannik M et al.: *Progress in Immunology II*, Vol. 5. Brent and Holborrow, (Eds.). North Holland, Amsterdam, 1974.

chapter 17

Clinical Interest in the Detection of Immune Complexes in the Aqueous Humor

J. P. Dernouchamps

The presence of immune complexes (IC) has been recently demonstrated in the aqueous humor (AH) of patients suffering from various ocular diseases.[1-4] The aim of this paper will be to discuss the possible clinical importance of the presence of immune complexes in cases of uveitis, ocular perforating injuries, and choroidal melanoma.

Material and Methods

Among the numerous techniques now available for the detection of IC, we used the agglutination inhibition test.[5] This test is based on the ability of the complexes to inhibit the agglutinating activity of rheumatoid factor (RF) or of subunit q of the first component of complement (Clq) on IgG-coated particles (latex). The agglutination inhibition tests were always performed with both agglutinating agents, because IC, depending on their size and the class or subclass of immunoglobulins involved, do not necessarily react with RF and Clq in the same manner.[5] Before testing the inhibitory properties, each sample was always checked for the possible presence of spontaneous agglutinating activity caused by en-

dogenous RF, which is often associated with the presence of IC.

Results and Discussion

Uveitis

In cases of uveitis, IC or RF was detected in 63 AH samples of the 119 examined (Table I). In general (57/63), these factors were restricted to the AH or reached about the same titer in the AH and in the serum. This suggests intraocular formation of these factors. Moreover, our tests were positive in the serum of 60 patients; in 24 of these 60 patients, AH was free of IC or RF. Detection of IC or RF was especially common in the AH of patients suffering from Fuchs' heterochromic cyclitis (26 of 34) and in the serum of patients with Behçet's disease or Vogt–Koyanagi–Harada's syndrome (11 of 16). In clinical practice, the presence of IC or RF in the AH has not been shown to be specific for a given etiology of uveitis. When considering the anatomical type of uveitis, the same conclusion may be drawn (Table II). However, it must be noted that the detection of IC or RF in the AH and in the serum was less frequently observed in cases of intermediate uveitis.

With regard to the prognosis, the occurrence of IC or RF in the AH could indicate that a given case of uveitis had become

University Eye Clinic, Université Catholique de Louvain, Brussels, Belgium.

TABLE I

Frequency of the Detection of Immune Complexes or Rheumatoid Factor in Aqueous Humor (AH) from Patients with Uveitis

	No. of cases	Immune complexes or rheumatoid factor		
		in AH only	in serum only	in AH and serum
Fuchs' cyclitis	34	12	1	14
Ankylosing spondylitis	4	1	0	2
Crohn's disease	2	0	1	0
Collagenosis	1	0	0	1
Herpetic keratouveitis	2	0	0	1
Hodgkin's disease	1	0	1	0
Phacoantigenic uveitis	1	0	0	1
Toxoplasmosis	3	0	2	0
Placoid epitheliopathy	1	0	0	1
Eales' disease	1	1	0	0
Atopy	2	0	0	0
Behçet's disease	13	1	7	2
Vogt–Koyanagi–Harada's syndrome	3	0	0	2
Sarcoidosis	1	0	1	0
Retinal detachment	2	1	0	1
Unknown origin	48	11	11	11
TOTAL	119	27	24	36

TABLE II

Frequency (Percentage) of the Presence in the Aqueous Humor (AH) of Immune Complexes or Rheumatoid Factor in Cases of Uveitis According to their Anatomical Type

Anatomical type	Number of cases	Immune complexes or rheumatoid factor		
		in AH only	in serum only	in AH and serum
Anterior uveitis				
Fuchs' cyclitis	34	35%	1%	41%
Other origins	29	21%	24%	38%
Intermediate uveitis	16	6%	12%	12%
Posterior uveitis	10	40%	30%	20%
Panuveitis	30	13%	33%	23%

chronic. Since posterior subcapsular cataract generally develops in cases of chronic uveitis, we searched for a possible correlation between this form of cataract and the presence of IC or RF in the AH of 54 patients suffering from uveitis, usually of unknown origin. This correlation was found to be at the limit of significance ($0.05 > p > 0.02$).[3,4] However, Fuchs' cyclitis, which is associated in 80% of the cases with the presence of IC or RF in the AH, is also characterized by the same posterior subcapsular cataract.

Ocular perforating injuries

In cases of ocular perforating injury, IC or RF were found in the AH of four patients of nine examined. Interestingly, these four patients developed sympathetic ophthalmia, which was confirmed histopathologically. The five patients in whom the tests were negative in the AH developed atrophy of the eye (two cases), ocular siderosis, ossification of the choroid, and traumatic cataract, respectively, but no inflammatory reaction threatening the other eye. The probability

TABLE III

Presence of Immune Complexes (IC) or Rheumatoid Factor (RF) in Aqueous Humor (AH) of Patients with Choroidal Melanoma

Patient No.	Histological type	Risk of metastasis	Presence of IC or RF in AH
1	Spindle cell A	0	0
2	Spindle cell B	0	+
3	Spindle cell A	0	0
4	Spindle cell A	0	0
5	Epithelioid	+	+
6	Spindle cell B	+	+
7	Spindle cell A	+	0
8	?	0	0
9	Spindle cell	0	0
10	Spindle cell	0	0
11	Epithelioid	0	+
12	Spindle cell B	0	0
13	Spindle cell	+	+

that the association between sympathetic ophthalmia and the presence in the AH of IC or RF could have arisen by chance was 0.008 (exact probability test of Fisher).

Choroidal melanoma

In cases of choroidal melanoma, IC or RF were detected in the AH of five of 13 patients examined (Table III). In three of these five patients, the histopathological analysis disclosed the presence of tumor cells along intrascleral vessels, suggesting a risk of metastasis. However, the number of patients was too small to allow statistically significant conclusions.

In summary, the presence of immune complexes or of rheumatoid factor in the aqueous humor has not been shown to be specific for a given ocular disease. However, the presence of immune complexes was found to precede (1) the development of secondary cataract in cases of uveitis, (2) sympathetic ophthalmia after ocular perforating injury and, (3) perhaps, metastases in cases of choroidal melanoma.

References

1. Dernouchamps JP et al.: *Am J Ophthalmol* **84**:24–31, 1977.
2. Dernouchamps JP and Michiels J: In *Immunology and Immunopathology of the Eye*. Silverstein AM and O'Connor GR (Eds.). Masson Publishing USA, New York 1979, pp. 40–45.
3. Dernouchamps JP: *Les protéines de l'humeur aqueuse*. Thèse. Université Catholique de Louvain, 1981.
4. Dernouchamps JP: *Doc Ophthalmol*, **53**:193–248, 1982.
5. Lurhuma AZ et al.: *Clin Exp Immunol* **25**:212–226, 1976.

chapter 18

Clinical and Experimental Studies of Immune Complexes in Uveitis

Kaoru Mizuno, Kiyoshi Watanabe, and Yasuo Mimura

It has been proved that in systemic autoimmune disorders such as systemic lupus erythematosus (SLE) and rheumatoid arthritis (RA), levels of circulating immune complexes (CIC) frequently increase. This may be one of the causes of the disorder. The role of CIC in ocular tissues has been investigated by various authors who found that lesions in the limbus, iris root, and capillaries of the ciliary body could be caused by CIC.[1–3] However, all of those studies were limited to rabbits; the role of CIC in humans remains unknown. This study reports the role of CIC on intraocular inflammation in patients with various forms of uveitis.

Materials and Methods

CIC was determined by the polyethylene glycol precipitation (PEG) method as proposed by Digeon et al.[4]

Determination of serum immunoglobulins and complement. In CIC-positive sera, IgG, IgA, IgM, and complement C′3 and C′4 were determined using a single radial immunodiffusion (SRID) method (Partigen, Behring Co., Marburg). Only for the measurement of IgG was serum diluted to 1:10. IgM was determined after 80 hours of incubation, and measurement of the other immunoglobulins was made after 48 hours.

Department of Ophthalmology, Osaka University Medical School, Osaka, Japan.

Determination of precipitated immunoglobulins and complement: CIC in CIC-positive serum were precipitated by the PEG method, and the sediment was dissolved in 50 μl of borate-buffered saline (0.1 M, pH 8.4, 0.8% NaCl). Serum IgG, IgA, IgM, C′3 and C′4 were investigated by the SRID method.

Animal experiments. To examine the effects of immune complexes (IC) on the ocular tissues, homologous aggregated IgG obtained by the Dickler method[5] was prepared in concentrations of 5 mg/ml, 0.5 mg/ml and 0.05 mg/ml. White rabbits weighing 3 kg were injected with 0.2 ml of each preparation into the vitreous, while 0.2 ml of 0.85% saline was used as a control. After 24 hours, the eyes were enucleated and studied histologically.

Results

Preliminary experiments. While the precipitation of aggregated IgG was hardly affected by changes in the PEG concentration, PEG, 3.5%–10%, progressively precipitated unaggregated IgG. Therefore, PEG was used at a concentration of 3.5%. The absorbance was directly proportional to the concentration of aggregated IgG, but independent of the concentration of normal human IgG when determined in 3.5% PEG.

CIC levels in various uveitides. It was only in uveitis of unknown origin that a significant increase of CIC level was found in

TABLE I
C. I. C. in Uveitis

Kinds of uveitis	No. of cases	CIC level	
		C.I.C. level (O.D.)	Cases of PEG positive sera (mean + 2S.D.)
Behçet's disease	58	0.119 ± 0.045	6 (10.3%)
Harada's disease	44	0.130 ± 0.075	8 (18.2%)
Sarcoidosis	6	0.108 ± 0.014	0 (0 %)
Peripheral uveitis	13	0.127 ± 0.050	2 (15.4%)
Uveitis of unknown origin	36	0.141 ± 0.076^a	9 (24.3%)[b]
Healthy controls	38	0.107 ± 0.035^a	2 (5.3%)[b]

[a] $t = 2.45, p < 0.02$.
[b] $x^2 = 4.03, p < 0.05$.

comparison with a group of healthy controls. Shown in Table I are positive cases in which CIC level was never lower than 0.177. This value was obtained by adding twice the standard deviation to the mean value of healthy controls. A chi square test showed that only in cases of uveitis of unknown origin could significant increases be demonstrated, as compared with healthy controls ($X^2 = 4.03$, $p<0.05$) (Table I).

Uveitis of unknown origin. In this group of cases significant differences were found from other groups of uveitis by the t and chi square test. A comparison was made between a group with anterior uveitis without lesions in the fundus, a group with posterior uveitis, and healthly controls. In the anterior uveitis group, a significant increase in CIC, as compared with healthy controls, was observed ($t = 2.99$, $p < 0.05$) (Fig. 1).

In addition, cases of anterior uveitis could be divided into an active phase and an inactive phase, and these could be compared with healthy controls. CIC showed a sig-

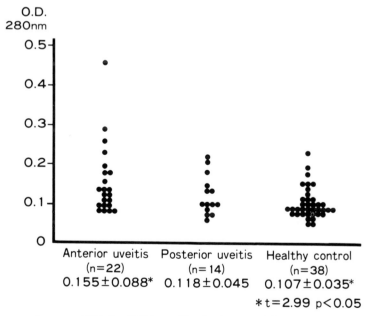

FIG. 1.   CIC in uveitis of unknown origin.

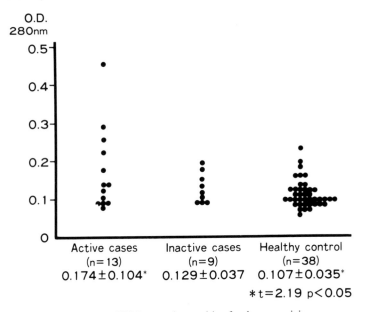

FIG. 2.   CIC in anterior uveitis of unknown origin.

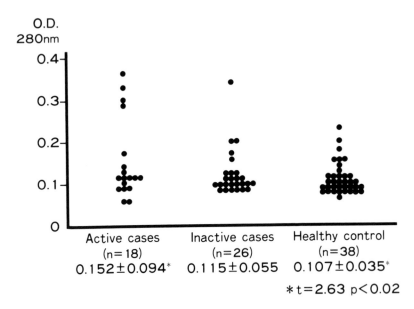

FIG. 3.   CIC in Harada's disease.

nificant increase in the active phase ($t=2.19$, $p<0.05$) (Fig. 2).

Harada's disease. On the whole, no significant differences were found when the cases were compared with healthy controls. The cases of Harada's disease were divided into two groups: active and inactive. The former group comprised fresh cases with active lesions and chronic cases with recurrent iritis; the latter group contained cases that were in remission or had healed lesions. When compared with healthy controls, the group of active cases showed a significant increase of CIC ($t=2.63$, $p<0.02$) (Fig. 3).

Behçet's disease. According to clinical

TABLE II
PP Index of PEG-Positive Sera

Kinds of uveitis	No. of cases	IgG	IgA	IgM	C3	C4
Behçet's disease	15	0.36 ± 0.12^a	—	9.65 ± 3.92	1.53 ± 0.92	12.33 ± 4.41
Harada's disease	12	0.33 ± 0.11^b	—	9.28 ± 5.95	1.88 ± 0.85	8.00 ± 5.10
Uveitis of unknown origin	10	0.39 ± 0.14^c	—	12.93 ± 4.21^d	1.95 ± 1.16	11.65 ± 5.51
Healthy controls	12	$0.24 \pm 0.07^{a,b,c}$,	—	9.10 ± 2.90^d	—	11.64 ± 5.22

[a] $t = 3.26$, $p < 0.005$.
[b] $t = 2.39$, $p < 0.05$.
[c] $t = 3.16$, $p < 0.01$.
[d] $t = 2.31$, $p < 0.05$.

$$\text{PP index} = \frac{\text{Precipitated immunoglobulin (complement)}}{\text{Serum immunoglobulin (complement)}} \times 100.$$

features of the ocular and systemic conditions of a given patient, the cases of Behçet's disease were classified into (1) the ocular attack phase, (2) the ocular remission phase, (3) the systemic attack phase (including ocular signs), and (4) the systemic remission phase. These patients were compared with healthy controls. No significant differences were observed in any phase.

Immunoglobulin and complement components of CIC. IgG, IgA, IgM, C′3 and C′4 were examined in the serum, and their corresponding proteins were precipitated by the PEG method. In the CIC no IgA was detected in either the cases or the healthy controls. No C′3 was found in the CIC of the healthy controls. As to the IgG component in CIC, the polyethylene glycol precipitates index (PP index) was significantly elevated in cases of Behçet's disease ($p<0.005$), Harada's disease ($p<0.05$), and uveitis of unknown origin ($p<0.01$) as compared with that of healthy controls. Cases of anterior uveitis of unknown origin showed a significantly high PP index in IgM ($p<0.05$). With regard to C′4, no significant differences were observed when compared with healthy controls, except for the cases of Harada's disease, which showed a slightly lower ratio (Table II).

Animal experiments. No sign of inflammation was observed in the eyes of controls in which saline was injected. In the eye in which 0.2 ml of aggregated IgG (5 mg/ml) was injected, however, infiltration of polymorphonuclear leukocytes was found in the vitreous, posterior, and anterior chambers.

Discussion

For the determination of CIC, various methods are currently used, but no ideal method has been found.[2] It is said that the values of CIC obtained vary according to the methods employed. In the present study, the authors used a method that utilized the precipitation of CIC in PEG solution.

This method is simple and requires no isotope or cultured cells, but the sediment contains not only IC, but also C1q, C′3 and C′4. Meanwhile, Digeon[4] reported that the quantity of sediment obtained by the PEG method was proportional to total content of IgG, IgM, and IgA in the sediment as well as to the content of IgG. From the results obtained, the volume of sediment obtained with PEG reflected mainly the serum level of CIC or aggregated IgG. As indicated, the level of CIC was high in the active phase of anterior uveitis of unknown origin. Since iridocyclitis can be produced by this injection of aggregated IgG in the vitreous body of rabbits,[1] it is tempting to think that there is an analogy between anterior uveitis of unknown origin and the experimental lesions produced in rabbits. In both cases there may be elevated levels of CIC. Although high levels of CIC can be detected in serum sickness and SLE, iridocyclitis is only rarely observed. Thus, we suppose that a high CIC level alone may not be sufficient to cause inflammation.

Shimada[6] reported that in immunogenic uveitis, only 20% of the antibody forming cells in the ocular tissues produced antibodies against the corresponding antigens, while the remaining 80% of these cells produced antibodies against other antigens. This may suggest that other precipitating events, such as previous inflammation, are necessary in order to produce ocular inflammation by circulating immune complexes.

With regard to the mechanisms of Harada's disease, Mimura[7] and Yuasa[8] considered the main pathological mechanism of this disease to be due to delayed hypersensitivity against uveal melanocytes. Antiuveal antibodies may be found in the serum of the patients with the disease. However, these antibodies presumably cause little damage to ocular tissues; the same antibodies were also detected in other kinds of uveitis. For this reason, these antibodies seem to have low specificity. Recently, however, the results of an immunoadherence study by Yuasa[8] and a study of ADCC by Tagawa[11] showed that antiuveal antibodies could be detected in fresh and recurrent cases of uveitis. These facts mean that several serum antibodies may play a role in the development of this disease. In the present study, also, CIC showed a significant increase in fresh and recurrent cases of Harada's. This suggests that IC may play a role in this ocular inflammation. Since CIC may be increased in other types of uveitis, the specificity of IC in Harada's disease is uncertain.

With regard to Behçet's disease, the results of the present authors' study showed that a CIC level above the mean value plus 2 SD was seen in six of 58 cases (10.3%); no significant differences were observed between these cases and healthy controls. As to types and stages of the disease as well as kinds of treatment, there was no statistical disparity. From these results it would seem that IC probably did not participate in the disease.

It is reported [9,10] that the constituents of immunoglobulins and complement in IC were different in various diseases and in various stages of these diseases. In the present study, an increase of IgG and C'3 in Behçet's disease and in Harada's disease was observed in CIC-positive sera. IgM as well as IgG and C'3 where elevated in anterior uveitis of unknown origin. It is unknown how the differences in the components of CIC relate to the pathogenesis of each disease.

References

1. Hohki T et al.: *Acta Soc Ophthalmol Jpn* **75**:1565, 1971.
2. Theofilopoulos AN and Dixon FJ: *Hospital Practice* **15**:107–121, 1980.
3. Gamble CN et al.: *Arch Ophthalmol* **84**:331:341, 1970.
4. Digeon M et al.: *Immunol Methods* **16**:165–183, 1977.
5. Dickler HB: *J Exp Med* **140**:508–522, 1974.
6. Shimada K: *Acta Soc Ophthalmol Jpn* **83**:1861, 1979.
7. Mimura Y: *Acta Soc Ophthalmol Jpn* **83**:1909–1975, 1979.
8. Yuasa T: *Folia Ophthalmol Jpn* **27**:992–998, 1976.
9. Abe C: *Saishin-Igaku* **33**:1360–1365, 1978.
10. Levinsky RJ: *Immunol Today*, 94–97, May 1981.

chapter 19

An Experimental Study of the Cytotoxic Effect of Antiserum against Retinal S-Antigen on Cultured Retinal Pigment Epithelium

M. Usui, Y. Furuse, K. Takamura, and M. Mitsuhashi

It has been proved histologically that retinal pigment epithelium (RPE) damage occurs at an early stage of experimental autoimmune uveoretinitis (EAU) induced by retinal antigens,[1-3] especially in those cases in which high doses of S antigen are administered.[4,5] This RPE damage may impair the blood-retinal barrier.[6] If this is correct, antibodies produced approximately 7–14 days after inoculation of S-antigen[7] would be able to penetrate from the choroid to the outer segments of the photoreceptor cells where S-antigen is located on the surface of the cytoplasmic membrane.[6,8] Subsequently, an immunological reaction could well ensue between the antibody and S antigen forming an immune complex on the outer segment membrane. This series of immunologic reactions is thought to play an important role in the pathogenesis of EAU. However, the pathogenesis of the damage to the RPE during the early stage of this disease has not been clarified.

In order to elucidate the pathogenic mechanisms related to RPE cell damage, the antigenicity of swine RPE was examined by immunoelectrophoresis and by immunofluorescent methods with anti-S and anti-A serum. Also, the cytotoxicity of these sera for cultured swine RPE as the target cell was investigated.

Materials and Methods

Antigen preparation

Retinal S antigen and A antigen were made from extracts of swine retina by the method described by Takano.[9,10] The antigens of RPE were extracted from homogenates of swine RPE collected from 50 swine eyes washed 5 times with PBS. A final suspension of 1×10^7 cells in 1 ml of PBS was prepared. Antigens extracted from these cells were examined by PAG disc electrophoresis to detect their protein constituents.

Immunization and antiserum

Purified S antigen, highly purified A antigen free of contamination by S antigen, and RPE antigen were separately inoculated into the hind foot pads of rabbits with complete Freund's adjuvant (CFA Difco Laboratories). Each antigen was inoculated four times at intervals of 2 weeks. Ten days after the final inoculation, the animals were sacrificed and bled. Their serum was stored at $-20°C$. The serum was analysed by im-

Department of Ophthalmology, Tokyo Medical College Hospital, Tokyo, Japan.

munoelectrophoresis (1.5% agarose, veronal buffer pH 8.6, 6 mA/cm, 1 hour) and counter immunoelectrophoresis (0.8% agar, veronal buffer pH 8.6, 2mA/cm, 1 hour).

Cultured cells

The following cells were investigated for their content of retinal antigens by the indirect immunofluorescent method, and were used as target cells for a cytotoxicity test: (1) swine RPE cells 14–21 days after primary culture; (2) swine corneal endothelial cells (CE) 14–21 days after primary culture; and (3) successively cultured human fibroblast cells (HF).

The culture method for the swine RPE cells and CE cells was as follows: After careful removal of the vitreous and the sensory retina from the eye cup, EDTA (0.02%) solution was poured into the cup and left for 30 minutes. Then the EDTA solution was replaced by Hanks' solution. Subsequently, the RPE was freed from the eye wall and floated in Hanks' solution. After pipetting, the RPE suspension was centrifuged for 5 minutes (1100 g). After the centrifugation, RPE was divided into 1 ml lots containing 2×10^5 RPE with TCM 199 medium containing 15% fetal bovine serum. RPE was cultured in 2 ml chambers (Lab-Tek) in 5% CO_2, 37°C for 2–3 weeks.

The CE was also removed from the swine corneal cup by applying 0.25% trypsin solution for 5 minutes. After the removal of the trypsin solution by aspiration, it was replaced by Hanks' solution, and the CE was collected by scrubbing gently with a spatula. The CE suspension was centrifuged for 5 minutes (1100 g), and CE was divided into 1 ml lots containing 2×10^4 of cells in the same medium used for the RPE. The duration and the culture conditions were the same as for RPE. Successively cultured HF cells were obtained from the Department of Immunology at our institution.

Immunofluorescent techniques

To detect the antigenicity of RPE, the indirect immunofluorescent technique was employed using serum containing antiretinal antibodies. The cells were washed with PBS

three times and fixed with 95% cold ethanol for 1 minute. After washing three times with PBS, antiserum diluted to 1:5 or 1:10 was poured into a Lab-Tek chamber. The cells and antiserum were incubated at 37°C for 1 hour; then the antiserum was aspirated, and the cells were washed with PBS three times for 5 minutes. Fluorescein-conjugated goat antirabbit γ-globulin (Behring-Werke AG, West Germany) diluted to 1:15, was employed on the RPE specimen in the chamber, and the RPE was incubated at 37°C for 1 hour. After the final incubation, the RPE was washed, dried, and covered with glycerol buffer.

Cytotoxicity test

All antisera used in the cytotoxicity test had been treated previously to inactivate complement. Three kinds of cells such as

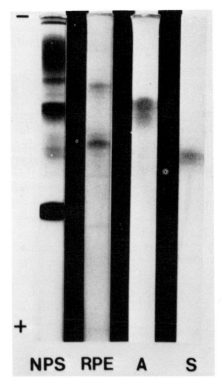

FIG. 1. Analysis of retinal antigens on PAG disc electrophoresis. NPS (normal porcine serum), RPE (extract of swine retinal pigment epithelium), A (swine retinal A antigen), S (swine retinal S antigen).

RPE, CE, and HF were used as target cells in this experiment. The cytotoxicity test was conducted as follows: 1 ml of antiserum was incubated with one of the above-mentioned cells at 37°C for 60 minutes. After incubation, 1 ml of rabbit complement diluted 1:15 was placed in the chamber, then incubated again at 37°C for 60 minutes. After the second incubation, the cultured RPE was stained with PBS containing 0.1% trypan blue. The number of dead cells per 200 RPE were counted by phase-contrast microscopy, and the percentage was calculated.

Results

Antigens

The purified S antigen showed two fractions on PAG disc electrophoresis. In highly purified A antigen, there was no evidence of contamination by S antigen as demonstrated by immunoelectrophoresis or PAG disc electrophoresis. More than five protein bands were detected in the RPE antigen mixture (Fig. 1).

Antisera

The antisera obtained by hyperimmunization were titrated by a ring type of precipitation reaction as follows: anti-S serum diluted 1:256; anti-A serum diluted 1:128; anti-RPE diluted 1:512. On immunoelectrophoresis, S antigen formed a precipitation arc line in the α-globulin region against anti-S serum. A antigen showed a precipitation arc line in the albumin region against anti-A serum. RPE antigen formed some reactive precipitation lines in α-, β-, and γ-globulin regions against anti-RPE antibody, especially in the γ region. However, the precipitation in the γ region was subsequently shown to be derived from a substance other than swine serum globulin. Also, the anti-RPE produced a faint precipitation line with S antigen. The anti-A and anti-S sera formed pairs of reaction lines with RPE antigen on counter-immunoelectrophoresis.

Immunofluorescent technique

A strong, specific fluorescent reaction with anti-RPE serum was observed in the

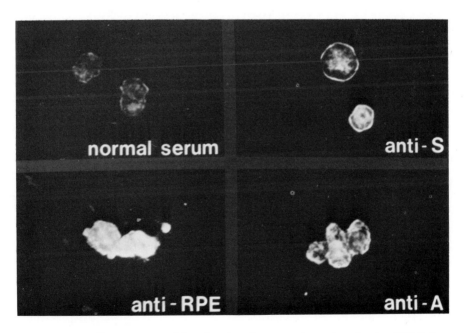

FIG. 2. Immunofluorescent reactivity of each antiserum against cultured swine RPE 8 days after the primary culture (× 160).

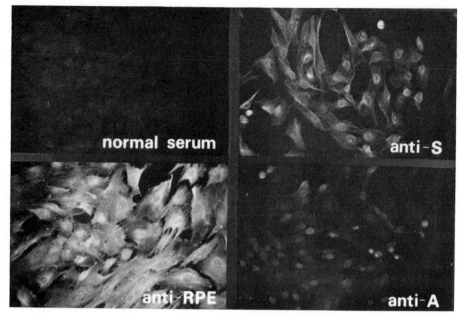

FIG. 3. Immunofluorescent reactivity of each antiserum against swine cultured RPE 8 weeks after primary cultivation. Immunofluorescent staining with anti-S and anti-A sera on the cultured RPE is weaker than that of RPE cultured for 8 days (× 160).

cultured swine RPE at multiple stages of culture, and RPE cells were totally stained with fluorescein. Moderate specific fluorescein staining with anti-S serum in RPE was noted to be present in whole-cell configurations (except for the pigment granules) and especially in the cytoplasmic membrane. By staining with anti-A serum, RPE showed a specific fluorescent binding of nuclear membranes. (Fig. 2). In RPE tested more than 6 weeks after primary cultivation, the specific fluorescent reactions with anti-S and anti-A were noted to be weaker, compared with the RPE cultured 3 weeks or less after primary culture (Fig. 3) but overall staining of all substrates was weaker at this time. As for the control, RPE stained with normal rabbit serum showed no evidence of specific fluorescence, nor did RPE stained by the direct method with fluorescein-conjugated goat antirabbit γ globulin.

Cytotoxicity test

Antibody control was performed with each decomplemented serum against cultured RPE, cultured CE, and HF as the target cells. No cytotoxic effect was seen in the control groups when treated serum was tested against cultured RPE, CE, or HF target cells. Rabbit complement control serum diluted 1:15 did not reveal any cytotoxic reaction to the various target cells. In a test of RPE cultured with normal rabbit serum and complement, a dead-cell percentage of 11.5 ± 5.9 was noted. Under the same testing conditions, the CE death rate was noted to be 6.3 ± 1.5%. The death rate of RPE incubated with anti-RPE serum and complement was constantly 100%. With anti-S serum and complement, the RPE death rate was 58.6 ± 15.4%. However, with anti-A serum, the dead cell rate was noted to be lower than that of anti-S serum, i.e., 30 ± 16.4%. The CE was affected by these three antisera with complement, but the death rate was less than 15%. There was no evidence of cytotoxicity among all antisera used in this study against HF (Fig. 4, Table I).

Discussion

It has been reported that from the earliest stages of EAU, there are various types of

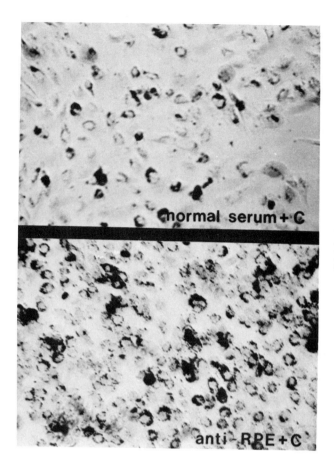

normal serum + C

anti-RPE + C

FIG. 4. Swine cultured RPE following incubation with normal rabbit serum and complement (upper). Note several staining cells and 7% dead cells. In the lower plate RPE incubated with anti-RPE and complement shows only dead cells ($\times$ 50).

RPE damage including vacuolization of the cytoplasm,[1] loss of microvilli, and destruction of the cells. A widening of the intercellular spaces and hole formation in the cell membranes also have been reported.[11] These types of damage are thought to be due to humoral immune responses, which appear approximately 7 days after inoculation of S antigen, rather than to be caused by cellular immunity.[1,11]

However, the mechanisms of the process of cell damage attributable to anti-S serum have not been elucidated. Although the antigenicity of RPE is a matter of controversy, Wacker reported that the U antigen (uveal antigen) appears to be deeply involved in the antigenicity,[12,13] and he proved by an immunofluorescent technique that S antigen does not exist in RPE.[14] Reich et al. proved by an immunofluorescent staining technique that antiserum produced by hyperimmuni-

zation of guinea pigs with injections of cultured rabbit RPE showed specificity for the cell membrane and the cytoplasm of cultured RPE of rabbits and mice. However, Reich's antiserum did not show any immunologic reaction to the sensory retina, including the photoreceptor cells.

On the other hand, it has been reported that EAU can be produced by inoculation of RPE itself with adjuvant.[16] Meyers reported that anti-outer segment serum produces cross-reacting precipitation lines between the outer segment antigen and RPE antigen.[17] Therefore, it cannot be said that there is no relationship between the RPE and the antigens of photoreceptor origin.

In nature, RPE phagocytizes the outer segments. Moreover, there has always been the possibility that the S antigen is located on the microvilli of the RPE. Also, phagocytized outer segments may occasionally be

TABLE I
Results of Cytotoxicity Test

Culture medium	Target cells in		
	Retinal pigment epithelium	Corneal endothelium	Human fibroblast
Medium (M) TCM 199	0	0	0
Medium + complement (C)	0	0	0
Normal rabbit serum + M	0	0	0
Anti-S serum + M	0	0	0
Anti-A serum + M	0	0	0
Anti-RPE serum + M	0	0	0
Normal rabbit serum + C	$11.5 \pm 5.9\%$	$6.3 \pm 1.5\%$	0
Anti-S serum + C	$58.6 \pm 15.4\%$	$8.3 \pm 2.4\%$	0
Anti-A serum + C	$30.3 \pm 16.4\%$	$7.5 \pm 2.1\%$	0
Anti-RPE serum + C	100%	$9.6 \pm 3.4\%$	0

Dead cell percentage = %.

noted in the RPE. Therefore, RPE antigens prepared by collection of fresh RPE naturally contain a small amount of S antigen, no matter how extensively the material is purified.

In our immunoelectrophoretic study, RPE antigen did not show any visible precipitation line against anti-S serum or anti-A serum. However, by counter-immunoelectrophoresis, there was formation of precipitation lines between RPE and anti-S serum and between RPE and anti-A serum. This evidence may suggest that there is at least minimal antigenicity in extracts of RPE against anti-S and anti-A sera. Furthermore, in cultured RPE up to 14–21 days after the primary culture, it was noted by immunofluorescent techniques that antigenic activity can be detected by anti-S and anti-A antibody. However, in RPE cultured more than 6 weeks, only weak responses against anti-S and anti-A antibody were noted. This result is in accordance with the report of Reich

et al.[15] If the antigens made from cultured RPE were used for experimental production of antibody, the antibody produced would show no response to the sensory retina because the antigenicity of the retinal antigen would be reduced in such cultured RPE. One can conversely apply this reasoning to the antigenicity of RPE, i.e., that the RPE itself, in nature, lacks the S antigen. However, the RPE which we cultured and which morphologically closely resembled that seen *in vivo,* showed evidence of S antigen. In a recent publication concerning the localization of S antigen by a direct enzyme-labeled antibody method on cryostat sections of the retina, the authors reported that this antigen was located in the microvilli of RPE.[8] Based on these results, it can be postulated that the RPE contains some antigenic entity, and the results of cytotoxicity tests in our study may serve to support the hypothetical presence of such antigenicity.

The cytotoxicity test suggests that the RPE may be destroyed by anti-S serum in the presence of complement. This is important in explaining the pathogenesis of EAU. Marak reported that the severity of EAU was mitigated by the addition of a C'3 complement inactivating substance.[18] His report supports the results of the cytotoxicity test of the present study, because if complement were inactivated the RPE would not be damaged even in the presence of antibody. The blood–retinal barrier at the level of the choriocapillaris-pigment epithelium complex is normally maintained intact; this barrier does not allow the passage of higher molecular weight proteins such as immunoglobulins.

In conclusion, our investigation demonstrated that the cytotoxic effects of anti-S antibody in the presence of complement may be responsible for RPE damage at the early stages of EAU.

References

1. De Kozak Y et al.: *Arch Ophtalmol (Paris)* **36**:231–248, 1976.
2. Wong VG et al.: *Arch Opthalmol* **93**:509–513, 1975.
3. De Kozak Y et al.: *Albrecht Von Graefes Arch Klin Exp Ophthalmol* **208**:135–142, 1978.
4. De Kozak Y et al.: *Cur Eye Res* **1**:327–337, 1981.

5. Faure JP et al.: *J Fr Ophthalmol* **4**:465–472, 1981.

6. Usui M et al.: *Acta Soc Ophthalmol Jpn* **84**:1064–1074, 1980.

7. De Kozak Y, et al.: *Eur J Immunol* **11**:612–617, 1981.

8. Yajima S et al.: *Acta Soc Ophthamol Jpn* **86**:1553–1566, 1982.

9. Takano S et al.: *Folia Ophthalmol Jpn* **32**:491–499, 1981.

10. Takano S: *Folia Ophthalmol Jpn* **33**:968–975, 1982.

11. Renard G et al.: *Arch Ophtalmol* (Paris)**36**:327–340, 1976.

12. Wacker WB and Kalsow C: *Mod Probl Ophthalmol* **16**:12–22, 1976.

13. Kalsow CM, and Wacker WB: *Int Arch Allergy Appl Immunol* **48**:287–293, 1975.

14. Wacker WB et al.: *J Immunol* **119**:1944–1958, 1977.

15. Reich-D'Almeida F and Rahi AHS: *Nature* **252**:307–308, 1974.

16. Faure JP et al.: *Arch Ophtalmol (Paris)* **37**:47–60, 1977.

17. Meyers RL and Pettit TH: *J Immunol* **114**:1269–1274, 1975.

18. Marak GE Jr et al.: *Ophthalmic Res* **11**:97–107, 1979.

Growth Inhibition of Cultured Corneal Cells by Mononuclear Leucocytes

Elaine Young and Walter J. Stark

Maumenee was the first investigator to ascribe immunologic rejection as the cause of corneal graft failure in humans.[1] This mechanism is now recognized as the single most important cause of delayed graft failure in man.[2,3] The importance of immunological events in rejection, and their sequence, are still not well understood. Studies by MacDonald[4] and others, for example, suggest that corneal allograft rejection is mediated by cellular immune responses. We have previously demonstrated the development of lymphocytotoxic antibodies in association with corneal graft rejection,[5] but it is not clear whether these antibodies are a cause of rejection. In an attempt to define the effects of the cellular and serological aspects of graft rejection, we have developed an *in vitro* assay that correlates temporally with rejection episodes. The assay employs the inhibition of growth of cultured stromal cells by mononuclear cells from the peripheral blood of sensitized individuals. This *in vitro* model for corneal graft rejection provides a system in which specific immunologic mechanisms may be analyzed both separately and in conjunction with one another.

Materials and Methods

Subjects

Nineteen patients who underwent penetrating keratoplasty at the Wilmer Institute

The Wilmer Institute, Johns Hopkins University School of Medicine, Baltimore, Maryland.

of Johns Hopkins in Baltimore were studied. Nine of these patients were experiencing clinically defined corneal rejection reactions (keratic precipitates, an advancing line of lymphocytes on the endothelial surface, stromal infiltrates, and/or linear epithelial staining) and 10 had clear grafts at the time of assay.

Blood samples

Peripheral blood mononuclear leucocytes (PBL) were isolated from the defibrinated blood of each patient by the technique of ficoll-hypaque density gradient centrifugation.[6] The PBL were washed, counted, and resuspended at $1.0–1.5 \times 10^6$/ml in complete RPMI 1640 medium (Gibco) supplemented with 25 mM HEPES buffer, 2 mM glutamine, 100 U/ml penicillin, 100 μg/ml streptomycin, and 10% fetal calf serum (FCS). Where indicated, PBL were first depleted of adherent cells by incubation on plastic for 2 hours at 37°C. The harvested, nonadherent population was then washed and resuspended to $1.0–1.5 \times 10^6$/ml for assay. Serum was prepared from an additional sample of blood drawn at the same time as for PBL. All sera were screened for antilymphocyte antibodies against a panel of 50 typed cells by D. Kappus at Johns Hopkins. Antilymphocyte antibodies were recorded as a percentage of the panel with which each serum reacted and expressed as "Panel Reactive Antibody" (PRA).

Inhibition Assay

Human corneal stromal cells maintained in Nutrient Mixture F-12 (HAM) medium

(Gibco), supplemented with 2mM glutamine, 100 U/ml penicillin, 100µg/ml streptomycin and 10% FCS, were used as targets of growth inhibition. A monolayer culture of stromal cells was trypsinized from a culture flask (Corning, 75 cm^2), resuspended in fresh growth medium at a 1:10 or 1:20 dilution, and 0.2 ml were plated into an 8-well tissue culture chamber/slide (Lab-Tek, 4838). After 2–4 hours at 37°C in 5% CO_2, the stromal cells had adhered, and the medium was replaced with 0.1 ml PBL suspension in 10% FCS or autologous serum in a final volume of 0.4ml. Control wells contained an equal volume of complete RPMI medium alone or medium plus patient or normal serum. Cultures were incubated for four days at 37°C in 5% CO_2 and examined daily for growth inhibition with an inverted microscope. Growth was rated as positive (those cultures growing as well as the control) or negative (those cultures with little or no detectable increase in stromal cell number).

Immunofluorescent staining

Leu 1 (antihuman T-cell), Leu 2a (antihuman suppressor T-cell), Leu 3a (antihuman helper T-cell), HLA-DR (antihuman DR), Leu M1 (antihuman monocyte–granulocyte), and Leu M2 (antihuman monocyte subset) were purchased from Becton Dickinson (Mountain View, CA). MO2 (antihuman adherent macrophage) and B1 (antihuman B-cell) were kindly provided by Dr. K. Kortright (Coulter Immunology, Hialeah, FL). Four-day assay cultures containing stromal targets and patient PBL were washed 3 times in 1% FCS/phosphate buffered saline (PBS) to remove nonadherent cells and debris. One-hundred microliters of cold 1% FCS/PBS were added to each well, and the appropriate amount of each monoclonal reagent was added to each sample, incubated 30 minutes on ice, and washed twice in cold 1% FCS/PBS. One-hundred microliters FITC goat antimouse IgG (Tago, Inc., Burlingame, CA) was then added to the wells at a final concentration of 20 µg/ml, the samples were incubated 40 minutes more on ice, washed twice in cold 1% FCS/PBS, and fixed in 4% formaldehyde in PBS containing 0.25 M glucose. Negative controls contained second-step reagent only or no antibody whatever. Individual wells were examined for cells bound to stromal targets and photographed by light and by fluorescent microscopy as previously described[7] to determine the lineage of the attached cell.

Results

The PBL of 19 patients who had undergone penetrating keratoplasty were examined for their capacity to inhibit growth of human corneal stromal cell targets during co-culture. Representative cultures illustrated in Figure 1 show that human stromal cells co-cultured with PBL from a patient undergoing a rejection reaction (1a) failed to grow to monolayer after 4 days in culture. In contrast, stromal cells co-cultured with PBL from a patient with a clear graft (1b) formed a monolayer under the same conditions. This observation of inhibition correlated well with the uptake of ^{3}H-thymidine by the stromal cultures after incubation with irradiated PBL from the same patients (not shown).

Of 19 patients tested, nine were experiencing rejection episodes (keratic precipitates or a "rejection line") at the time of testing, while 10 had clear grafts. Of the nine experiencing reactions, six had circulating antibodies that reacted with 30% or more of a typed lymphocyte panel and the serum from three reacted with less than 30% of the panel (Table I). The PBL of eight out of nine (89%) rejecting individuals inhibited stromal growth in the absence of autologous serum, and all inhibited in the presence of serum. Of the 10 patients with clear grafts, none inhibited stromal growth in the absence of autologous serum and only 1 inhibited in its presence. In all cases, patient serum alone did not inhibit stromal cultures, and in many cases, it actually enhanced monolayer formation.

In experiments in which adherent cells were depleted from PBL prior to the assay, the remaining nonadherent population did not possess the capacity to inhibit stromal cell proliferation (Table I). Close examination of inhibited stromal cultures revealed a cell that was tightly bound to the stromal target (Fig. 2). To determine the lineage of

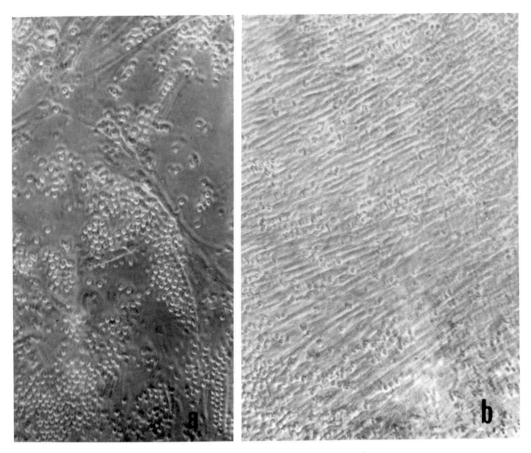

FIG. 1. Human stromal cells co-cultured for 4 days in the presence of 1X10⁶/ml peripheral blood mononuclear leucocytes (PBL) from *(a)* a patient undergoing a corneal rejection reaction, and *(b)* a grafted patient with a clear graft.

TABLE I
Growth Inhibition of Cultured Human Corneal Cells

Serum % PRA[a]	Patient graft status	Cultures inhibited when incubated with			
		1X10⁶/ml PBL alone	1X10⁶/ml Macrophage depleted PBL alone[b]	1X10⁶ml PBL plus autologous serum	Serum alone
<30	Rejecting	3/3	0/2	3/3	0
≥30	Rejecting	5/6	0/1	6/6	0
<30	Nonrejecting	0/6	n.d.	0.6	0
≥30	Nonrejecting	0/4	n.d.	1/4	0
Total Rejecting		8/9	0/3	9/9	
Total Nonrejecting		0/10	n.d.	1/10	

[a] "Panel Reactive Antibody." Lymphocytotoxic antibody against a panel of 50 typed lymphocytes.
[b] Depletion of adherent cells prior to assay, see Materials and Methods.

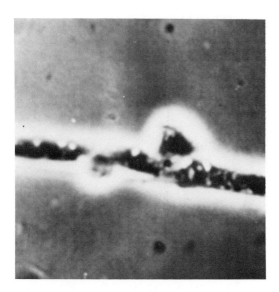

FIG. 2. Growth-inhibited stromal cell showing binding by a population of cells from the peripheral blood.

this bound cell, these cultures were washed several times to remove unbound cells and debris. The cultures were then stained by indirect immunofluorescence using a panel of monoclonal reagents with specificities for T- and B-lymphocytes, monocytes/macrophages, and granulocytes. Figure 3 depicts a stromal target with bound peripheral blood cells, photographed under conditions of *(a)* phase microscopy, and *(b)* after fluorescent staining for HLA-DR. While the attached cells are difficult to see in the light micrograph, they readily become apparent after fluorescent staining. Results shown in Table II indicate that the bound cell is negative for T- and B-cell surface antigens and positive for macrophage/monocyte markers.

Discussion

These data demonstrate that peripheral blood cells of persons undergoing corneal graft rejection reactions are capable of inhibiting the growth of cultured human stromal cells *in vitro*. The PBL inhibited stromal growth in eight of nine (89%) of the rejecting patients and in none of the nonrejecting patients. While the number of individuals tested was small, there was good correlation between clinical rejection epi-

sodes and the capacity to inhibit stromal growth. We are presently testing the specificity of this assay in patients with ocular bacterial infections and nonspecific ocular inflammation.

Immunofluorescent staining of peripheral blood cells bound to inhibited stromal targets suggests that a macrophage-like cell may be the effector in this phenomenon. The bound cell stained brightly for HLA-DR, MO2, Leu M1, and Leu M2, all of which identify cells of the monocyte lineage. Reagents specific for B- and T-cells failed to stain the attached population. The importance of an adherent cell in this assay was further demonstrated by a loss of growth inhibition in those cultures which contained a macrophage-depleted PBL population only. Previous studies have shown that macrophages isolated from Maloney sarcoma virus-tumor bearing mice have strong cytostatic activity for a lymphoma cell line.[8] In those studies, the macrophage effects were nonspecific, with some antigenically unrelated cell lines strongly inhibited. Preliminary experiments in our laboratory indicate all human stromal cultures serve equally well as targets, whereas the growth of rabbit stromal cells is unaffected by co-culture with PBL from rejecting patients.[9] We are currently studying the capability of the adherent cell alone in inhibition.

The effect of serum in this assay is not clear. In general, autologous serum had no effect on the results. The PBL from one rejector and one nonrejector inhibited stromal growth only in the presence of autologous serum. In both patients, the level of lymphocytotoxic antibody (PRA) was high (93% and 76% in the rejector and nonrejector, respectively). Most patients develop lymphocytotoxic antibody following surgery regardless of the clinical outcome of the graft.[5] In general, PRA levels less than 5% are not consistently reproducible. PRA greater than 60% may represent multiple antibodies or sensitization to some public specificity on the panel of lymphocytes, such as BW4 or BW6, shared by a majority of the population (D. Kappus, unpublished observations). Therefore, anti-HLA antibodies may have led to an antibody-dependent cellular cy-

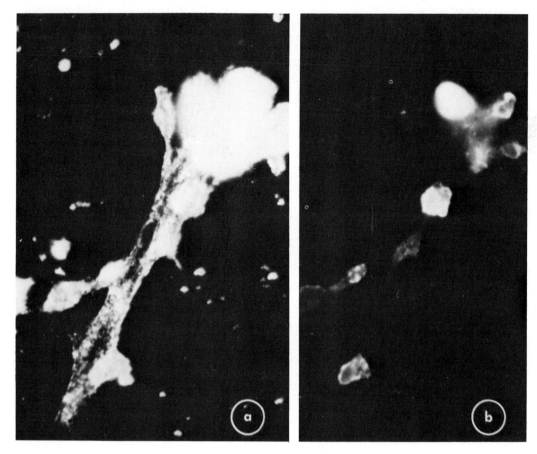

FIG. 3. Growth-inhibited stromal cell with bound peripheral blood cell, (a) light microscopy, and (b) fluorescent microscopy after staining for HLA-DR.

TABLE II
Surface Characteristics of Peripheral Blood Cell Binding to Stromal Targets

Monoclonal reagent	Human cell specificity	Fluorescence of bound cell
Leu 1	≥ 95% Circulating T-cells	−
Leu 2a	Cytotoxic/suppressor T-subset	−
Leu 3a	Helper/inducer T-subset	−
B1	100% B-cells	−
HLA-DR	Monocytes, macrophages, B-cells, activated T-cells	+
Leu M1	80–100% monocytes, 95% mature granulocytes	+
Leu M2	60–95% monocytes	+
MO2	Adherent monocytes	+

totoxic reaction (ADCC) similar to that seen in studies of renal transplantation.[10] This could account for the serum effect displayed by these two patients.

These results suggest a central role for the macrophage in rejection and the assay provides an *in vitro* model, which can be manipulated to analyze the important immunologic events occurring during rejection.

Acknowledgments

Supported in part by grants EY 03999, EY 01302, and EY 01765 from the National Institutes of Health.

The authors thank D. Kappus for serum screening and R.A. Prendergast and J. Donnelly for helpful suggestions with the manuscript.

References

1. Maumenee AE: *Am J Ophthalmol* **34**:142–152, 1951.
2. Khodadoust, AA: In *Corneal Graft Failure*, Ciba Foundation Symposium. Elsevier, Amsterdam, 1973, pp. 151–163.
3. Stark WJ et al.: *Am J Ophthalmol* **79**:795–802, 1975.
4. MacDonald AL et al.: *Transplantation* **23**:431–436, 1977.
5. Stark WJ et al.: *Invest Ophthalmol* **12**:639–645, 1973.
6. Boyum A: *Scand J Clin Lab Invest* **21**(Suppl 97):77, 1968.
7. Young E. and Roth FJ: *J Invest Dermatol* **72**:46–51, 1979.
8. Kirchner H et al.: *J Natl Cancer Inst* **55**:971–975, 1975.
9. Young E and Stark WJ: *Invest Ophthalmol* **22** (Suppl 3):210, 1982.
10. Thomas JM et al.: *Transplantation* **22**:94–100, 1976.

chapter 21

Inhibition of Locally Mediated Corneal Allograft Rejection by Topically Applied Cyclosporin-A

Paul A. Hunter

Cyclosporin-A (CS-A) is a potent new immunosuppressive agent which is known to act preferentially on T-lymphocytes and has been shown to be effective in prolonging allograft survival in many experimental and clinical transplantation studies.[1] A number of studies have stressed that the timing in the administration of CS-A early in the allograft reaction is critical for successful immunosuppression.[2,3] It would also appear that, in view of the current theories of its mechanism of action,[1] CS-A must be available in sites where host lymphocytes become sensitized to foreign antigen.

The first experiments[4] to test the efficacy of CS-A in corneal allografts used a well-established rabbit model in which corneal rejection was brought about by the additional transfer of donor skin 2 weeks following keratoplasty. In this model, corneal allograft survival was prolonged when CS-A was administered systemically for 2 weeks following sensitization. However, while these experiments were being carried out, workers who were involved in clinical trials of CS-A in other transplantation fields reported a number of side effects,[5] the most important of which is nephrotoxicity. Other recorded side effects include hepatotoxicity, hirsutism, gum hypertrophy, tremors, neurasthenia, benign breast tumor, and depres-

sive psychosis. Three lymphomas have been detected in patients receiving additional immunosuppression. Although the nephrotoxicity may be dose-dependent and partially reversible,[6] the exposure of patients undergoing corneal surgery to such risks might not be acceptable if a safer route of administration could be found.

In an attempt to avoid these potential side effects, it was suggested that CS-A might be applied topically. This experiment was carried out by Shepherd et al[4] using the same model that had been earlier used to test the effects of systemically administered CS-A. As this is a second set model, the application of CS-A to the recipient eye predictably had no effect on corneal graft rejection.

A single set model of corneal graft rejection was therefore devised in which rejection resulted through the transfer of corneal antigens only in order that topically applied CS-A might be tested.[7]

Materials and Methods

Corneal allografts 6 mm in size were exchanged between albino and Dutch rabbits with the grafts placed eccentrically near the limbus. This encouraged the rapid ingrowth of vessels and at the same time it ensured iris adhesions that further enhanced vascularization. A detailed description of this model has been published in the literature elsewhere.[8] Rejection was readily diagnosed by the appearance of epithelial or endothe-

Department of Clinical Ophthalmology, Moorfields Eye Hospital, London, England.

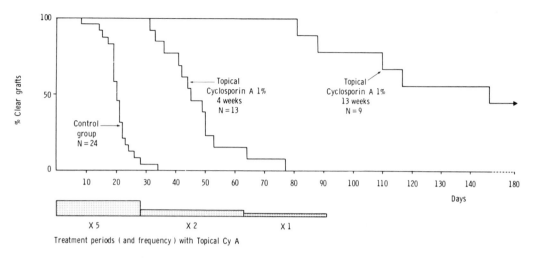

FIG. 1. Survival of rabbit corneal allografts treated with topical cyclosporin A.

lial rejection lines, or the appearance of clouding in a previously clear graft stroma. Cyclosporin-A drops were administered to the recipient eye as a 1% solution dissolved in arachis oil.

Results

In a control series of untreated corneal allografts, the onset of rejection occurred between the 8th and 34th day postoperatively. No graft in this group failed to be rejected. (Figure 1).

A group of rabbit corneal allografts was then treated with either topically applied Cyclosporin-A 1% in arachis oil five times daily or with arachis oil alone in a fully masked trial. Treatment was continued for 28 days or until rejection occurred. None of the grafts in the CS-A treated group were rejected while treatment was maintained. The first day on which any graft was rejected was day 31, and one graft remained clear for a period of 77 days (Fig. 1). The mean survival time of grafts in this group was $47.31 \pm SD$ 12.56 days, more than double that of the control group ($20.78 \pm SD$ 4.80 days) which was treated with arachis oil alone ($p < 0.001$ Student t test).

Thus, topically applied cyclosporin-A was seen to inhibit, but not prevent, corneal allograft rejection in rabbits. Therefore, a further experiment was performed in which

treatment was maintained for a period of 3 months following surgery. After 1 month the frequency of drops was reduced to twice daily and during the third month each rabbit received only 1 drop daily to the grafted eye. In this experiment, no graft was rejected during the first 2 months, but two grafts were rejected 2–3 weeks after the dosage was reduced from twice to once daily. After stopping treatment, three of the remaining seven grafts were rejected, but four grafts remained clear for more than 180 days (Fig. 1).

When these four grafts were revascularized 3 months after stopping treatment with CS-A, three developed typical signs of rejection within 30 days. This would suggest that the prolonged graft tolerance, seen after stopping treatment, is likely to be due to the return of the immunological privilege of the cornea as the central vessels regressed, rather than the presence of immune tolerance.

Discussion

The mechanism by which CS-A produces immunosuppression is still largely unknown. A recent study[9] has suggested that it inhibits the generation of cytotoxic effector T-lymphocytes without inhibiting the expression of suppressor T-lymphocytes. Another experiment[10] using the human mixed lymphocyte reaction has shown that CS-A exerts its effect on responding T-cells by preventing their proliferative response to interleukin-2 (IL2). It does not, however, inhibit the response of activated T-cells to IL2. It has been postulated that as T-cells become sensitive to IL2 through signals from the HLA-DR of stimulator cells, then CS-A may block receptors for DR molecules on the responding T-cells, thereby inhibiting the expression of IL2 receptors on the same cells.

Although the routes of absorption, distribution, and metabolism of topically applied CS-A are not known, it may be possible in the light of the experiments described to make some general observations about the role of locally mediated immune responses involved in corneal allograft rejection.

Where graft rejection was the result of a second set reaction,[4] topical CS-A was ineffective presumably because sensitization to donor antigen occurred at sites that were not exposed to sufficient amounts of the drug, that is, the skin allograft and its draining lymph nodes. However, using the single set model described here, topically applied CS-A was highly effective in inhibiting the allograft reaction. Comparison of control groups in the two models would seem to indicate that this effect is unlikely to be due to a reduced degree of immunological challenge in the single set model as has recently been suggested.[11] The untreated controls in the single set model were all rejected between 1 and 5 weeks, whereas rejection did not occur in the second set model until a period of 2–8 weeks had elapsed following sensitization. From these experiments, and from what is currently known of CS-A's mechanism of action, it is possible to infer that in the single set model the sensitization and subsequent proliferation of T-lymphocytes may occur as local phenomena within that part of the eye that is accessible to topically applied CS-A.

In a separate experiment using topical CS-A, Kana et al,[11] were able to demonstrate delayed corneal allograft rejection in a second set model. Although this finding appears to contradict the earlier results of Shepherd et al,[4] there were a number of differences between the two experiments most obviously in the strength of the CS-A and the vehicle used. Kana's group also used larger grafts and allowed the sutures to remain in place indefinitely. These diferences between the two models may offer an explanation for the apparent discrepancy in the results. The larger graft and persistent sutures may have invoked a greater local immunological response which would, therefore, be accessible to topically applied CS-A although the systemic sensitization eventually brought about rejection in all cases.

The concept of local ocular immunoresponsiveness is not new. Ten years ago, Jones[10] summarized a number of possible local mechanisms by which corneal allograft rejection might be brought about. I should like to submit that the experiments I have described, quite apart from suggesting a possibly safer clinical use for CS-A, lend further support for such a hypothesis.

References

1. Morris PJ: *Transplantation* **32(5)**:349–354, 1981.
2. Marwick JR et al.: *Lancet* **2**:1037–1039, 1979.
3. Homan WP et. al.: *Transplantation* **29**:361–366, 1980.
4. Shepherd WFI et al.: *Br J Ophthalmol* **64**:148–155, 1980.
5. Editorial. *Lancet* **2**:779–780, 1979.
6. Sweny P et al.: *Lancet* **1**:663, 1981.
7. Hunter PA et al.: *Clin Exp Immunol* **45**:173–177, 1981.
8. Hunter PA et al.: *Br J Ophthalmol* **66**:282–302, 1982.
9. Hess AD et al.: *J Immunol* **126**:961–965, 1981.
10. Palacios R and Moller G: *Nature* **290**:792, 1981.
11. Kana JS et al.: *Invest Ophthalmol Vis Sci* **22**:686–690, 1982.
12. Jones BR: In *Corneal Graft Failure*, CIBA Foundation Symposium, Associated Scientific Publishers, Amsterdam, 1973, pp. 349–354.

chapter 22

Lymphocytes in Allergic Conjunctival Disorders

**Takenosuke Yuasa, Yasunori Yamamoto, Rei Tada,
Yoshikazu Shimomura, Yayoi Nakagawa, Yasuo Mimura**

The conjunctival disorders caused by an immediate type of allergic reaction, such as vernal conjunctivitis and allergic conjunctivitis, may result from the production of IgE in the conjunctival tissue. However, the mechanism of excess production of IgE is still unknown. When an allergen makes contact with a sensitized individual, the first reaction to the allergen is the blastoid transformation of lymphocytes[1-3] that recognize the allergen. Therefore, the blastoid transformation of lymphocytes was examined in patients with allergic disorders of the conjunctiva.

Materials and Methods

Peripheral blood lymphocytes (PBL) from allergic patients and healthy controls were washed 3 times, and suspended in RPMI 1640 medium containing streptomycin, penicillin, and 10% human type AB serum. The final concentration of lymphocytes was 1×10^6/ml. PHA-P (final concentration, 10 μg/ml) or 1·10·100μg/ml of a mite antigen was added to the suspension, and it was incubated in a Falcon microplate (#3040) under 5% CO_2 atmosphere at 37°C for 7 days (3 days when PHA-P was added). ^{3}H-thymidine in a dose of 0.2 μCi (20 μl) was added to the suspension in each well 18 hours prior to the recovery of the lymphocytes. The

Department of Ophthalmology, Osaka University Medical School, Fukushima, Osaka, Japan.

uptake of ^{3}H-thymidine was determined by a scintillation counter after recovery of the lymphocytes from the suspension; a stimulation index (S.I. = thymidine uptake with allergen/without allergen) was calculated.

Results

Our studies showed no significant difference in the S.I. to PHA-P between healthy subjects and patients with vernal conjunctivitis or allergic conjunctivitis. Thus, in the nonspecific proliferative responses of lymphocytes no peculiar reaction pattern was seen in the patients with allergic conjunctival disorders, and controls were similar.

In the reaction to the mite antigen, the antigen levels necessary to elicit the maximal S.I. were different in various individuals. The maximal S.I in each case is compared in Figure 1; there were significant differences between healthy subjects and patients with vernal or allergic conjunctivitis. All patients with vernal or allergic conjunctivitis showed positive skin tests to a house dust allergen. Lymphocytes in the patients with vernal conjunctivitis showed the greatest reactivity to the mite antigen. Only lymphocytes from vernal conjunctivitis patients showed significantly elevated S.I. as compared with the other two groups even when the concentration of the mite antigen was 1μg/ml, as shown in Figure 2.

The relation between the S.I of lymphocytes in the peripheral blood and the RAST

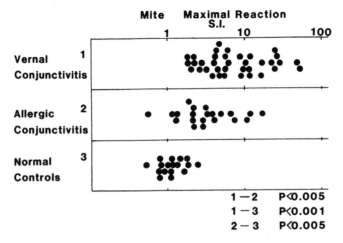

FIG. 1. Maximal stimulation index (S.I.) in lymphocyte transformation test with mite antigen.

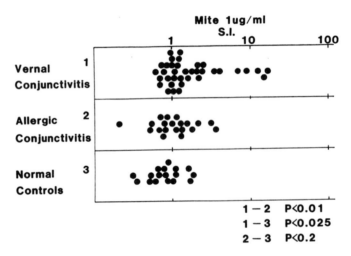

FIG. 2. Stimulation index (S.I.) in lymphocyte transformation test with mite antigen (1 μg/ml).

score of sera from patients with vernal conjunctivitis to the mite antigen is shown in Figure 3. There were significant differences between the cases showing a score of 3 or 4, and those with a score of 0 ($p<0.02$). A stronger tendency toward blastogenic transformation of lymphocytes was seen among the cases with higher RAST scores. However, there was no significant relation between the total IgE level in serum, the severity of the exacerbation, and the S.I. The ocular symptoms appear and resolve

within relatively short periods of time, but the S.I. value of peripheral lymphocytes does not change in the same way as do ocular symptoms.

Increase in the serum IgE level and in the production of IgE in the conjunctiva are observed in patients with conjunctival allergy.[4] Therefore, IgE production by lymphocytes was examined *in vitro* in order to discern the mechanism of excess production of IgE by lymphocytes.

The PBL obtained from patients with al-

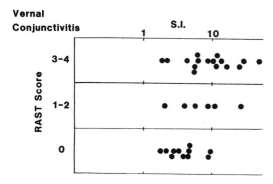

FIG. 3. RAST score and maximal S.I. to mite antigen in vernal conjuctivitis.

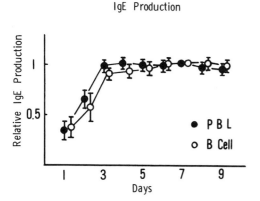

FIG. 4. IgE production *in vitro* during the culture.

lergic conjunctival disorders were suspended at a final concentration of 1×10^6/ml, in RPMI 1640 containing 10% fetal calf serum (FCS), and incubated at 37° under 5% CO_2 atmosphere. T-cells were isolated by the formation of E-rosettes, and separated B-cells (non-T cells) were incubated under the same conditions. An aliquot of the incubation mixture was harvested every day, and the IgE level of the supernatant was determined by a PRIST method as shown in Figure 4. The maximum IgE level was obtained on the 7th day after the initiation of incubation, though the IgE level had already reached a plateau level by the 4th day of incubation. The following information was obtained by assaying the IgE levels of the supernatant fluid.

An increase in IgE production was observed in lymphocytes obtained from patients with vernal conjunctivitis, but not in lymphocytes obtained from patients with allergic conjunctivitis or from healthy controls (Fig. 5). The ratio of IgE production by PBL and by B-cells was not directly proportional to the serum IgE levels. Therefore, it may assumed that the amount of IgE produced by B-cells might be greater when incubated with PBL. An increase in IgE production was not achieved by adding pokeweed mitogen (PWM), whereas an increased production of immunoglobulin is

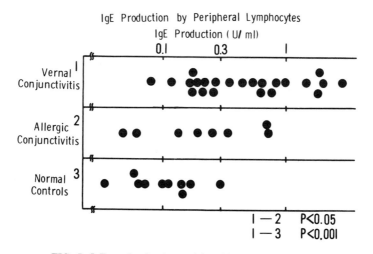

FIG. 5. IgE production by peripheral blood lymphocytes (PBL).

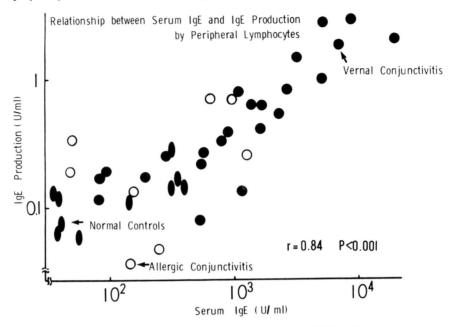

FIG. 6. Produced IgE *in vitro* and serum IgE level.

seen in IgG- or IgM-producing cells under these conditions.[5,6] The phenomenon may be the result of the activity of T-helper cells. However, in controls a correlation between the IgE level in the serum and the amount of IgE production by the PBL are shown (Fig. 6); a close correlation was observed between both specimens.[6]

The cause of the excess production of IgE in patients with atopic diseases, including conjunctival allergy, might be an abnormal helper function or suppressor function of T-lymphocytes. The helper function[7] was examined as follows: B-cells, 1×10^6/ml, were incubated alone or with an equal amount of autologous or heterologous T-cells, in the presence of PWM ($\times$ 100). On the 7th day of incubation, the IgE levels in the incubation aliquots were determined by the PRIST method. The suppressor function was examined as follows: T-cells of donors, 1×10^6 ml, were cultured for 3 days with 10 μg/ml of concanavalin A (Con A) for induction of suppressor T-cells. After this incubation, cultured T-cells in concentrations of 0.5, 1, and 2 $\times$ 10^5 in 100 μl of the medium, were mixed with a constant amount of PBL, (1×10^5 of responders) in 100 μl.

The incubation was continued for an additional 7 days, and the IgE level in the medium was determined. The responders were patients with allergic diseases.

An increase in IgE production by B-cells was demonstrated in many cases (as shown in the upper column of Figure 7) when autologous T-lymphocytes were incubated with PWM. A comparable result is obtained by the use of PBL. On the other hand as shown in the lower column in Figure 7, when B-cells obtained from patients with vernal conjunctivitis were added to the T-cells obtained from healthy subjects, IgE production was lower in most cases than when autologous T-cells were used. Therefore, it may be suggested that T-cells in the healthy subject have lesser helper function than T cells in allergic patients.

T-cells obtained from a pair of patients with vernal conjunctivitis, having more than 1,500 I.U./ml and less than 800 I.U./ml of serum IgE, were exchanged mutually in incubation with B-cells, and IgE production was examined. The ratio of IgE production, i.e., autologous T-cells/heterologous T-cells, is shown in Figure 8. When the autologous serum IgE level was lower than the heter-

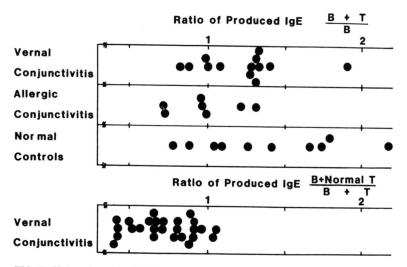

FIG. 7. Helper function of T-lymphocytes: The upper column shows the effects of autologous T-cells added to B-cells and the lower column shows the effects of normal T-cells added to atopic B-cells.

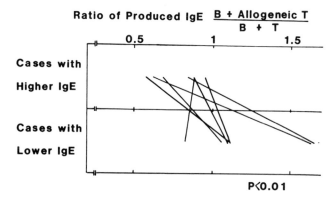

FIG. 8. T-cell exchange effects in helper function of lymphocytes from cases with vernal conjunctivitis.

ologous level, IgE production increased when heterologous T-cells were added, and vice versa. Thus, the helper function of T-cells from the donor showing the higher circulatory IgE level was the greater of the two.

On the other hand, the above results might be caused by lowered suppressor function of T-cells in the allergic patients. Thus, the suppressor function[8,9] of T-cells induced by Con A was examined. As shown in Table I, IgE production was suppressed

TABLE I

Dose Dependent Suppressor Function of Concanavalin A-induced T-lymphocytes in Allergic Conjuntival Disorders

	Suppression (+)	Average Suppression	Suppression (−)
Vernal conjunctivitis	7 cases	74.9%	15 cases
Allergic conjunctivitis	1	73.9	7
Normal controls	4	79.1	8

proportionally to the added amount of Con A-stimulated T-cells in 12 cases (27.9%). However, the supposed decrease in suppressor function could not be demonstrated in patients with vernal conjunctivitis as compared with controls. The IgE production was suppressed approximately 75%, whether the suppressor lymphocytes came from vernal conjunctivitis patients or from healthy controls.

Discussion

These findings suggest that an increase in the helper function of T-cells in allergic patients might be established, but an impairment of suppressor function could not be demonstrated. Serum IgE levels in the patients with allergic disorders may be 10–100 times higher than that in normal subjects, but the helper functions, as demonstrated in the present studies, cannot be the sole explanation for the excess production of IgE.

References

1. Zietz SJ et al.: *J Allergy* **38**:321, 1966.
2. Richter M and Naspitz CK: *J Allergy* **41**:140, 1968.
3. Rocklin RE et al.: *J Clin Invest* **53**:735, 1974.
4. Yuasa T: *Excep Med Int Congress Series* **450**:947, 1978.
5. Romagnani S et al.: *Clin Exp Immunol* **42**:167, 1980.
6. Ohta K et al.: *Int Arch Allergy Appl Immunol* **63**:129, 1980.
7. Hirano T et al.: *J Immunol* **119**:1235, 1977.
8. Too TJ: *J Allergy Clin Immunol* **63**:143, 1979.
9. Delafuente JC et al.: *Ann Allergy* **44**:331, 1980.

chapter 23

Pathology of Corneal Plaque in Vernal Keratoconjunctivitis

A.H.S. Rahi,[a] R.J. Buckley,[b] and I. Grierson[a]

Vernal disease shows a wide spectrum of clinical, immunological and histopathological features. It affects, in various combinations and with varying severity, the conjunctiva, the limbus, and the cornea.[1,2] The conjunctival lesions vary from a fine papillary reaction to a more severe and persistent inflammation with formation of the characteristic ''cobblestones'' (giant papillae) in the upper tarsal plate. There may be associated transient limbal hyperemia with swelling and cellular infiltration. In severe cases, however, limbitis with vegetations and Tranta's dots may develop and lead to neovascularization (pannus) and subepithelial fibrosis (lace scarring). Corneal vernal disease appears to develop as a complication of the conjunctival disease and consists at first of superficial punctate erosions involving the upper half of the cornea. The micro-erosions may coalesce to give rise to massive epithelial defects to the base of which mucus may adhere, resulting in the formation of persistent plaques (Fig. 1) that prevent regeneration of the epithelium. The ''indolent ulcer'' leads to permanent corneal scarring, which fortunately does not involve the visual axis.

In an analysis of 100 patients with vernal disease attending Moorfields, it was found that in 82% of the cases the onset was in the first 10 years of life.[1] The disease was nearly always bilateral, and in about half of the patients it was strictly seasonal. Two-fifths of the patients had mild disease with palpebral and some limbal pathology. In another two fifths the disease was of moderate severity in which, in addition to marked papillary conjunctivitis, a large proportion showed corneal disease. In the remaining 20% of the patients the disease was severe and the cornea was always involved. Corneal plaque is an uncommon but severe manifestation of vernal keratopathy. It is likely to develop in badly managed cases and in patients who respond poorly to conventional therapy.

Since the precise nature of vernal plaque is unknown, a detailed light and electron microscopic, histochemical, and immunological examination was undertaken. Its aims were to understand the resistance to mucolytic agents, to identify the constituents of the greyish amorphous material of which these plaques are composed, and to explain why there is no surgical plane between Bowman's zone and the plaque.

Materials and Methods

The present communication concerns pathological examination of 25 specimens of vernal plaque. The tissues were processed for routine light microscopic examination. Larger specimens were cut into several pieces for scanning and transmission electron microscopy. Frozen sections were used

[a]Institute of Ophthalmology and [b]Moorfields Eye Hospital, London, England.

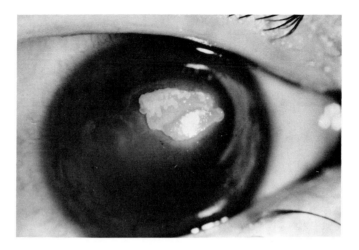

FIG. 1. Clinical photograph of a plaque in the upper cornea of a patient with severe vernal keratoconjunctivitis.

for immunological and histochemical studies. Sections were stained with fluorescein-labeled antibodies to demonstrate the presence of fibrin, immunoglobulins, and other serum proteins. Histochemical analysis included demonstration of fibrin, tyrosine, sulphur-containing amino acids, and glycoproteins.

Results

Light microscopic appearance

The plaques are formed of amorphous material which often appears to be finely granular and eosinophilic. The staining intensity, however, is variable and patchy. In more superficial parts, the plaque appears laminated, suggesting repeated episodes of deposition of material from tear film. The material is intimately adherent to Bowman's zone which appears normal. The superficial stroma shows infiltration by lymphocytes and by occasional plasma cells and eosinophils.

Histochemical findings

The plaque is PAS and Alcian blue (pH 2.5) positive and consists of a mixture of proteins and mucopolysaccharides. It contains large amounts of tyrosine and sulphur-containing amino acids. Sections stained by the MSB, PicroMallory, and Masson 44/41 technique showed the presence of both young and old fibrin, which provides evidence that plaque formation occurs in repeated episodes.

Ultrastructure (Fig. 2)

Transmission electron microscopy shows large deposits on Bowman's layer of electron-dense fibrillar material interspersed with more electron-lucent amorphous aggregates. The fibrillar material is unbanded but its appearance is not inconsistent with that of fibrin. The electron-lucent, amorphous material is somewhat similar to the mucoid material in goblet cells. The individual microfibrils in Bowman's layer are less clearly demonstrated largely because the superficial interfibrilar space is filled with electron-lucent material similar in appearance to that present in the plaque. The transition between the plaque and Bowman's layer is gradual and irregular, and collagen fibrils can be seen sporadically within the deeper layers of plaque immediately above Bowman's layer. The amount of mucoid material and the degree of penetration of the Bowman's layer is variable. The plaque appears to contain, in addition, cellular debris, vesicular elements and membranous structures originating from

FIG. 2. Electron micrograph of vernal plaque overlying Bowman's layer (BZ). Note the stratified appearance of the deposits; the electron-dense (fibrin) and electron-lucent (mucus) material appear to have accumulated in repeated episodes.

necrotic corneal epithelium and possibly inflammatory cells.

Scanning electron microscopy

The anterior (i.e., superficial) surface is continuous but undulating and devoid of epithelial covering. There are distinct nodular extrusions which probably correspond to the surface aggregates of mucoid material seen by electron microscopy.

Immunological findings

The plaque contains most of the plasma proteins including albumin, acute phase proteins, and transferrin. Immunofluorescent examination shows the presence of IgG, IgA and complement within the plaque. Antibodies to fibrinogen show stratified deposits of fibrin in the intermediate layers of the plaque. There is some fluorescence in the superficial layers, but it is absent in the region immediately in contact with the Bowman's layer.

Discussion

This communication contains only the preliminary findings of an extensive study that we have undertaken to understand the pathobiology of corneal plaque in vernal disease. The stratification noted in this study confirms an earlier report from our institution[3] that was based only on a limited number of cases. The presence of fibrin, however, as a major component of vernal plaque is reported here for the first time. This explains the failure of mucolytic agents to dissolve the corneal deposits and allow the epithelium to cover the defect. Furthermore, we were able to demonstrate frayed collagen fibers from Bowman's layer penetrating the plaque, and the mucoid material in the plaque, in turn, packing the interfibrillar space in the superficial Bowman's zone. These findings not only clarify the absence of a demonstrable surgical plane between the plaque and Bowman's layer, but also explain why superficial keratectomy becomes necessary to remove the plaque *in toto*. The survival of substantial amounts of fibrin within the vernal plaques, for what appears to be an extended period of time, may be explained to some extent by an absence of a significant neutrophilic infiltrate into this tissue. Neutrophils are primarily responsible for clot dissolution at other sites in the body.

It is known that histamine and prostaglandins are present in the tears of patients with vernal disease. In a recent study[4] it was shown that the major basic protein derived from granules of eosinophils is also markedly elevated in these patients. Since this protein is toxic in tissue culture to respiratory epithelial cells, it is possible that corneal erosion and keratitis are in part induced by the products of eosinophils dissolved in the mucous discharge. We have observed free eosinophil granules in tissue sections in vernal disease. It is possible, therefore, that corneal disease develops following the release of these granules from a limbal lesion.

The presence of immunoglobulins, complement, and other serum proteins suggests that the corneal plaque may develop following inspissation of the plasmoid tears formed during acute exacerbations. It is possible that the IgG and C3 are actually responsible for the cytotoxic changes in the cornea and might indicate an immune-complex mediated inflammation. Whether the vernal plaque represents an end stage of an IgE-mediated allergic inflammation or acts as a depot for the toxic factors released from inflammatory cells is unclear, but the presence of large amounts of fibrin is reminiscent of a cell-mediated tissue allergy[5] of which the Mantoux reaction is the prototype. It is of relevance in this context that peripheral blood lymphocytes (PBL) from patients with vernal disease show increased DNA synthesis when challenged with mite antigens[6] suggesting a pathogenetic role of delayed hypersensitivity (to common allergens) in vernal keratoconjunctivitis. Further studies are required, however, to clarify these conjectures.

References

1. Buckley RJ: In *The Mast Cells*, Pepys and Edwards (Eds.). Pitman, London, 1979, p. 518.
2. Easty DL et al.: In *The Mast Cells*, Pepys and Edwards (Eds.). Pitman, London, 1979, p. 493.
3. Rice, NSC et al.: *Trans Ophthalmol Soc UK* **91**:483, 1971.
4. Gleich GJ, Udell, IJ, Ackerman, SJ and Abelson MB: *ARVO Exports* p. 100, 1981.
5. Gogi R et al.: *Histopathology* **3**:51, 1979.
6. Yuasa T et al.: *Acta Soc Ophthalmol Jpn* **85**:1060, 1981.

chapter 24

HSV Infection of the Trigeminal Ganglia Prevents Subsequent Superinfection by Another HSV Strain

Ysolina M. Centifanto-Fitzgerald

One striking characteristic of herpes simplex virus (HSV) infections is the recurrent nature of the lesions. After the primary episode of disease runs its course, the virus remains in the ganglia in a latent state. This provides a source of virus for subsequent episodes of disease.

Previous work from our laboratory has demonstrated that initial colonization of the ganglia by one strain of HSV precludes superinfection of the same ganglia by other HSV strains[1]. Although the eyes could be infected by the second strain, the disease was less severe. The virus was shed in the tears for several days, but only the initial infecting HSV strain was found in the ganglia. The mechanisms involved in this protection are not clear, and they warrant investigation. Specifically, we wanted to know the role of humoral and secretory antibodies and whether this protection could be related to the amount of latent virus in the ganglia.

Methods and Materials

Experimental design

New Zealand white rabbits were infected in both eyes with the E-43 strain of herpes simplex virus Type 1 (HSV-1). Six months later, the animals were challenged with the McKrae strain. Cultures were taken from the tear film of these animals with a moistened cotton swab three times a week for 60 days. The animals were sacrificed, and both left and right ganglia were processed for virus recovery. The day before challenge, the animals were bled and tears were collected to ascertain immune status.

Cells

RK-13 cells (Microbiological Associates, Rockville, MD) were grown in basal minimal medium supplemented with 10% fetal calf serum, 1% glutamine, sodium bicarbonate, and antibiotics. Vero cells grown in Medium 199, supplemented with 5% fetal calf serum, were used for the DNA work.

Antibodies

Rabbits were immunized with the E-43 strain of HSV-1 grown in RK-13 cells. One ml of a 1:1 mixture of immunogen: Freund's Complete Adjuvant (Cappel Laboratories, Inc., Cochranville, PA) was injected intramuscularly into the hind legs of four New Zealand white rabbits. Each rabbit was immunized weekly for three weeks, boosted 10 days later, and bled 15 days following the final immunization. The potency of the sera was tested in a microneutralization assay.[2]

Lions Eye Research Laboratories, LSU Eye Center, Louisiana State University Medical Center School of Medicine, New Orleans, Louisiana.

Restriction endonuclease analysis

Confluent monolayer cultures of Vero cells were infected with each virus isolate at an input multiplicity of one plaque-forming unit (PFU)/cell and were allowed to absorb at 37°C for 1 hour. Phosphate-free Eagle's minimal essential medium supplemented with 1% calf serum and containing 50 μCi/ml of ^{32}P-orthophosphoric acid (New England Nuclear, Boston, MA) was added, and the cultures were incubated for 24–30 hours at 37°C.

The infected cells were shaken off the glass and centrifuged for 10 minutes at 900 rpm in an International Centrifuge. The cell pellet was gently resuspended in 0.5 ml of 10 mM EDTA. Ten μl of a 20% SDS solution and 25 μl of pronase (20 mg/ml) were added, and the samples were incubated at 37°C for an additional 24 hours.

The virus DNA was then extracted with one volume of 80% phenol (pH 7.0 with 50 mM Tris base, 0.2 M NaCl) for five minutes on a rocker table. One-third volume of chloroform was added, and the sample was agitated for 5 minutes. The samples were centrifuged at 2000 rpm for 5 minutes to separate the phases. The DNA-containing layer was re-extracted with one volume of phenol/chloroform (2:1) and dialyzed for 36 hours against Tris-EDTA buffer (10 mM Tris, 1 mM EDTA).[3]

Results

Effect of immunization

The titers of the HSV-1 immunized animals were high, ranging from 1:128 to 1:256. These immune animals were infected in both eyes with the McKrae strain of HSV and sacrificed at 7 days postinfection, and the ganglia were processed for virus recovery. Virus was recovered from all eight ganglia. The data suggest that although antibodies to HSV may have modified the ocular infection, they did not prevent colonization of the ganglia.

The animals that were initially infected with the E-43 strain were challenged with the McKrae strain exactly 6 months postinfection. All the eyes showed infection, and

the virus was shed from these animals at a rate comparable to that of the control. After two months, the animals were sacrificed and the ganglia processed for virus recovery. The virus was recovered in RK-13 cells and subsequently cultured on Vero cells for DNA analysis.

All of the virus recovered from the 18 ganglia removed from the doubly-infected animals was shown to be the E-43 strain (Fig. 1). In one case, the left ganglion showed a variant of the E-43 strain. This variant strain has been observed in our E-43 stock. It is our understanding that the vi-

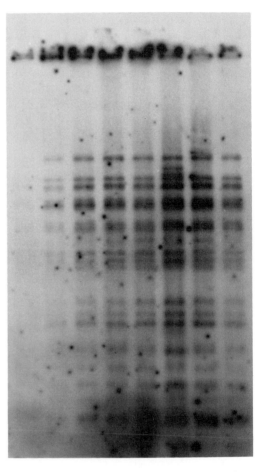

FIG. 1. Cleavage patterns of DNA from isolates from the trigeminal ganglia of doubly-infected rabbits, obtained by means of restriction endonuclease analysis with Bam HI. DNA fragments were separated by electrophoresis on 0.8% agarose. All isolates have the characteristic pattern of the E-43 strain.

rus stock is a heterogeneous mixture and that in some cases this variant is selected out. It is not clear whether the selection takes place at the level of acute infection in the ganglia or during *in vitro* growth after processing of the ganglia.

Discussion

The results of this study indicate that colonization of the trigeminal ganglia by HSV after initial ocular infection precludes later ganglionic superinfection with another HSV strain. We have shown that this protection is not mediated by antibodies to HSV, because the hyperimmune animals were readily infected, and virus was recovered from the trigeminal ganglia of all the animals.

This observation raises questions about our current understanding of herpetic disease. It is known that the trigeminal ganglia act as a reservoir of virus for recurrent ocular disease[4] and that even in severe ocular infections, only a small percentage of the neuronal cells are infected. Yet in our study, superinfection of the same ganglia did not occur.

In mice, it has been shown that immunization reduces the incidence of latent infections, compared to nonimmunized controls.[5] Also, Walz and co-workers have shown that the immune response restricts the number of neurons that acquire latent virus. But at the same time, latent ganglionic infection can occur in immunized animals, with a high titer of neutralizing antibodies present.[6] We have confirmed this finding, inasmuch as, in this work, neither humoral nor secretory antibodies prevented infection of the trigeminal ganglia. These data imply that the protective mechanism(s) involved in ganglionic superinfection are related more to the physiology of the neuronal cells or the nature of the trigeminal ganglia as a privileged site than to the characteristics of the port of entry, i.e., the cornea cells. An investigation of this hypothesis is now in progress. In any case, this work is of importance in furthering our understanding of recurrent ocular herpetic disease.

Acknowledgment

This work was supported in part by USPHS grants EY02389 and EY02377 from the National Eye Institute, National Institutes of Health, Bethesda, Maryland.

References

1. Centifanto-Fitzgerald, YM et al.: *Infect Immun* **35**(3):1125–1132, 1982.
2. Rawls WE et al.: *J Immunol* **104**:599–606, 1970.
3. Buchman TB et al.: *J Infect Dis* **138**:488–489, 1978.
4. Nesburn AB et al.: *Invest Ophthalmol* **15**:726–731, 1976.
5. McKendall RR: *Infect Immun* **16**:717–719, 1977.
6. Walz MA et al.: *Nature* **264**:554–556, 1976.

chapter 25

Cell-Mediated Immunity in HSV Keratitis:

Modulating Effect of Serum Antibodies

Bryan M. Gebhardt, Candice Cutrone, and Stephen Kaufman

Cell-mediated immunity is necessary for resistance to a variety of viral infections.[1] Since cell-mediated immunity is a function of lymphocytes, particularly thymus-derived lymphocytes (T-lymphocytes or T-cells), most immunologists have focused their attention on the functional activities of these cells in resistance to viral infection. A major thrust centers on the role of host resistance in the prevention and cure of infection. This study is part of a detailed analysis of lymphocyte-mediated (cell-mediated) immunity in HSV-1 ocular infection.

Several reports have indicated that lymphocytes are required to resolve HSV-1 infections of the cornea of the eye.[2-6] Despite the fact that these investigations demonstrate the requirement of lymphocyte immunity in "curing" HSV-1 ocular infection, many features and details of this immunity have not been precisely defined. The experiments reported in this paper form part of a continuing analysis of the stimulation and regulation of lymphocyte immunity during corneal infection by HSV-1.

It was found that experimental ocular infection by the RE-6 strain of HSV-1 results in specific lymphocyte immunity 3 weeks after primary infection and that this immunity can be modulated by a serum substance present in infected, but not in control, animals.

Lions Eye Research Laboratories, LSU Eye Center, Louisiana State University Medical Center School of Medicine, New Orleans, Louisiana.

Materials and Methods

Infection with HSV

New Zealand white (NZW) rabbits (2–3 kg) were used. RE-6 strain (HSV-1) at a multiplicity of infection of 10^7 per ml was used to infect both corneas of each rabbit. One ml of chlorpromazine (Thorazine) was administered intramuscularly one half hour before inoculation and two drops of .5% proparacaine hydrochloride (Ophthaine) were used as a topical anesthetic.

Lymphocyte stimulation

The rabbits were killed by exsanguination via cardiac puncture. Approximately 30 ml of blood was retained for serum; the remainder was collected in heparin at a concentration of 1 ml heparin per 30 ml peripheral blood. The peripheral blood lymphocytes (PBL) were separated using Ficoll-Hypaque. This process consists of layering 4 ml of PBL over 3 ml of Ficoll-Hypaque. The mixture was then centrifuged at 1100 rpm for 40 minutes. The lymphocytes at the interface were collected and washed twice in Hanks' balanced salt solution (HBSS).

The spleen and preauricular lymph nodes were removed and each was minced in HBSS. Then each was separately reduced to a single cell suspension using an 80-mesh wire screen and a series of fine gauge needles. The suspensions were washed twice in HBSS. Cell suspensions at a concentration of 1.25×10^6 cells per ml were made of PBLs and of spleen or preauricular lymph node lymphocytes, using a culture medium of 90% RPMI 1640 and 10% rabbit serum from either a control rabbit or an infected rabbit. Six cell suspensions were made: (1) spleen lymphocytes from the infected rabbit in control serum or (2) in infected serum, (3) preauricular lymph node lymphocytes in control serum or (4) in infected serum, (5) PBL in control serum and (6) PBL in infected serum.

Each cell suspension was cultured in round bottom microculture plates, then each suspension was divided into three groups for harvesting at days 4, 5, and 6. Each group consisted of 28 cultures. The 28 wells were subdivided into seven sets of quadruplicate samples. The cells were cultured alone or with 10 μl of full-strength, 1:10, or 1:100 dilutions of HSV antigen or equivalent volumes and dilutions of control antigen.

The cultures were incubated at 37°C in an atmosphere of 5% CO_2/95% air. The three groups from each suspension were harvested at 4, 5, and 6 days. Eighteen hours before harvesting, 10 μl of tritiated thymidine (^{3}H-TdR) was added to each group. The amount of ^{3}H-TdR incorporated by the cells into each cell culture was determined in a liquid scintillation spectrometer.

HSV antigen

Antigen was produced by infecting a rabbit kidney (RK) cell line with RE-6 strain of HSV-1 for 24 hours. The cells were harvested, frozen, and thawed three times, collected in a pellet after centrifugation, and resuspended in a small volume (3–5 ml) of phosphate buffer saline (PBS). The cell fragments were reduced in size by sonication; the resulting suspension was considered to be "full strength." The control antigen was produced similarly from uninfected cells.

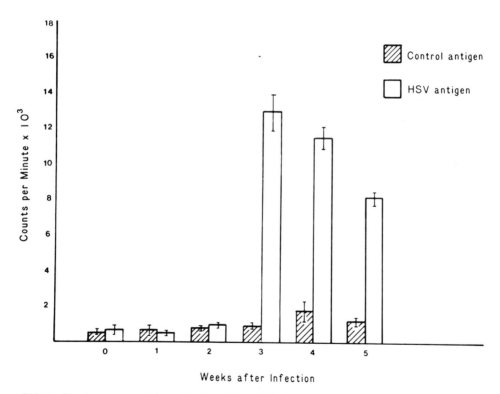

FIG. 1. The time-course of the maturation of the cellular immune response to HSV-1 antigens. Open bars indicate the lymphocyte response to HSV antigen; lined bars depict the lymphocyte response to control antigen.

Results

Clinical evidence of infection

Five to seven days after corneal infection, the animals had inflamed, reddened eyes. This inflammation began to resolve by day 14 and cleared completely by day 21.

Lymphocyte stimulation by RE-6 antigens

PBL proved exceedingly difficult to isolate in sufficient numbers and purity for use in these experiments. The PBLs that were tested did not give any evidence of responsiveness, and these negative results are not included in this report.

Lymphocytes from the spleen and preauricular lymph nodes of infected rabbits were maximally responsive to HSV antigen 3 weeks after infection. No significant response was seen at 1 and 2 weeks after infection, and the response began to drop off by 4 and 5 weeks (Fig. 1). No response to control antigen was noted at any of the times tested.

Routinely, lymphocytes were cultured in either autologous or pooled homologous serum collected from uninfected animals. Subsequent attempts to culture lymphocytes in the serum of infected animals revealed that such sera inhibited the lymphocyte response to HSV antigens. Further study of this inhibitory influence revealed that serum collected from rabbits 3 weeks after infection was maximally inhibitory (Fig. 2). This inhibitory effect was seen in cultures incubated for 4, 5, or 6 days. In addition, the suppressive effect was observed in lymphocyte cultures incubated with various concentrations of antigen (data not shown).

Further characterization of the serum inhibitory factor revealed that it could be isolated with the immunoglobulin fraction following ammonium sulphate precipitation and ion-exchange chromatography on DEAE cellulose (Table I). Thus, serum immunoglobulin from infected rabbits dramatically suppressed the response of immune lymphocytes as compared with immunoglobulin obtained from uninfected rabbit serum.

In an effort to test whether or not specific anti-HSV antibodies were responsible for the inhibition of lymphocyte reactivity to

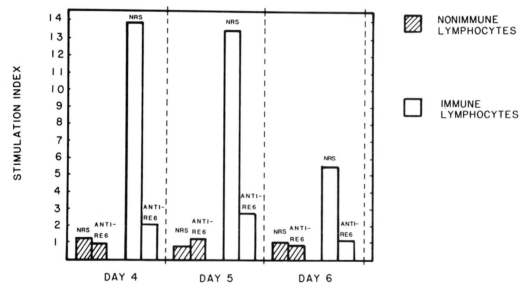

FIG. 2. The effect of serum from infected and uninfected rabbits on the lymphocyte response to HSV antigens. Open bars depict immune lymphocyte response to normal (uninfected) rabbit serum (NRS) or serum from an infected rabbits (ANTI-RE6). Lined bars depict the response of nonimmune lymphocytes to NRS or ANTI-RE6.

TABLE I

Serum Immunoglobulin-Mediated Suppression of the Lymphocyte Response to HSV *in vitro*

	^{3}H-TdR[a] Incorporation CPM ± SD[b]		SI[c]	% Suppression[d]
	− HSV antigen	+ HSV antigen		
Normal serum	465 ± 82	11,556 ± 864	25	—
Immune serum	624 ± 97	2,041 ± 506	3	83
Normal Ig	337 ± 77	9,681 ± 412	29	87
Immune Ig	489 ± 112	1,240 ± 207	3	—

[a] Tritiated thymidine

[b] Counts per minute ± standard deviation

[c] Stimulation Index $= \dfrac{\text{CPM (+ HSV antigen)}}{\text{CPM (− HSV antigen)}}$

[d] Percent suppression $= \dfrac{\text{CPM immune serum or Ig}}{\text{CPM normal serum or Ig}}$

TABLE II

Antibody-Mediated Suppression of the Lymphocyte Response to HSV *in vitro*

	^{3}H-TdR[a] Incorporation CPM ± SD[b]		SI[c]	% Suppression[d]
	− HSV antigen	+ HSV antigen		
Normal Ig	235 ± 112	5,091 ± 525	22	—
Anti-HSV antibody	609 ± 207	1,158 ± 144	2	77

[a] Tritated thymidine.

[b] Counts per minute ± standard deviation.

[c] Stimulation Index $= \dfrac{\text{CPM (+ HSV antigen)}}{\text{CPM (− HSV antigen)}}$.

[d] Percent suppression $= \dfrac{\text{CPM (anti-HSV)}}{\text{CPM (normal Ig)}}$.

HSV antigens, serum antibodies were isolated by affinity chromatography. Although only very small quantities of specific antibody have been obtained, an inhibitory effect of these antibodies on the lymphocyte response to HSV antigen has been recorded (Table II). Additional testing at various antigen:antibody ratios is needed to characterize more fully the action of the antibodies on cell-mediated immunity to HSV.

Discussion

In the present study, lymphocyte reactivity to HSV-1 antigens was not evident until 3 weeks after primary corneal infection. At that time, specifically reactive lymphocytes were present in the lymph nodes and spleens of HSV infected rabbits; similarly responsive lymphocytes were not present in control, uninfected rabbits.

A particularly interesting observation made in the course of these experiments was the suppressive effect of autologous serum on the lymphocyte response to HSV antigens. Lymphocytes from HSV-1 infected rabbits responded to HSV-1 antigens when cultured in the presence of serum obtained from uninfected rabbits, but this response was greatly reduced when the cells were cultured in serum from an HSV infected animal. The serum factor responsible for the suppression is apparently anti-HSV antibody, as suggested by these experiments. Meyers-Elliott and Chitjian[4] suggested that anti-HSV serum antibody might inhibit lymphocyte reactivity to HSV-1 antigens. In fact, these authors did not prove that anti-

HSV antibody was actually the serum substance responsible for the suppression. The mechanism by which anti-HSV antibody inhibits lymphocyte recognition of viral antigen is of considerable interest and may be of great relevance to the pathogenesis of herpetic keratitis. At certain stages of the infection, specific antibody might prevent lymphocyte attack on virus-infected cells, permitting the virus to survive, replicate, and spread to uninfected cells.

It is somewhat curious that peripheral blood lymphocytes did not respond to HSV-1 antigens. Since these cells are thought of as frontline defensive cells, it would be expected that they would be poised and ready to react to viral antigens. Further studies of blood lymphocytes are warranted in order to define the reasons for this lack of response by these cells. Perhaps the presence of specific antibodies in the serum causes the unresponsiveness of the blood lymphocytes.

Acknowledgments

This work was supported in part by USPHS grants EY03150 and EY02377 from the National Eye Institute, National Institutes of Health, Bethesda, Maryland.

References

1. Dulbecco R and Ginsberg HS: *Virology.* Harper and Row, Hagerstown, Maryland, 1980, pp. 1017–1030.
2. Steele RW et al.: *J Infect Dis* **131**:528–533, 1975.
3. Ilonen J: *Acta Pathol Microbiol Scand* **87**:151–157, 1979.
4. Meyers-Elliott RH and Chitjian PA: *Invest Ophthalmol Vis Sci* **19**:920–929, 1980.
5. Rasmussen L and Merigan TC: *Proc Natl Acad Sci USA* **75**:3957–3961, 1978.
6. Carter C and Easty DL: *Br J Ophthalmol* **65**:77–81, 1981.

chapter 26

Genetic Regulation of Autoimmunity

Noel R. Rose

This essay is presented with two goals in mind. The first is to describe our mounting understanding of the importance of the major histocompatibility complex *(MHC)* in regulating normal immunological responses. The second purpose is to apply these new concepts to the genetic regulation of autoimmunity, using thyroid autoimmunity as a specific example. I suggest that the lessons we have learned over the past several years in studying thyroiditis are pertinent to other organ-localized autoimmune diseases, including those affecting the eye.

MHC and Immunological Regulation

The immunological response is based upon two modalities of recognition. The first is the familiar immunoglobulin system. The circulating immunoglobulins, or antibodies, are the secreted products of the recognition structures of the B-lymphocytes. Their existence begins in the cytoplasm of the maturing B-cell and they are incorporated into the cell membrane of the mature B-cell as antigen-specific receptors.[1] Based on their complementary stereochemical structure, the receptors are able to bind selectively to individual antigenic determinants, called epitopes. Each B-cell carries only one such receptor, but there is a B-cell receptor for every antigenic determinant. In order to achieve a sufficient degree of diversity of receptor molecules with maximal parsimony

Department of Immunology and Infectious Diseases, The Johns Hopkins University, School of Hygiene and Public Health, Baltimore, Maryland.

of genetic material, the maturing B-cell employs special genetic mechanisms. In brief, three clusters of genes are employed by the B-cell to direct the synthesis of immunoglobulin molecules: one cluster for the heavy chain (σ, μ, γ, α, or ϵ) and two for the two types of light chain (λ and κ).[2,3] A variable number of *V* genes encoding for the hundred or so amino acids at the N terminus of the heavy or light chain couple with a small number of *J* genes and, in the case of the heavy chain, *D* genes, to determine the sequence of the variable portion of the immunoglobulin molecule. These *VJ* and *VDJ* gene combinations interact successively with one of the *C*-region genes which encode the class- and subclass-specific constant regions of the immunoglobulin peptides. With the added contribution from somatic mutation, these genetic rearrangements permit the immunological apparatus to produce B-cells capable of responding to essentially any given antigen determinant.[4] When the B-cell encounters its corresponding epitope, it is triggered to undergo blastogenic transformation, to proliferate, and eventually to develop into a clone of antibody-secreting plasma cells.

The second modality of recognition is represented by the T-cell network. T-cells also possess the ability to recognize individual antigenic determinants, although the structure of the receptor on the surface is not yet worked out. In addition, the T-cell recognition system involves a different family of genes, that of the *MHC*.[5] This gene cluster codes for two kinds of cell surface determinants, Class I and Class II. Class I determinants are expressed on virtually all cells of the body. They comprise two peptide chains, one with a molecular weight of approximately 45,000; the other is a low molecular weight β_2 microglobulin of approximately 12,000 daltons. As in the immunoglobulin system, the genes controlling the larger peptides are highly polymorphic, so that there are great antigenic differences in the surface constituents distinguishing members of the same species. These major alloantigens interfere with the transplantation of tissues from one to another member of the same species, giving rise to the name histocompatibility antigens. The major histocompatibility antigens are also remarkable because a great proportion of T-cells is capable of recognizing them, as shown in mixed lymphocyte cultures or graft-versus-host reactions. It has been estimated that as many as 37% of T-cells are able to recognize particular *MHC* antigens, which is at least a hundred times more than the number of T-cells that can recognize a conventional antigenic determinant. In the mouse, there are two main regions controlling the synthesis of Class I antigens, the *K* and *D* regions.[6] The analogous regions in the human *MHC* are designated *B*, *A*, and *C*.

A second series of cell surface determinants is encoded by other genes of the *MHC*. These Class II determinants are expressed mainly on immunologically active cells, such as lymphocytes, a subpopulation of macrophages, Langerhans' cells, and dendritic cells. They are surface glycoproteins with a two-chain structure—one chain having a molecular weight of approximately 32,000 daltons, and the second approximately 28,000 daltons.[7] In the mouse, they are controlled by genes of the *I* region of the *MHC*, called *H-2*, which is found between *K* and *D*; the analogous region in the human is referred to as *D/DR*. The *I* region houses the *Ir* genes that regulate the immunologic response. They determine the vigor of the response of mice to individual antigenic determinants. In addition, genes that determine lymphoid cellular interactions, mixed lymphocyte reactions *in vitro,* or graft-versus-host reactions *in vivo* are found in the *I* region. The antigenic products associated with the *I* region, referred to as *Ia,* are highly polymorphic.

There are two striking features about the *MHC*. First is its evolutionary conservatism. *MHC*'s have been identified in many different vertebrate species, including amphibians, birds, and mammals, and all of them carry essentially the same major genetic and biochemical features. Second, there is great similarity between the immunoglobulin system and the *MHC*. It is seen not only in the high degree of polymorphism that characterizes both systems, but also in shared sequences of amino acids. For example, the β_2 microglobulin portion of the Class I determinants is homologous to one of the constant domains of the immunoglobulin heavy chain. There are even analogies in the amino acid structure of human *D/DR* antigens and immunoglobulin molecules. It is reasonable to propose, therefore, that both immunological recognition systems had a common evolutionary origin in primitive cell-to-cell recognition and that both are important in maintaining individual integrity in multicellular animals.

The importance of immunoglobulins in maintaining health and protecting against disease has been known for nearly a century. However, the role of *MHC* products has only recently come to the attention of immunologists. Evidence of their impact has accumulated in both clinical and experimental work. In the realm of clinical immunology, the importance of the human *MHC*, called human leukocyte antigen *(HLA),* has been underlined by its association with a great variety of diseases, particularly malignant disorders of the lymphoid system and autoimmune diseases.[8] As yet, association is based only on statistical considerations and provides little information about the function of *MHC*-encoded cellular antigens.

From experimental studies, it has been possible to learn that *MHC* determinants control cellular interactions, including the interaction of T-lymphocytes with antigen-presenting cells[9] and the interaction of T-cells with B-cells in the process of antibody formation.[10] These interactions depend upon products of the *I* region. Products of the *K* and *D* regions restrict the interactions of cytotoxic T-cells with virus-infected or malignant target cells, and thereby regulate cell-mediated immunity.[11,12] The importance of this *H-2* restriction becomes clear when one considers that T-cells provide the principal protection against viruses harbored by host cells. The *K* and *D* regions also play a restrictive role in the severity of tissue lesions inflicted by cytotoxic T-cells in experimental autoimmune diseases, as will be described later.

Heredity in Human Thyroiditis

The evidence that chronic lymphocytic thyroiditis has a hereditary component was derived first from studies of familial aggregation. It was found that several thyroid diseases, including Graves' disease and thyroiditis, tended to cluster in certain families.[13] Other organ-specific autoimmune diseases, such as pernicious anemia, atrophic gastritis, idiopathic adrenal insufficiency, and diabetes mellitus, are often found in the same families. The second line of evidence came from studies of genetic markers, especially those of the *HLA* system. It was found that many of the human organ-specific autoimmune diseases are associated with particular *HLA* haplotypes, particularly *HLA-A-1/B-8/Dw3* (or *DRw-3*). Interestingly, only the atrophic form of chronic thyroiditis has been associated with this *HLA* haplotype.[14] Nevertheless, these findings add to the evidence that there is a genetic predisposition toward the development of thyroiditis.

Unfortunately, these studies can carry us only so far. Although human families share a gene pool, they also inhabit a common environment. Infectious agents, dietary habits, and exposure to environmental toxicants are also shared by the same families. It is sometimes difficult to distinguish genetic heredity, therefore, from ''cultural'' heredity based on shared environmental factors (including infections).

We have undertaken a study of juvenile thyroiditis in patients with two purposes in mind.[15] The first is to obtain a quantitative estimate of the importance of genetic factors in the initiation of thyroid autoimmunity. The second is to gain some idea of what chromosomes may be involved. These studies were carried out in children and adolescents, because we felt that they are most likely to have a strong genetic predisposition. In addition, it seems possible that accumulated environmental influences are less in the juvenile population.

In each family, one juvenile propositus had either chronic lymphocytic thyroiditis or thyrotoxicosis. Antibody determinations were performed on the parents in each family, as well as on each clinically normal sibling. The occurrence of thyroid autoantibodies in the clinically normal children of thyroiditis families was much greater than in control families with no evident thyroiditis. From these data, it was possible to obtain some estimate of the importance of genetic versus environmental influences in precipitating thyroiditis. This was done by identifying families in which both parents had evidence of thyroid autoimmunity based on the presence of circulating antibodies to thyroglobulin, or thyroid microsomes; families in which only one parent had evidence of thyroid autoimmunity; and families in which neither parent had evidence of thyroid autoimmunity. The unaffected children in each family could then be divided into those with and without antibodies. The clinically affected proband was excluded from these totals.

Strikingly, the distribution of antibodies among the three groups was significantly different. A clinically normal child in a family in which both parents had evidence of thyroid autoimmunity had an approximately 70% chance of having thyroid autoantibodies. A child from a family in which only one parent had thyroid autoimmunity had about a 50% chance of developing autoantibodies, whereas a child in a family in which neither parent had evident thyroid autoimmunity had about a 30% chance of developing autoantibodies. Since we can assume that environmental factors are equally powerful in these three groups, these differences must be due to genetics. It is clear that genetic predisposition plays a major role in the development of thyroid autoimmunity. On the other hand, it is not possible from these data to determine how many genes, and what kinds of genes, are involved in the inheritance of thyroid autoimmunity.

As a first step in identifying particular genes that may influence the development of thyroid autoimmunity, a large number of genetic markers was examined in these families. Correlations of these markers with thyroid disease or thyroid autoimmunity were sought. First attention was given to the *MHC* which, in the human, is found on chromosome 6. As earlier investigations of populations indicated, there is a close association, in the families we have studied, between thyrotoxicosis and a particular *HLA* haplotype, A-1/B-8, suggesting linkage disequilibrium with this particular haplotype. No such linkage could be found with chronic lymphocytic thyroiditis, making these negative findings particularly striking. Nevertheless, children who share one *HLA* haplotype with the affected sibling are themselves positive for thyroid autoantibodies in about 70% of cases, and those who share two haplotypes were positive in 90%.[16] While the interpretation of this finding is still uncertain, it seems likely that genes located on chromosome 6 with *HLA* are involved in the predisposition to thyroiditis, but these genes are not in linkage disequilibrium with particular *HLA* haplotypes, at least those recognizable with our presently available reagents.

These studies show that genetic predisposition plays a major role in the initiation of human chronic thyroiditis. On the other hand, we still have little information about the nature or number of the genes involved. For that purpose, it is necessary to turn to experimental models.

Spontaneous Autoimmune Thyroiditis in the Obese Strain (OS) Chicken

OS chickens represent a family of white Leghorns that is genetically predisposed to autoimmune thyroiditis.[17] The development of disease is accompanied by the appearance of autoantibodies to thyroglobulin. B-cells are required for the development of thyroiditis, since bursectomized chickens do not develop the disease. Autoantibodies are able to transfer disease within the OS family.

Analysis of OS chickens has shown that there are at least three major traits that influence the development of thyroiditis.[18] Spontaneous disease is most likely to occur in birds that possess all three of these traits. Animals that have only one or two of the genetic lesions have a lower probability of developing disease.

The most important gene predisposing to development of spontaneous thyroiditis in the OS chicken is genetically linked to the avian *MHC*, the *B* alloantigen. There is a close correlation between the *B* haplotype and the severity of disease and titer of thyroglobulin autoantibodies.[19] The *B*-associated proportion is expressed mainly by thymus-derived T-lymphocytes. Cell transfer experiments have shown that poor responder birds can be converted to good responders by providing them with good responder T-cells.[20,21] It seems likely that these *MHC*-associated traits represent immune response genes, determining the vigor of response of chickens to one or more epitopes on the thyroglobulin molecule.

A second, independently inherited trait in the obese chicken affects the maturation of the thymus.[22] In normal chickens, the preponderance of cells migrating from the thymus during early embryonic development has primarily suppressor function. As long as suppressor cells predominate, normal animals do not respond to self-antigens. In the OS birds, however, there is evidence that helper cells predominate among the earliest emigrants from the thymus. These cells can be inactivated by whole-body irradiation. This defect in thymus maturation probably predisposes OS chickens to the development of a variety of organ-specific autoimmunities, as seen by the multitude of autoantibodies found in these birds. However, they do not develop any significant autoimmune disease other than thyroiditis.

A third abnormality is found in OS chickens in the intrinsic function of the thyroid epithelial cells.[17] Independent of autoimmune damage, these cells seem to take up larger quantities of iodine than normal thyroid cells, even in organ culture or during growth on the chorioallantoic membrane. On the other hand, their production of thyroid hormones, T4 and T3, is reduced and the mono-iodo thyronine:di-iodo thyronine (MIT:DIT) ratio is abnormal. Functionally, the thyroids of OS chickens behave autonomously, suggesting that their cells are not susceptible to pituitary regulation.

It is likely that the confluence of these three different traits is responsible for the spontaneous occurrence of thyroiditis in OS chickens. However, there are other genetic contributions.[23] For example, an X-linked gene has been found that adds to susceptibility to thyroiditis. In addition, female sex hormones themselves add to the proclivity toward thyroiditis, particularly in animals with intermediate genetic susceptibility.

Autoimmune Thyroiditis in the Mouse

Experimental autoimmune thyroiditis can be produced in the mouse by immunization with mouse thyroglobulin plus an appropriate adjuvant. The response of different strains of mice is genetically controlled.[24] The major gene determining response has been localized to the *H-2* complex. Mice with *H-2* haplotypes such as *k, s* and *q* respond promptly to injections of thyroglobulin by producing relatively high titers of autoantibody and severe lesions in their thyroids. Other strains with haplotypes *b* or *d* produce antibody at a slower rate and develop only minor thyroid lesions. Tests of congenically matched strains of mice have shown definitively that the control is localized in the *I* region of the *H-2* complex.

Further studies using mouse strains in which recombination has taken place within *H-2* have shown that there are two levels of genetic control.[25] A gene at the *I* region determines the initial steps in recognition of thyroglobulin.[26] This gene has been termed *Ir-Tg*. It probably encodes the recognition of a limited number of epitopes on the thyroglobulin molecule, possibly by governing the ratio of helper T-cells to suppressor T-cells activated by the particular determinant. The effect of *Ir-Tg* can be measured *in vitro* by means of the lymphocyte proliferation test.[24] The stimulation index differs in different strains of mice, correlating with their *H-2* haplotype.

A second level of genetic control is found at the *D* or *K* regions of *H-2*.[27–29] *In vivo*, it determines the severity of thyroid lesions in mouse strains that are good immunological responders. Some strains develop severe thyroiditis, while others show only moderate disease. These changes are reflected *in vitro* by a lymphocyte-mediated cytotoxicity reaction, in which damage to thyroid epithelial cells is produced by T-cells. Cytotoxicity is diminished by treatment of the target cells with antisera to *H-2K* and *H-2D* determinants.[29].

In brief, both Class I and Class II histocompatibility antigens are involved in the regulation of the autoimmune response of mice to thyroglobulin. Class II antigens are involved in the initial steps of recognition of thyroglobulin; Class I antigens restrict the T-cell-mediated cytotoxic reaction traced *in vitro*.

In summary, the autoimmune response in thyroiditis is controlled by a number of genes. Some genes act on the target organ, the thyroid, and render it vulnerable to immunological attack; other genes affect the function of the thymus and alter the tempo of production of T-cell populations with helper and suppressor functions. Finally, genes within the *MHC* of the species regulate the response to antigenic determinants of the thyroglobulin molecule. There are at least two critical steps in the response. Genes of the *I* region regulate early steps in T-cell recognition and proliferation, whereas genes at *K* or *D* of the mouse *MHC* regulate cytotoxic actions of effector T-cells. Studies of experimental animals help us to understand the genes that may be involved in the predisposition to autoimmune thyroid disease in the human.

Since this article was written, much of the structure and genetic control of the T-cell receptor has been solved (Siu G et al: *Nature* 311:344–350, 1984). The receptor is a heterodimer composed of two disulfide-linked peptide chains, α and β, each of which has a variable and constant region. The variable region genes undergo rearrangement much like that described for immunoglobulin *V* genes.

Acknowledgments

This research has been supported by PHS-NIH grants AM 20023 and AM 20029, and by a grant from the William Beaumont Research Foundation.

References

1. Klinman NR et al.: *Ann Immunol* **172**:489–502, 1976.
2. Croce CM et al.: *Proc Natl Acad Sci USA* **76**:3416–3419, 1979.
3. Shander M et al.: *Transplant Proc* **12**:417–420, 1980.
4. Gearhart PJ: *Immunology Today* **3**:107–112, 1982.
5. Katz DH and Benacerraf B: *Trans Rev* **22**:175–195, 1975.
6. Nathanson SG et al.: *Ann Rev Biochem* **50**:1025–1052, 1981.
7. Silver J: *CRC Crit Rev Immunol* **2**:225–257, 1981.
8. Vladutiu GD and Rose NR: *Immunogenetics* **1**:305–328, 1974.
9. Sonderstrup-Hansen G et al.: *Eur J Immunol* **8**:520–524, 1978.
10. Wylie, DE et al.: *J Exp Med* **155**:403–414, 1982.
11. Zinkernagel R and Doherty P: *Nature* **248**:701–702, 1974.
12. Hunig T and Bevan MJ: *Nature* **294**:460–462, 1981.
13. Rose NR: In *HLA in Endocrine and Metabolic Disorders*. Farid NR (Ed.). Academic Press, New York, 1981, pp. 1–10.
14. Farid NR et al.: *Tissue antigens* **17**:265–268, 1981.
15. Burek CL et al.: *Clin Immunol Immunopathol* **25**:1982, (in press).
16. Rose NR and Burek CL: In *Endocrinology Metabolism*. Cohen MP and Foa PP (Eds.). Alan R. Liss, New York, 1982, pp. 139–176.
17. Sundick RS and Rose NR: Autoimmune thyroiditis in obese-strain chickens. In *Immunologic Defects in Laboratory Animals*. Gershwin ME and Merchant B (Eds.). Plenum Press, New York, 1981, pp. 3–15.
18. Rose NR et al.: *Clin Exp Immunol* **39**:545–550, 1980.
19. Bacon LD and Rose NR: *Proc Natl Acad Sci USA* **76**:1435–1437, 1979.
20. Polley C et al.: *J Immunol* **127**:1465–1472, 1981.
21. Livezey M et al.: *J Immunol* **127**:1465–1472, 1981.
22. Jakobisiak M et al.: *Proc Natl Acad Sci USA* **73**:2877–2880, 1976.
23. Bacon L et al.: *Immunogenetics* **12**:339–350, 1981.
24. Okayasu I et al.: *Cell Immunol* **61**:32–39, 1981.
25. Kong, YM et al.: In *Genetic Control of Autoimmune Disease*. Rose NR, Bigazzi PE and Warner NL (Eds.). Elsevier North-Holland, New York, 1978, pp. 433–442.
26. Beisel KW et al.: *Immunogenetics* **15**:427–430, 1982.
27. Kong YM et al.: *J Immunol* **123**:15–18, 1979.
28. Maron R et al.: *Immunogenetics* **15**:625–627, 1982.
29. Creemers P et al.: *J Exp Med* **157**:559–571, 1983.

chapter 27

Immunostimulation as a Therapeutic Alternative for Uveitis

An Experimental Study

A.C. Martenet, H. Cornelis, A. Fontana

Uveitis, at least in its severe forms, still remains a therapeutic challenge. Not only Behçet's disease, but many cases of chronic anterior and intermediate uveitis often resist all classical forms of treatment. Immunosuppression by means of cytostatic drugs has certainly been of great help in numerous desperate cases.[1] However at least 13.5% of the patients do not show an improvement when treated with cytostatics. Appropriate treatment is hindered by insufficient knowledge of the pathogenesis of uveitis and of the immune reactions implicated in this inflammatory disease. While many uveitis patients are quite healthy except for their eyes, some forms of uveitis, such as Candida hyalitis, cytomegalovirus chorioretinitis, and various other forms of retinal vasculitis can occur in immunodeficient subjects. In these patients the disease is likely to respond to drugs that stimulate the host's own defense system.

In this context one might refer to the studies using transfer factor in uveitis.[2] Levamisole, one of the first immunopotentiating drugs,[3] has also been tested in isolated cases of uveitis.[4] More recently Iso-prinosine, a compound which seems to enhance certain cell-mediated immune responses,[5] has also been used with satisfactory results in rheumatoid arthritis,[6] a disease whose pathogenesis might share at least some similarities with uveitis. To investigate certain immunomodulating drugs for their influence on uveitis, we have chosen a model experimental autoimmune uveitis (EAU) as described by Faure et al.[7] The clinical signs of the experimental disease correlate fairly well with intermediate uveitis in humans.

Together with this therapeutic trial, we also looked for the presence of interleukins (IL), factors that are likely to be involved in EAU in induction phase of the immune response to bovine S antigen.

Materials and Methods

Animals

Both Kunath rabbits (mostly males, 2.5–3.7 kg) and female guinea pigs (mostly piebald, some pigmented, 250–350 g except for 13 animals weighing more than 800 g in group IIa, see below) were used. Although rabbits are known to respond less vigorously to retinal S antigen than guinea pigs, the rabbit eye is convenient for uveitis studies because its size allows not only clinical and histological examination, but also studies on aqueous humor. All animals in each group were of about the same age and weight and

University Eye Clinic (Prof. R. Witmer), Zürich, Switzerland and the Section of Clinical Immunology of the Department of Medicine of the University of Zürich (Prof. P.J. Grob), Zürich, Switzerland.

Dr. H. Cornelis is a fellow of the Schweizerischen Fonds zur Verhütung and Bekämpfung der Blindheit.

had normal eyes, except for one rabbit with slight vascularization of both corneas. This animal was used as a control.

Induction of experimental autoimmune uveitis

Animals were immunized by a single injection of bovine retinal S antigen, kindly provided by Dr. J.P. Faure and Miss C. Dorey of the Hôtel-Dieu in Paris. The antigen, mixed with complete Freund's adjuvant, was injected in a dose of 25–50 μg in one hind footpad of rabbits, or in a dose of 20–30 μg, half in each hind footpad of guinea pigs. Control animals (#1, #2, and #3 of each species in each group) received NaCl 0.9% with Freund's adjuvant.

Treatment

The animals were divided into three groups. Animals of group I (12 rabbits and 23 guinea pigs) were not treated and served as controls for the natural course of the disease.

In group II (19 rabbits and 36 guinea pigs) half of the animals were treated with Levamisole (Janssen Pharmaceutica, Baar, Switzerland) as follows:

In subgroup IIa Levamisole was administered to three of six rabbits and to six of 12 guinea pigs 2 days after inoculation.

In subgroup IIb Levamisole was administered to four of eight rabbits and to eight of 16 guinea pigs, 2–6 days after onset of signs of uveitis. For that purpose, the animals were divided according to their signs into two comparable series with equal numbers of severe, moderate, and slight cases in each group. Levamisole was dissolved in distilled water to make a 1% solution, and a dose of 2.5 mg/kg was injected subcutaneously in the neck of the animal.

In the course of the third month, when the uveitis had become very chronic, some animals of group IIa received Levamisole, either as a first or as a second treatment.

In group III (13 rabbits and 37 guinea pigs) half of the animals were treated with Isoprinosine (Max Ritter, Pharma, Zurich) as follows:

In subgroup IIIa three of six rabbits and

seven of 14 guinea pigs received Isoprinosine for five days starting on day 2 after inoculation.

In subgroup IIIb: three of six rabbits and nine of 18 guinea pigs received Isoprinosine for five days immediately after onset of the clinical signs of disease. Isoprinosine was diluted to a 10% solution in tap water, and administered orally at a dose of 50 mg/kg/day twice daily for 5 days. A comparable series was observed as in subgroup IIb.

In groups IIb and IIIb animals with minimal inflammatory reactions were not included.

Clinical studies

All animals were examined 1–2 times every week at the slit lamp (even more often before onset of inflammatory signs) and sometimes with the ophthalmoscope. The intensity of uveitis was estimated as between 0 and + + +. These observations were made independently by at least two persons.

Histological studies

Some of the animals were sacrificed after 1–5 months. Their eyes were enucleated, fixed in formalin, and embedded in celloidin; 12 μ sections were then stained with hematoxylin and eosin or Van Gieson stain for histological examination. This study is still in progress with the remaining animals.

Interleukin assays

The aqueous humors of eight rabbits (paracenteses done on day 0, 15, 33, 42, 55 and 132) and the vitreous humors of four rabbits, taken at the time of sacrifice (1, 3, or 5 months after immunization) were tested for their activity on mouse lymphocytes. The specimens were ultrafiltered on YM-10 (Amicon) membranes, concentrated tenfold, and tested on mouse thymocytes stimulated with phytohemagglutinin (PHA). Some samples were also added to an ovalbumin-specific mouse T-cell line growing in the presence of interleukin 2 only. The assays used are described in detail by Fontana et al.[8,9]

An interleukin-1 standard, prepared by lipopolysacharide stimulation of mouse peritoneal cells, and an interleukin-2 standard, obtained by concanavalin A stimulation of mouse spleen cells, were used as positive controls.

Results

Clinical course

None of the control animals showed any signs of uveitis during the whole period of examination.

After inoculation with S antigen, practically all of the guinea pigs and 60–70% of the rabbits developed a uveitis of the intermediate type, with mainly cells and flare in the vitreous, as well as inflammation of the anterior segment in the severe cases. This uveitis started 10–20 days after inoculation and appeared a little earlier in guinea pigs than in rabbits. The disease reached its maximal intensity rapidly, remained stable

for 2–4 weeks, and then diminished. The disease then became chronic and was sometimes complicated by cataract formation. In the mild cases the clinical signs disappeared completely after 6–10 weeks.

Levamisole appeared to influence the clinical course of the disease when administered after onset of the clinical signs of inflammation. Compared to untreated animals, animals treated with Levamisole appeared to show a more rapid abatement of clinical signs. This was especially noteworthy among animals that manifested strong initial reactions (Figs. 1, 2). By contrast, administration of the drug 2 days after inoculation seemed to enhance the disease in rabbits, (Fig. 3), while it did not show any effect in guinea pigs (Fig. 4). Finally administration of the drug in the very chronic states of the disease (after the third month) appeared ineffective.

Isoprinosine did not appear to influence

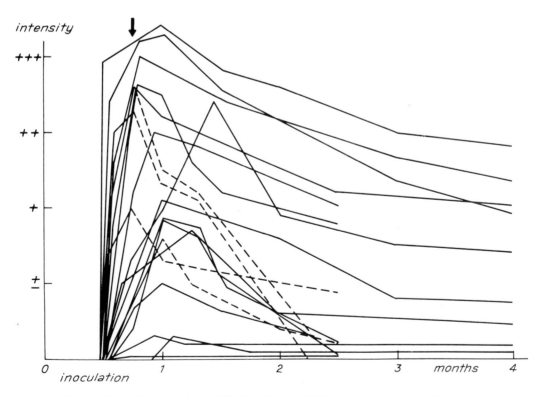

FIG. 1. Clinical effect of Levamisole on EAU in rabbits (solid lines: untreated; broken lines: treated; arrow: treatments begin).

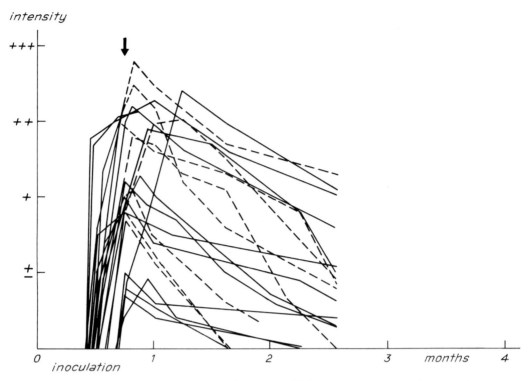

FIG. 2. Clinical effect of Levamisole on EAU guinea pigs (solid lines: untreated; broken lines: treated; arrow: treatments begin).

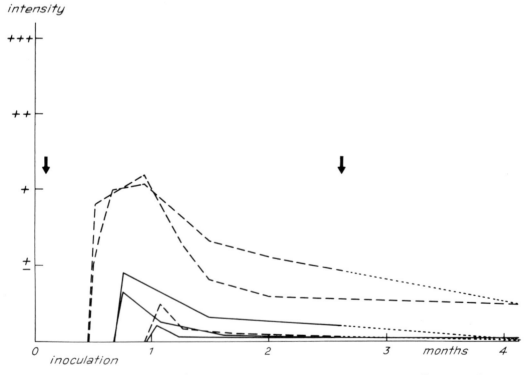

FIG. 3. Clinical effect of Levamisole ↓ on EAU in rabbits (solid lines: untreated; broken lines: treated; arrow: treatments begin).

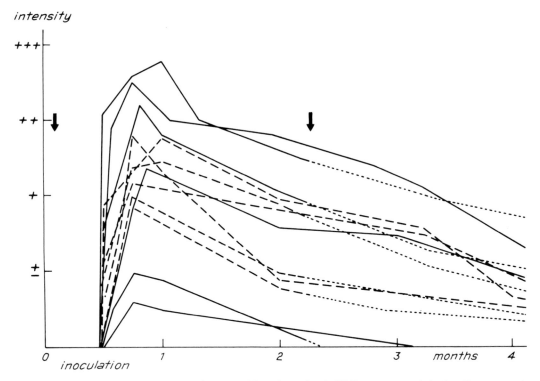

intensity

FIG. 4. Clinical effect of Levamisole ↓ on EAU in guinea pigs (solid lines: untreated; broken lines: treated; arrow: treatments begin).

FIG. 5. EAU, rabbit, one month. Patchy infiltration of choroid and retina.

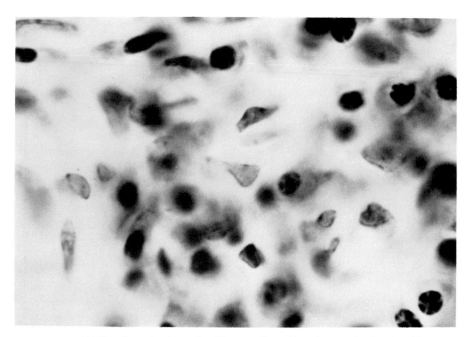

FIG. 6. EAU, guinea pig, 4 months. Plasma cells and lymphocytes in the choroid.

the clinical course of EAU in rabbits or in guinea pigs. It did not matter whether it was administered immediately after inoculation or after onset of the uveitis.

Histological findings

In the animals sacrificed at 4 weeks there was a patchy infiltration of the choroid and of the retina with polymorphonuclear, cells are well as lymphocytes (Fig. 5). There were also many cells in the vitreous, while the iris and ciliary body remained almost free of lesions. The extent of the infiltration correlated well with the clinical signs. After 2–3 months, there were more lymphocytes and plasma cells in the infiltrates (Fig. 6), while after 4 months there was mostly a fibrotic organization of choroid, retina, and even the vitreous. There was no apparent difference in the histological picture of the treated versus the nontreated animals. The controls showed no histological changes.

Interleukin assays

Whereas the standard interleukin-1 and interleukin-2 preparations augmented the PHA response of mouse thymocytes, and the interleukin-2 preparation supported the growth of an ovalbumin-specific mouse T-cell line, none of the aqueous humors or vitreous humors tested showed such an effect. In spite of ultrafiltration, some of the eye specimens inhibited the PHA response. (See examples in Table I.)

TABLE I
Activity of Aqueous Humor and Vitreous of EAU Animals on Thymocytes and IL-2 Dependent Cells

Culture medium supplemented with	^{3}H-Tdr incorporation (cpm) $\pm$ sD in cultures of	
	PHA stimulated thymocytes	Ova cells[a]
Nothing	5563 $\pm$ 690	221 $\pm$ 24
IL-1 standard	52078 $\pm$ 2687	289 $\pm$ 73
IL-2 standard	59138 $\pm$ 4124	31001 $\pm$ 3653
Aqueous humor:		
57/82 l	6201 $\pm$ 332	126 $\pm$ 34
59/82 r	4153 $\pm$ 518	183 $\pm$ 42
Vitreous:		
66/82	5103 $\pm$ 496	253 $\pm$ 46
17/82	2641 $\pm$ 87	144 $\pm$ 29

[a] Ovalbumin specific mouse T-cell line.

Discussion

The uveitis produced in rabbits and guinea pigs by inoculation of the footpad with retinal S antigen seems to be a satisfactory model for testing the effectiveness of drugs. Although the onset of EAU is generally rather acute, its prolonged, chronic evolution imitates rather closely the intermediate uveitis seen in humans. This is a form of uveitis whose treatment can be very difficult. In our series, Levamisole, given after the first appearance of clinical signs of inflammation, was found to decrease the disease activity in both of the animal species used in this experiment, but especially in the rabbits. However, it must be emphasized that this beneficial clinical effect was not apparent on histological examination. Furthermore, the drug had a rather adverse effect, at least in rabbits, when given before the onset of the uveitis, a situation which is evidently not seen in human cases. Finally, the lack of effectiveness of the drug when administered in the chronic stages might well have been expected. Such failure has been repeatedly observed in uveitis treated with other drugs, even with cytostatics.[10] This fact confirms once again the importance of starting therapy early in cases of severe uveitis. Isoprinosine gave rather disappointing results in this first series. After the reports of Hadden et al.[11] on the stimulation of lymphocytes and lymphokine-induced macrophage proliferation, of Renoux et al.[12] on enhanced suppressor cell activity, and of Wybran et al.[6] on satisfactory results with this drug in rheumatoid arthritis, this failure was rather unexpected. It might be related to the dosage or to the mode of administration of Isoprinosine in our animals. Since Isoprinosine is known to have very low toxicity[13] we think that further trials with higher doses or a more prolonged administration might be worthwhile.

Both in untreated animals and in EAU animals treated with Levamisole or Isoprinosine, no IL-1 or IL-2 was detected in the aqueous humor or in the vitreous humor. Interleukin 1 is a soluble macrophage-derived factor, whose interaction with antigen- or mitogen-activated T-cells, (possibly helper cells) releases interleukin 2. The latter stimulates the proliferation of lymphocytes.[14] According to Hadden,[15] Isoprinosine stimulates the interleukin 1 production *in vitro*. Induction first takes place in the lymphatic system located near the site of immunization. Later, at the effector site of the immune response in the eye, when sensitized lymphocytes attack cross-reacting antigens in experimental animals, the contribution of interleukins seems to be of less importance.

This could partially explain the negative results concerning interleukins in the EAU eyes. Furthermore, the interleukins could be present in only very low concentrations. Finally, the chronic stage of the disease at the time of sacrifice could contribute to these results. The lack of interleukin activity might also support the hypothesis that B-rather than T-lymphocytes are implicated in EAU, as well as in posterior or intermediate uveitis in humans. Indeed, our histological examinations showed many plasma cells. Witmer et al.[16] had made the same observation in a uveitis produced by ferritin, and recently Kaplan et al.[17] found only B-cells in the vitreous of eyes treated by vitrectomy for severe uveitis.

All of these observations are, however, in opposition to the report of Belfort et al.[18] who showed only T-cells in the aqueous humor of various forms of uveitis; Nussenblatt et al.[19] also attribute a major role to T-lymphocytes in S antigen-induced EAU. Finally, Nussenblatt et al.[20] noted good therapeutic results in experimental uveitis treated with cyclosporin A, a specific T-cell inhibitory drug.

Conclusion

The clinical course of EAU in rabbits and guinea pigs can be influenced by immunostimulation, especially with Levamisole. Isoprinosine did not alter the course of the disease. In Levamisole-treated animals the clinical signs of EAU disappeared faster. The mechanism of action of Levamisole on the inflammation, however, is not yet clear, especially since it did not appear to influence the histopathological findings. Detection of interleukin was not possible in these experiments. This might be due to technical problems or to the hypothesized theory that B- rather than T-lymphocytes are implicated in EAU. Further studies on such models as we have developed should lead to a better understanding of the pathogenesis of uveitis and of its treatment in humans.

Acknowledgments

This investigation was entirely supported by the Schweizerischen Fonds zur Verhütung and Bekämpfung der Blindheit.

We are much indebted to Mrs. O. Kradolfer for her help with the animal work, to Ms. E. Strickler for the preparation of the histological slides, and to Ms. E. Weber for excellent technical assistance in lymphocyte culture work.

References

1. Martenet AC: *Ber Dtsch Ophtalmol Ges* **78**:223–228, 1981.
2. Abramson A et al.: *Br J Ophthal* **64**:332–338, 1980.
3. Renoux G: *Drugs* **19**:88–89, 1980.
4. James DG: *Trans Ophthalmol Soc UK* **97**:468–473, 1977.
5. Simon LN and Glasky AJ: *Cancer Treat Rep* **62**:1963–1969, 1978.
6. Wybran J et al.:*J Rheumatol* **8**:643–646, 1981.
7. Faure JP et al.: *Mod Probl Ophthalmol* **16**:21–29, 1976.
8. Fontana A et al.: *Rheumat Int* **2**:49–53, 1982.
9. Fontana A et al.: Astrocytes and C6 glioma cells secrete interlukin I-like factors. *J Immunol* (in press).
10. Martenet AC: *Klin Mbl Augenhk* **176**:648–651, 1980.
11. Hadden JW and Giner. Sorolla A: Presented at National Cancer Institute Conference on Biological Response Modifiers, Bethesda, Md., March 1980.
12. Renoux G et al.: *Int J Immunopharmacol* **1**:239–241, 1979.
13. Ginsberg T, Simon LN, and Glasky AJ: Presented at Seventh International Congress of Pharmacology, Paris, France, July 1978.
14. Smith KA et al.: *J Exp Med* **151**:1551–1556, 1980.
15. Hadden JW: In *Proceedings First International Conference on Immunopharmacology*, Hadden JW et al. (Eds.). Pergamon Press, New York, 1980, pp. 327–340.
16. Witmer R et al.: *Ophthalmologica* **154**:301–309, 1967.
17. Kaplan HJ et al.: *Arch Ophthal* **100**:585–587, 1982.
18. Belfort R et al.: *Arch Ophthal* **100**:456–467, 1982.
19. Nussenblatt RB et al.: *Invest Ophthalmol Vis Sci* **19**:686–690, 1980.
20. Nussenblatt RB et al.: *Arch Ophthal* **100**:1149, 1982.

chapter 28

Skin Testing in Uveitis

E. Bloch-Michel,[a,b] **J. Dry,**[a] **R. Campinchi,**[b] **M.P. Le Corvec,**[a,b,] **and F. Niessen**[a,b,]

Intradermal tests using various antigens have long been employed for the etiological diagnosis of uveitis. In general, it has been stated that once the uvea has been sensitized to one antigen, an anamnestic inflammatory response will occur in the same eye after new contact with the same antigen.

Skin testing was proposed some time ago[1] as the best means of detecting a state of hypersensitivity to microbial allergens. Later, this method was criticized[2] and generally abandoned except for a few proponents.[3]

Over the past 14 years we have developed a standardized technique for skin testing patients with uveitis; this technique was performed on 178 of 1400 patients with the disease. The results reported here indicate that skin testing is a potentially profitable method for studying patients with uveitis. It may detect a state of hypersensitivity to one particular allergen; it may reveal a peculiar immunological background; or both.

Patients and Methods

Patients

Of 1400 patients with uveitis, 178 were selected for skin testing according to the following criteria: (1) Clinical history indicating a hypersensitivity state (atopy, recurrent inflammatory diseases, etc.). (2) No other etiology found, according to our criteria.[4,5] (3) Skin testing to be performed at the end of the second attack in recurrent forms of uveitis or after three months' duration of the chronic forms. (4) All tests to be performed on outpatients. (5) Contraindications included: (a) local disease: hemorrhagic forms of uveitis, juxtamacular lesions; (b) general disease: suspected collagen-vascular and/or Behçet's disease. (6) No corticosteroid employed, or current corticosteroid therapy less than 10 mg of Prednisone and/or 2 drops per day of Dexamethasone, the ocular inflammation remaining stable for at least 15 days.

Methods

Intradermal tests. A sequential series of three different allergens, (1) bacterial: Streptococcus, Staphylococcus, tuberculin; (2) fungal: Candida; and (3) airborne allergens were inoculated weekly for a period of three weeks. The concentrations of allergens injected were those generally used in allergy testing.[6] The skin tests were evaluated immediately and 48 hours after injection.

Ocular reaction. The manifestations of ocular reaction, including visual acuity, intensity of the flare (in aqueous and vitreous), and state of the fundus were carefully evaluated by the same ophthalmologist on the day of injection of the antigen and on the day of reading the skin tests. In case a frank modification of the uveitis (transient increase of inflammation or paradoxical improvement) was noted, the same antigen was reinjected 15 days later via a subcutaneous route to confirm the existence of a local hypersensitivity reaction of this particular antigen, and not to others.

[a]From the Centre d'Allergie, Hôpital Rothschild, Paris and the [b]Service d'Ophtalmologie, Hôpital Lariboisiere, Paris.

Our 178 patients included 109 females and 69 males; 19 children under age 15 and 159 adults. Regarding the anatomic type, the sample comprised 58 cases of anterior uveitis, 35 cases of posterior uveitis, 36 cases of panuveitis, and 49 cases of intermediate uveitis. Of the cases, 83 were unilateral and 95 bilateral; 99 patients were not receiving corticotherapy of any kind; 79 were receiving local (35) and/or systemic steroids (44).

Results

In connection with the demonstration of a local inflammatory reaction to the test antigen, the diagnosis of "allergic uveitis" was made in 64 of 178 cases (32 certain, 32 presumptive). The antigens thought to be responsible for the disease were, in order of numerical frequency, Candida (26), tuberculin (16), other microbial (14), and airborne allergens (10).

Among these 178 cases, an atopic background (personal history of asthma, eczema, hay fever) and/or a history of allergy in the family was found in 41 patients (24.1%).

Among the 64 cases considered to be allergic in origin, antiallergic treatment was undertaken in 39 of them with the aim of eliminating reactivity to a particular allergen or of desensitizing the individual to various allergens. Such treatment led to improvement in 23, no effect in 10, and aggravation in 6. In general, improvement was noticed when candidin or tuberculin was used as a desensitizing antigen. On the other hand, aggravation occurred mainly among atopic patients who were generally treated with airborne allergen extracts.

Discussion

The high incidence (24%) of atopy among patients selected for skin testing indicates that these patients could more reasonably be expected to give positive results than others with uveitis. Uveitis and hay fever often occur simultaneously; skin tests to pollen extracts may be hazardous for the eye condition, and desensitization to pollen is contraindicated in such patients if a recurrence or an aggravation of the eye condition is to be prevented. Likewise, some cases of uveitis among patients sensitized to airborne allergens, like house dusts or mites, may be exacerbated by skin testing or by desensitization with high doses of allergens. In general, the role of an atopic background is not clear in the occurrence of uveitis or in its course.

It is remarkable that of 64 patients considered to be allergic, another explanation for the uveitis was later found in four cases (2 HLA B27 positive, 1 Behçet, 1 toxoplasmosis). Also 114 patients with negative tests 26 (23%) were in the later years recognized as HLA B27 positive (11), Behçet or Behçetlike (5), toxoplasmosis (7), Fuchs (2), and multiocular (1).

The difference suggests that if no ocular reaction appears after skin testing, another etiology for the uveitis will be found. Further, it indicates that the local reaction that occurs in the eye after skin testing may be due to an immunologic epiphenomenon.

On the other hand, the improvement of the eye condition with injections of certain allergens such as candidin or tuberculin, used either for testing or for treatment, is not clear-cut. This might indicate that certain subpopulations of T-cells (of the suppressive type) could be selectively stimulated by skin testing. Further investigation is required in the attempt to elucidate this issue.

References

1. Woods AC: *Endogenous Inflammations of the Uveal Tract*. Williams & Wilkins, Baltimore, 1961.
2. Coles RS and Nathaniel A: *Arch Ophthalmol* **61**:45–49, 1959.
3. Kastler M: In *Uveitis: Immunologic and Allergic Phenomena*, Campinchi R et al.(Eds.) Thomas, Springfield, 1973, p. 322.
4. Campinchi R et al.: *XXIII eme Concilium Ophthalmologicum Data*, Excerpta Medica, Amsterdam, 1979, p. 95.
5. Niessen F et al.: *Ber Deutsch Ophthalmol Ges* **78**:209, 1981.
6. Bloch-Michel E: *L'Ouest Medical* **28**:869, 1975.

chapter 29

Idiopathic Insidious Bilateral Uveitis

Clinical and Autoimmune Characteristics

David BenEzra, Juan H. Paez, and Genia Maftzir

In most uveitis cases, the etiology is unknown or at best presumptive. In our uveitis clinic, more than 50% of the bilateral uveitides are labeled as idiopathic, since no clinical or routine laboratory examination demonstrates a systemic manifestation of the disease or an infectious agent that could be implicated in the etiology. During the last decade, 85 patients were classified as suffering from idiopathic bilateral uveitis—pars planitis. According to their presenting clinical symptoms, these patients were regrouped under "benign," "chronic," or idiopathic insidious bilateral uveitis (IIBU) whose course was relatively "malignant" with severe visual sequelae. Forty-eight of these patients underwent immunological tests aimed at the detection of cellular or humoral immune responses toward ocular antigens.

Materials and Methods

Patients

These were referred to our clinic for further work and evaluation. Only patients were included who presented with ophthalmic manifestations of uveitis as their only complaint and in whom routine laboratory examinations did not disclose any systemic involvement. Eighty-five patients

The Immuno-Ophthalmology and Pediatric Ophthalmology Units, Department of Ophthalmology, Hadassah Hebrew University Hospital, Jerusalem, Israel.

were recruited and followed for a minimum of 2 years.

Tests

Blast transformation. From each patient 20 ml of blood were drawn from the cubital vein. Two ml were used for serology and 18 ml were transferred into heparinized tubes and allowed to sediment. The leukocytes were separated and their blast transformation responses to various ocular antigens and to nonspecific stimulants in microcultures were carried out as previously described.[1-3] One microgram, 10 micrograms, and 100 micrograms of crude ocular tissue antigens[2] were used to detect specific anamnestic responses. Two criteria for a positive specific response were examined. "Nonrigid criterion": A stimulation index of 1.5 or more to one concentration of allogeneic or xenogeneic antigen. "Rigid criterion": a stimulation index of 2.0 or more to two different concentrations of the allogeneic and xenogeneic antigens. For the nonspecific stimulation with phytohemagglutinin (PHA), concanavalin A (Con A), and pokeweed mitogen (PWM), one microgram per culture was used. In each set of experiments, at least two patients with nonuveitic ophthalmic conditions were included as controls.[3] Forty-eight uveitis patients underwent immunologic testing on their initial visit.

Migration inhibitory factor (MIF). This test, carried out with nonpurified peripheral leukocytes, was essentially based on the method reported by McCoy et al.[4]

Serum antibodies. A passive hemagglutination test was used as described by Gery and Davies.[5]

Results

According to the severity of the ocular findings, the clinical course among the 85 patients followed one of three distinct courses: benign (52), chronic (24), or idiopathic insidious bilateral uveitis (IIBU)[9]. In 48 of the cases, tests were done to determine specific and nonspecific lymphocyte blast transformation response and to check for the presence of humoral antibodies in the sera. Table I illustrates the blast transformation response to PHA, Con A, and PWM. When compared with the results obtained from nonuveitis patients, none of the groups showed any characteristic pattern of response. Only the mean response of the IIBU group to Con A was slightly higher than that recorded in the other groups. The specific blast transformation response obtained when lymphocytes were challenged *in vitro* by allogeneic and xenogeneic retinal and uveal crude extract antigens is shown in Table II.

When nonrigid criteria for specific blast transformation were used, eight of 25 patients with a benign disease, six of 14 with chronic disease, and all patients with IIBU demonstrated a positive response to the ocular antigens. However, when more rigid criteria of positivity were used, none of the patients in the benign and chronic groups yielded positive results, while five of nine

TABLE I
Nonspecific Lymphocyte Blast Transformation Response[a]

Group	No.	Stimulating mitogen		
		PHA	Con A	PWM
Benign	25	1.1 ± 0.3	1.0 ± 0.3	0.9 ± 0.3
Chronic	14	0.9 ± 0.4	1.3 ± 0.5	0.8 ± 0.4
IIBU	9	1.3 ± 0.3	1.6 ± 0.4	1.0 ± 0.3

[a] Assessed in comparison with the stimulation demonstrated by two nonuveitis patients counted as 100% and represented as a stimulation index (S.I.).

TABLE II
Specific Lymphocyte Blast Transformation Response

Group	S.I.	Challenging tissue extracts				
		H.Ret.	H.Uvea	B.Ret	B.Uvea	R.Ret
Benign	≥ 1.5	3/25	1/25	5/25	3/25	0/25
	≥ 2.0	1/25	1/25	1/25	0/25	0/25
Chronic	≥ 1.5	2/14	3/14	2/14	0/14	0/14
	≥ 2.0	0/14	1/14	2/14	0/14	0/14
IIBU	≥ 1.5	9/9	8/9	7/9	8/9	1/9
	≥ 2.0	6/9	6/9	5/9	2/9	0/9

H. Ret. = Human retinal extract; B. Ret. = Bovine retinal extract; R. Ret. = Rabbit retinal extract.

TABLE III
MIF Production in Presence of Bovine Antigen Extracts

Disease	No.	Positive[a] responses to challenging extract			
		Retina	Uvea	Cornea	Thyroid
IIBU	9	6	6	0	0
Benign	10	0	0	0	0
Chronic	6	0	0	0	0
Hashimoto's disease	2	0	0	0	2

[a] A positive indication for MIF production was recorded when an inhibition of 30% or more was observed in the presence of the challenging extract.

patients with IIBU were positive despite the application of rigid criteria (Table II).

The leukocytes of six patients with IIBU also demonstrated significant MIF production (Table III). Low titers of hemagglutinating antibodies toward lens, retina, or uveal antigens were observed in two patients with benign disease, two patients with chronic, and in all patients with IIBU. However, in dilutions of 1:9 or more of the sera, none of the benign, one of the chronic, and four of the IIBU patients gave positive responses.

Discussion

Grouping cases of idiopathic bilateral uveitis (pars planitis) under three types according to initial ocular symptoms and findings appears to be clinically satisfactory. However, the immunologic parameters tested in this study did not differentiate between the benign and the chronic types.

Both of these groups had cellular and humoral immune responses toward ocular antigens that were comparable with control (nonuveitis) patients. Only the IIBU group had clear-cut differentiating immunologic responses and could be accepted as a separate entity. Nonetheless, it is still possible that the subdivision into the three types is sound, as the clinical and laboratory findings are influenced by the state of the individual specific immunologic and nonspecific inflammatory regulatory mechanisms.

We have previously speculated on the possible underlying autoimmune phenomena associated with idiopathic insidious bilateral uveitis (IIBU) after the demonstration of lymphocytes sensitized to retinal and uveal antigens in the peripheral blood of a small group of patients.[6] The data as reported in the study demonstrate further the autoimmune nature of IIBU. All patients tested showed a significant cellular and/or humoral

immune reaction toward crude extracts of both allogeneic and xenogeneic retinal and uveal antigens. Marked cellular immune responses were obtained using specific blast transformation and secretion of the migration inhibition factor when lymphocytes from the peripheral blood of patients with IIBU were challenged *in vitro* by retinal or uveal antigens.

References

1. BenEzra D et al.: *Proc Soc Exp Biol Med* **125**:1305, 1967.
2. BenEzra D: *Arch Ophthalmol* **94**:661, 1976.
3. BenEzra D et al.: *Am J Ophthalmol* **82**:866, 1976.
4. McCoy JL et al.: In *In vitro Methods in Cell Mediated and Tumor Immunity*, Bloom BR and David JR (Eds.). Academic Press, New York, 1976, p. 607.
5. Gery I and Davies AM: *J Immunol* **87**:351, 1961.
6. BenEzra D: In *Michaelson's Textbook of the Fundus of the Eye*, 3rd ed. Livingston, 1980.

chapter 30

Immediate Hypersensitivity in Human Retinal Autoimmunity

**Jean Sainte-Laudy, Jean-Pierre Faure, Yvonne de Kozak,
Etienne Bloch-Michel, Phuc Lê Hoang, and Jacques Benveniste**

The human basophil degranulation test (HBDT) allows one to detect immediate hypersensitivity at the cellular level through reagin-dependent sensitization of patient's basophils against specific antigen. Blood basophils with specific (Ig E) antibody fixed on the cell surface lose their ability to be stained by toluidine blue when exposed *in vitro* to the corresponding antigen. Counting the basophils with or without the antigen allows one to determine the percentage of degranulation, i.e., the percentage of basophils able to degranulate for a given dose of antigen. HBDT has proven its high specificity for the diagnosis of immediate hypersensitivity to common allergens, food products, insect venoms, drugs and other chemicals, and parasitic antigens.[1] The test was recently extended to autoimmune diseases, confirming for systemic lupus erythematosus that reagin-mediated phenomena against autoantigens may be a component of autoimmunity.[2,3]

Evidence that mast cell sensitization to the retinal S antigen occurs in rats with experimental autoimmune uveoretinitis[4,5] prompted us to perform HBDT with ocular antigens in patients suspected of autoimmune ocular disorders. Using HBDT in the presence of lens proteins, we found a high incidence of positive tests in patients with presumed lens-induced uveitis.[6] In this report we examine 58 patients with uveitis or retinitis of various types for basophil reactivity to retinal S antigen.

Materials and Methods

S antigen was isolated from bovine retinas according to Dorey et al.[7] and lyophilized. HBDT was performed by the technique of Benveniste[1] slightly modified. Five dilutions of antigen ranging from 100 μg/ml to 1 ng/ml were distributed in microtitration plates and dried at 37°C. The plates could be stored several months at room temperature. Blood was enriched in leukocytes by simple sedimentation at 1 g. The supernatant was spun at 1000 rpm and the cells were washed twice. Ten μl of cell suspension were placed in wells of the plate, five containing S antigen and one control without antigen. After incubation 15 minutes at 37°C, 90 μl of an acidic solution of toluidine blue were added to each well. After 10 minutes the content of each well was transferred to counting chambers. The percentage of degranulation is expressed as follows:

$$\frac{\text{(basophil No. without antigen)} - \text{(basophil No. in the well corresponding to the maximum degranulation)}}{\text{(basophil No. without antigen)}} \times 100$$

Due to zonal phenomena there exists an optimal amount of antigen inducing maximal degranulation and varying for each patient. The threshold of positivity of the test is 25% degranulation.

Laboratoire d'Immunopathologie de l'Oeil, CNRS ER 227, INSERM U 86, Hôtel-Dieu, Paris; Laboratoire d'Immunologie 75014 Paris; Laboratoire d'Immunopharmacologie de l'Allergie et de l'Inflammation, INSERM U 200, Clamart/Paris, France.

TABLE I
Results of HBDT with S Antigen in Patients with Chorioretinal or Retinal Lesions

	No. of positive/total
Patients with chorioretinal or retinal lesions	16/46 (35%)
First attack	4/15 (27%)
Recurrence	12/25 (48%)
Recent onset (< 3 months)	5/16 (31%)
Long duration (> 3 months)	10/28 (36%)
Unilateral	2/9 (22%)
Bilateral	14/36 (39%)
With systemic corticosteroid therapy	6/18 (33%)
Without systemic corticosteroid therapy	8/23 (35%)
Retinal vasculitis	9/22 (41%)
Multifocal chorioretinitis	3/9 (33%)
Focal toxoplasmic chorioretinitis first eye + secondary cyclitis second eye (sympathy ?)	2/2
Post-traumatic sympathetic ophthalmia	1/2
Pigment placoid epitheliopathy	1/3
Retinitis pigmentosa	0/2

Results

HBDT with S antigen was negative in 14 controls without ocular disease. The test was positive in 16 of 58 patients (28%). All patients with positive test presented lesions of the posterior segment of the eye (16 positive/46 = 35%), whereas the test was negative in 10 patients with anterior and/or intermediate uveitis without posterior involvement. Reaginic sensitization to S antigen seems to be more frequent in recurrent diseases than in first attack, and in bilateral rather than unilateral uveitis. It was frequently associated with retinal vasculitis, multifocal chorioretinitis, and sympathetic uveitis (Table I).

We did not find that systemic corticosteroid therapy had any effect on the results of HBDT because the number of basophils was frequently very low (5 cells/μl) in patients receiving corticosteroids; therefore, the test is impracticable for such patients.

Discussion

HBDT provides a very accurate means for *in vitro* diagnosis of allergies. It explores not only the presence of specific Ig E antibodies but also their fixation on the cell membrane and the ability of the basophil to degranulate in the presence of allergen. The results of HBDT correlate well with the skin test for many common allergens and HBDT is particulary useful when the skin test is impracticable. HBDT does not require expensive apparatus and thus is especially convenient for large routine application. Its sensitivity allows it to demonstrate hypersensitivity phenomena in a number of pathological states other than classical atopy.

In this work we found basophil sensitization to S antigen in patients suffering from chorioretinal inflammation in a percentage of cases slightly lower than the tests for cell-mediated immunity.[8] Both lymphocyte transformation test (LTT) and HBDT were performed simultaneously in 40 of our patients with chorioretinal involvement. Results of LTT are given by Tanoé et al. (Chapter 61). We did not find any correlation between the results of both tests, the results being more often discordant (65%) than concordant (35%). Although a definitive conclusion cannot be drawn from this small series, reaginic sensitization to S antigen seems unrelated to lymphocyte responsiveness to the antigen in most patients. Pooling the results of LTT and HBDT leads to the conclusion that immune reactivity to the S antigen is very common (80%) in patients with posterior ocular inflammatory disease. Since both tests were found negative in normal people and in most patients with only anterior or intermediate uveitis, this finding seems significant.

As for drug hypersensitivity, specific anti-S antibodies were often undetectable by serological methods, whereas HBDT allowed their detection. We found circulating anti-S IgE by ELISA in six cases of 24 patients with S antigen-positive HBDT and did not find specific IgE in 16 patients with negative HBDT.

References

1. Benveniste J: *Clin Allergy* **11**:1–11, 1981.
2. Egido J et al.: *Ann Rheum Dis* **39**:312–317, 1980.

3. Camussi G et al.: *Lab Invest* **44**:241–251, 1981.
4. de Kozak Y et al.: *Eur J Immunol* **11**:612–617, 1981.
5. de Kozak Y: Chapter 31, This volume.
6. Sainte-Laudy J et al.: *Nouv Presse Med* **10**:992, 1981.
7. Dorey C et al.: *Ophthalmic Res* **14**:249–255, 1982.

chapter 31

Immediate Hypersensitivity in Experimental Retinal Autoimmunity

Yvonne de Kozak[a], Jean Sainte-Laudy[b], Junichi Sakai[a], Jacques Benveniste[c] and Jean-Pierre Faure[a]

Immediate hypersensitivity (IHS) has been shown to be implicated in several experimental[1-3] and human[4-7] autoimmune diseases. In the pathogenesis of experimental autoimmune uveoretinitis (EAU) induced in rats by immunization with the retinal S antigen, there is evidence of cell-mediated,[8,9] immune complex-mediated,[8,10] and reagin-mediated[11] mechanisms. It was suggested that the release of vasoactive mediators from choroidal mast cells (MC) sensitized to S antigen could trigger early changes in the permeability of the blood–retinal barrier, allowing the access of cytotoxic cells as well as free antibodies and/or immune complexes to the retinal target tissue.[11]

In this report we present new evidence for the concept that an IHS reaction to S antigen might initiate the induction of EAU. We investigated the inhibition of ocular inflammation by drugs (disodium cromoglycate (DSCG) and ketotifen) that impede MC degranulation or deplete MC of their vasoactive amines (compound 48/80).

The serum content of anti-S IgE and IgG antibodies was determined during the course of the response to S antigen immunization and compared with the sensitization of peritoneal MC. Blocking of in vitro MC de-

granulation by class-specific antisera was used to identify the class of the reagin involved.

Materials and Methods

Induction and assessment of EAU

Adult Lewis rats were immunized by a single injection of 5, 30, or 50 µg protein of bovine retinal S antigen[12] mixed with different adjuvants[10] as listed in Table I. The clinical and histological severity of the ocular inflammation was scored from 0 to 3.

Counting of choroidal MC

The total number of MC and the percentage of spontaneously degranulated MC were determined on flat-mounted preparations of choroidal tissue stained with toluidine blue as described by de Kozak et al.[11]

Degranulation of peritoneal MC

Peritoneal MC from S antigen-sensitized animals were incubated in vitro with S antigen. Ten µl containing about 2000 cells each were placed in six wells of microtiter plates in which solutions of S antigen ranging from 100 µg/ml to 1 ng/ml had been preevaporated, and also in two control wells.[13] The percentage of degranulated cells was determined as described by de Kozak et al.[11] The class of the reagin responsible for S antigen-induced MC degranulation was identified by the inhibitory effect of class-

[a]Laboratoire d'Immunopathologie de l'Oeil, CNRS ER 227, INSERM U 86, Hôtel-Dieu, Paris; [b]Laboratoire d'Immunologie 75014 Paris; [c]Laboratoire d'Immunopharmacologie de l'Allergie et de l'Inflammation, INSERM U 200, Clamart/Paris, France.

TABLE I

Immune Response of Rats to S Antigen 1 Month after Immunization

Immunization		No. of rats with EAU/total	Peritoneal MC degranulation with S antigen		Anti-S antibodies[a]	
S antigen (μg)	Adjuvants		No. of positive tests/total	Mean percentage of degranulated MC	IgG	IgE
50	*M. tuberculosis* 500 μg in IFA SC	5/5	5/5	43 ± 3	1.386	0.991
30	*M. butyricum* 50 μg in IFA (=CFA) SC	5/5	5/5	49 ± 2	1.446	1.086
5	"	5/5	5/5	34 ± 2	1.493	0.817
none	"	0/5	0/5	11 ± 4	0.016	0.002
50	*H. pertussis* 5.10⁹ in alum IP	3/5	5/5	39 ± 2	0.737	0.082
50	*M. tuberculosis* 500 μg + IFA SC + *H. pertussis* 10⁹ + alum IP	6/6	6/6	41 ± 3	1.415	0.808
5	"	4/4	4/4	37 ± 2[b]	1.181	0.14[b]
none	"	0/5	0/5	8 ± 1	0.082	0.005
50	IFA SC	0/5	0/5	6 ± 1	1.228	0.521

[a] Antibody contents of sera (diluted 1:200) determined by ELISA and expressed as the mean value in each series of absorbance at 490 nm.

[b] Samples taken 11–18 days after immunization.

IFA: incomplete Freund's adjuvant; CFA: complete Freund's adjuvant; SC: subcutaneously (foot pads); IP: intraperitoneally.

specific anti-rat Ig antibodies on the degranulation. The MC were incubated with various dilutions of either goat anti-rat IgE (GARa/IgE) or IgG (GARa/IgG) antibodies (Nordic) for 20 minutes at 37°C prior to incubation with S antigen.

Antibody determination in sera

IgG and IgE antibodies against S antigen were studied by an enzyme-linked immunosorbent assay (ELISA) on polystyrene microplates (Nunc) coated with S antigen.[14] The wells were filled successively with rat sera at dilutions of 1:100 or 1:200, with either GARa/IgG or GARa/IgE antibodies, and finally with peroxidase-conjugated rabbit anti-goat IgG antibody (Nordic). The intensity of the peroxidase reaction on an ortho-phenylenediamine substrate was read using an Autoreader ELISA photometer (Dynatech).

Protocols of drug administration

Disodiumcromoglycate (DSCG, Fisons Laboratories) (20 mg/ml) or Ketotifen fumarate (Sandoz Laboratories) (20 or 2 mg/ml) were instilled in both eyes 4 times a day from the fourth day to the 20th day after immunization with 30 μg of S antigen in CFA.

Solutions of compound 48/80 at various concentrations in saline were injected either subconjunctivally or intraperitoneally. A subconjunctival (S Conj) injection of 125 μg of the compound in 0.05 ml of saline was done 2 days before the immunization, followed by daily S Conj injections of 25 μg in 0.05 ml of saline. As an alternative, 150 μg of the compound in 0.5 ml of saline were injected intraperitoneally two days before immunization. The degranulating efficiency of compound 48/80 was assessed on flat mounts of choroid after S Conj injection and on both choroidal and peritoneal MC after intraperitoneal (IP) injection.

Results

Pathological features of EAU

EAU occurred in almost all rats immunized with S antigen in bacterial adjuvants. Controls injected with adjuvants alone and rats immunized with S antigen in incomplete Freund's adjuvant (IFA) showed no ocular reaction (Table I). Variations in the severity of EAU were observed according to the dose of S antigen and according to the adjuvant used. The severity of the ocular inflammation was maximal in the rats injected with 50 μg of S antigen together with both mycobacterial and pertussis adjuvants. In those animals, the first histological abnormalities were observed on the 10th–12th day as large areas of edema in the retina as well as inflammatory cells in the uvea and in the vicinity of retinal vessels. In a few days extensive infiltration by polymorphonuclear leukocytes (PMN) and lymphoid cells developed. At the end of the third week after immunization the visual cells were destroyed; the PMN had disappeared, but a granulomatous infiltrate with epithelioid cells persisted for months. Seven months or 1 year after immunization, the outer layers of the retina were completely destroyed, but some inflammatory cells were still present in the choroid, around internal retinal vessels, and in the vitreous, in seven of eight rats. The inflammation was less severe, but the main pathological features were similar, in rats injected with lower doses of S antigen with or without pertussis vaccine. In particular, early retinal edema, transient PMN infiltration, and extensive tissue damage were noted with all immunization protocols including bacterial adjuvants.

Modifications of choroidal MC during EAU

In normal rats, the number of choroidal MC/mm^2 was 141 ± 10, and the percentage of degranulated MC was 18 ± 2. Rats injected with 50 μg of S antigen and both mycobacterial and pertussis adjuvants were

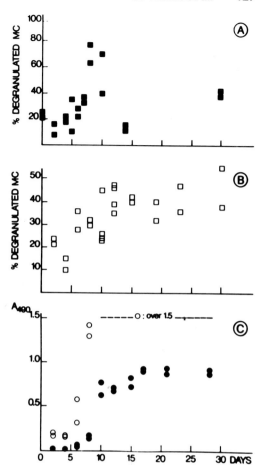

FIG. 1. Time course of MC sensitization and antibody levels after immunization with 50 μg of S antigen and both 500 μg of *M. tuberculosis* in IFA subcutaneously and *H. pertussis* 10^9 in alum intraperitoneally. *(A)* Percentage of spontaneously degranulated choroidal MC counted on flat-mounted preparations of choroid. *(B)* Percentage of *in vitro* peritoneal MC degranulation in the presence of S antigen. *(C)* Anti-S IgG (○) and IgE (●) antibody contents of 1:100 diluted sera.

killed on schedules varying from 2 days to 1 year after immunization. The number of choroidal MC increased at first with a maximum on day 6, reaching almost twice the normal value; then it suddenly decreased to a minimum on day 10. At this time, the percentage of degranulated MC increased, reached its maximum on day 8, when 70% of cells appeared degranulated, and then rapidly decreased (Fig. 1A). Choroidal MC degranulation preceded the beginning of the disease which occurred in all eyes on day 10–12. Seven months and 1 year after immunization, the MC count was found to be very high: 609 ± 77 in 5 eyes and 366 ± 104 in 4 eyes respectively with no evidence of degranulation.

Sensitization of peritoneal MC

In the same animals, anti-S reaginic sensitization of peritoneal MC was detected by an *in vitro* degranulation test with S antigen. The test was positive from the 6th day. The percentage of degranulation increased until the 12th day and then remained constant for 1 month (Fig. 1*B*). The test was still positive (mean percentage of degranulated cells [43 ± 4] in 10 rats killed at 7 months, but negative [9 ± 2] in six rats killed at 12 months).

The peritoneal MC degranulation test was also performed on rats killed 3–4 weeks after immunization with various doses of S antigen and with various adjuvants. The results in Table I show that the presence in the adjuvant of either Mycobacteria or *H. pertussis*, or both, had a similar effect on MC sensitization. In each case, the test was strongly positive, whereas it was negative with incomplete Freund's adjuvant. An immunizing dose of 5 μg of S antigen was sufficient for MC sensitization. The test was negative after immunization with bacterial adjuvants without S antigen.

Effect of anti-IgG and anti-IgE antibodies on peritoneal MC degranulation test

In five immunized rats the mean degranulation percentage without antisera was 48 ± 2. This value dropped to 23 ± 3 ($p < 0.01$) in the presence of GARa/IgE diluted 100 fold, and to 15 ± 3 ($p < 0.001$) at a 1000-fold dilution. No inhibition of degranulation was observed with GARa/IgG: 43 ± 4 at a 100-fold dilution; 43 ± 1 at a 1000-fold dilution.

IgG and IgE anti-S antibodies in serum

Antibodies against S antigen of the IgG and IgE classes were both detected from the 6th–8th day after immunization with 50 μg of S antigen and both Mycobacterial and pertussis adjuvants. The antibody content increased rapidly from the 8th to the 10th day and then remained high (Fig. 1*C*). Antibodies of both classes were still present in lower amounts after seven or 12 months, respectively.

One month after immunization (Table I) we did not find any significant difference in the serum antibody content as regards the immunizing dose (5, 30, or 50 μg) of S antigen. In a series in which blood was taken 11–18 days after immunization with 5 μg S antigen the anti-S IgE antibodies were very low. Both IgG and IgE antibodies were high when Mycobacteria in IFA, whether combined or not with pertussis vaccine, were used as adjuvant. On the other hand, IP immunization with S antigen and pertussis in alum led to moderate IgG and very low IgE antibody content of the serum. Immunization with S antigen in IFA, i.e., without bacteria, produced high levels of both IgG and IgE circulating anti-S antibodies.

Effects of inhibitors of MC degranulation

Repeated instillations of either DSCG or Ketotifen from the 4th day after immunization delayed the onset of EAU by about 2 days compared with untreated controls (Table II). Whereas all untreated rats presented EAU, no disease or mild disease was found clinically in three of 20 eyes in treated animals. Histologically, on the 21st day, two eyes showed no abnormality, and four eyes showed mild or moderate lesions. Results with DSCG and Ketotifen were similar. A significant decrease of the percentage of MC degranulated in the presence of S antigen was found in the treated animals. The IgG and IgE anti-S antibody content of sera was not modified by the treatments.

Effects of compound 48/80

One IP injection of 150 μg of compound 48/80 induced a general reaction of shock and a slight as well as transient ocular reaction. The choroidal MC were less numerous and less stainable than normal from the 6th hour to the 6th day after the injection. The number of peritoneal MC was very low until the 7th day; then it increased above

TABLE II
Effect of Drugs on Immune Response to S Antigen[a]

Treatment	Day of onset	EAU Histology			Peritoneal MC degranulation test[b]	Anti-S antibodies[c]	
		None or moderate	Severe	Mean score		IgG	IgE
DSCG	16.3 ± 0.5[d]	3/10	7/10	2.7 ± 0.2	27 ± 3[d]	1.385	0.524
Untreated controls	14.3 ± 0.4		10/10	3.0 ± 0	46 ± 8	1.257	0.552
Ketotifen	15.4 ± 2.4[d]	3/10	7/10	2.1 ± 0.4	33 ± 4[d]	1.310	0.543
Untreated controls	13.0 ± 0.4		10/10	3.0 ± 0	56 ± 7	1.328	0.564
Compound 48/80 150 µg IP	16.6 ± 0.5[d]	12/22	10/22	1.6 ± 0.3	24 ± 2[d]	1.563	0.519
Untreated controls	13.3 ± 0.2		10/10	3.0 ± 0	46 ± 8	1.468	0.571
125 then 25 µg S Conj	17.3 ± 1.7[d]	6/8	2/8	1.1 ± 0.4	n.d.	1.230	0.520
Untreated controls	12.8 ± 0.2		10/10	3.0 ± 0	n.d.	1.328	0.564

[a] Immunization was carried out with 30 µg of S antigen in CFA.

[b] Mean percentage of degranulated MC in the presence of S antigen.

[c] Antibody contents of sera determined by ELISA and expressed as the mean value in each series of absorbance at 490 nm. Sera were diluted 1:200.

[d] Significantly different from untreated controls.

IP: intraperitoneal; S Conj: subconjunctival; n.d.: not done.

the normal value. One S Conj injection of compound 48/80 produced immediate local anaphylactic phenomena varying in intensity according to the injected dose. Doses from 10 to 125 µg induced a complete degranulation of the choroidal MC of the injected eye after 30 minutes and 1 µg produced almost complete degranulation. On the following days, the choroidal MC were stainable again.

One IP injection of 150 µg of 48/80 given 2 days before the immunization with S antigen (30 µg in CFA) significantly inhibited EAU. The mean onset time was delayed; no disease was detected in eight eyes of 22, and the intensity of the inflammation and the extent of retinal destruction were markly reduced in four other eyes (Table II). An even better inhibition was observed when the first injection (125 µg) was given subconjunctivally and followed by daily 25 µg S Conj injections. With this last protocol, tested in four rats, the disease was completely suppressed in four eyes, mild in two eyes, and severe in two other eyes.

In the rats treated by 48/80 IP injection, the peritoneal MC degranulation test with S antigen was less positive than in untreated controls, but the anti-S IgG and IgE antibody content of sera was not modified.

Discussion

Soon after immunization of rats with S antigen, choroidal MC degranulation and sensitization of peritoneal MC to S antigen were demonstrated as the first manifestations of EAU. An early increase of the permeability of the blood–retinal barrier was evidenced by intraretinal exudation preceding cell infiltration. These findings suggest that the release of vasoactive mediators from reagin-sensitized choroidal MC would increase permeability and, therefore, favor the expression of the other components of the immune response against the retinal target. In addition, MC sensitization to S antigen persisted for a long period after the immunization.

Experiments with drugs that modulate the MC function suggest a causal link between MC activation and ocular disease. Drugs that block MC degranulation, DSCG and Ketotifen, were moderately effective in delaying and attenuating EAU. The depletion of MC from their mediators by the injection of the degranulating agent compound 48/80 before immunization effectively prevented or attenuated the ocular inflammation.

The isotype of the antibody responsible for the sensitization of peritoneal MC to S antigen was thought to be IgE, according to the blocking effect of anti-rat IgE antibody on MC degranulation, contrasting with the poor efficiency of anti-IgG antibody in modifying the test. Circulating anti-S antibodies of both IgG and IgE classes were detected in the sera of animals with EAU. Antibodies of both classes appeared before the clinical onset of the disease.

The following discrepancies were found between the presence of circulating anti-S IgE antibodies and the presence of anti-S IgE antibodies fixed on MC, as detected by the peritoneal MC degranulation test. MC sensitization was detected earlier after immunization than circulating anti-S IgE. In animals immunized with a low dose of S antigen, MC sensitization was effective on the 2nd–3rd week, although the serum did not contain high amounts of IgE antibodies. The same phenomenon was found when pertussis vaccine in alum, without IFA, was used as adjuvant. It seems that the cytophilic IgE antibodies are fixed on MC as soon as they are synthesized and that their presence in the serum reflects an excess of their production. Treatment by either MC antidegranulating or mediator-depleting drugs reduced in parallel the severity of EAU and the peritoneal MC sensitization, but did not modify the level of circulating antibodies. This finding confirms that the ocular inflammation is related to MC-bound anti-S IgE rather than to circulating IgE antibodies.

Furthermore, high amounts of IgE and IgG circulating anti-S antibodies were present in animals immunized with S antigen in IFA. These animals did not develop ocular disease, and their MC were not sensitized to the antigen.

These findings support the hypothesis of a causal link between the sensitization of mast cells to the autoantigen and the occurrence of the ocular disease.

Acknowledgment

This study was supported by grants from CNRS (ER 227) and INSERM (U 86, CRL 79.5.535.1).

References

1. Voisin GA and Toullet F: In *Immunology and Reproduction*, Edwards RG (Ed.). Collings, New York, 1968, pp. 93–105.
2. Moore MJ et al.: *Res Commun Chem Pathol Pharmacol* 9:119–132, 1974.
3. Paterson PY: In *Autoimmunity: Genetic, Immunologic, Virologic and Clinical Aspects*, Tatal N (Ed.). Academic Press, New York, 1977, pp. 643–692.
4. Permin H: *Lancet* 200–201, 1977.
5. Norn S et al.: In *The Mast Cell: Its Role in Health and Disease*, Pepys J and Edwards AM (Eds.), Pitman Medical Publishers, Tunbridge Wells, U.K., 1979, 53–60.
6. Egido J et al.: *Ann Rheum* 39:312–317, 1980.
7. Camussi G et al.: *Lab Invest* 44:241–251, 1981.
8. Faure JP and de Kozak Y: In *Immunology of the Eye, 2, Autoimmune Phenomena and Ocular Disorders*, Helmsen, RJ et al. (Eds.). Information Retrieval Inc, Washington DC, 1981, pp. 33–48.
9. Salinas-Carmona MC et al.: *Eur J Immunol* 12:480–484, 1982.
10. de Kozak Y et al.: *Curr Eye Res* 1:327–337, 1981.
11. de Kozak Y et al.: *Eur J Immunol* 11:612–617, 1981.
12. Dorey C et al.: *Ophthalmic Res* 14:249–255, 1982.
13. Benveniste J: *Clin Allergy* 11:1–11, 1981.
14. Tuyen VV et al.: *Curr Eye Res* 2:7–12, 1982.

chapter 32

Experimental Retinal Vascular Disease Model in Rats

Paul C. Stein and Devron H. Char

Autoimmune retinitis includes those diseases with retinal vascular infiltrates of unknown etiology. Retinal vasculitis can occur as an isolated entity or in association with systemic collagen vascular diseases. Immune complexes may be important in the pathophysiology of these disorders; we and others have demonstrated elevated levels of circulating immune complexes in various types of uveitis.[1,2]

We have studied retinal vascular changes associated with heterologous immune complex disease in a rat model, utilizing antibody against shared retinal endothelial antigens. We observed cross reactivity between retinal and renal antigens, and we have seen evidence of immunoglobulin and platelet deposition on the retinal vascular endothelium.

Materials and Methods

Antigen preparation

Retinas were dissected from enucleated bovine eyes. Purified retinal basement membranes (RBM) were obtained as described by Carlson et al.[3] The washed RBM suspension was divided equally. One portion was lyophilized (RBM); the other portion was digested with 2 mg of elastase (Sigma, Type 111, 70 U/mg), containing 0.1% so-dium azide, for 24 hours at 37°C.[4] The supernatant (RBM-S) was sterilized by filtration and 100 μl aliquots were frozen. A sample was assayed for total protein (Lowry) and analyzed by polyacrylamide gel electrophoresis (PAGE). Representative samples at different stages of purification were also fixed in glutaraldehyde for ultrastructural examination.

Immunization of rabbits

New Zealand white rabbits were immunized with approximately 0.5 ml of RBM-S (1.0–1.5 mg protein) or with 0.5 ml of RBM (10 mg lyophilized) mixed 1:1 with complete Freund's adjuvant (CFA), which was injected intramuscularly and subcutaneously. After resting 1 month, the rabbits were given a booster injection of antigen. Immune serum was obtained by cardiac puncture. An aliquot was absorbed with ox erythrocytes, glutaraldehyde cross-linked bovine serum, or by the addition of 50 μl of bovine serum to 250 μl of test serum.

Immunization of Experimental Rats

Lewis and Brown-Norway (BN) rats were injected in the hind footpads with 100 μl of RBM-S (0.2–0.3 mg protein) or with RBM (3 mg lyophilized) mixed 1:1 with CFA. A control group was injected with adjuvant alone. The animals received a booster injection of antigen in incomplete Freund's adjuvant after 30 days. Other rats were also injected intravenously with 1 ml of rabbit antiserum (IgG) prepared against RBM or

Ocular Oncology Unit, Department of Ophthalmology and the Francis I. Proctor Foundation, University of California, San Francisco California.

RBM-S. An IgG fraction from normal rabbit serum was injected into a control group. Animals were sacrificed 1, 3, 7, and 30 days after passive transfer of serum or 3 weeks after the last antigen injection.

Immunofluorescence and Electron Microscopy

The left eyes and a representative kidney wedge-biopsy from sacrificed animals were placed into isopentane cooled with dry ice. Cryostat sections (5-10 μ) were air dried and extensively washed in PBS. Bound immunoglobulins (rat IgG or rabbit IgG) were demonstrated with appropriate antisera. Indirect fluorescent assays were also performed using specific goat (Fab'2) antisera followed by incubation with FITC-labeled (rabbit) anti-goat IgG diluted 1:20 with PBS. Routine tissue preparation procedures were not adequate for the precise demonstration of retinal IgG deposits in passively immunized rats. It was necessary to utilize fixation with 4% paraformaldehyde–.2% picric acid and embed in water-miscible, methacrylate plastic. Sections (1μ) were digested with "Protease" (1mg/ml) (Sigma, St. Louis, Mo.) for 3 hrs at 37°C in order to "unmask" antigenic sites blocked by fixation or by embedment. They were then washed, incubated with fluorescein-labeled anti-rabbit IgG, and examined by fluorescence microscopy.

The specificity of the antisera produced in immunized rats and rabbits was determined by the application of 50 μl of diluted (1:5-1:2000) test serum (absorbed or unabsorbed) to a panel of cryostat sections (kidney, retina, intestine, muscle) and to cryostat sections from a mouse sarcoma (EHS-sarcoma) containing a connective tissue matrix of Type IV (basement membrane) collagen.[5] Rat IgG subclasses binding to substrate tissues were identified by incubation with fluorescein-labeled antisera. Rabbit IgG was identified with fluorescein-labeled goat and anti-rabbit IgG.

In vivo complement fixation in experimental rats and rabbits was examined by direct immunofluorescence using labeled anti-rat C3 or anti-rabbit C3 (Cappel Laboratories) respectively. *In vitro* activation of complement was ascertained by incubating cryostat sections with guinea pig sera (complement source) or guinea pig sera depleted of C3 (control). Slides were then incubated with anti-guinea pig C3-FITC, washed, and examined for fluorescence.

The right eyes were immediately dissected from sacrificed animals, taking care not to exert any pressure on the ocular tissues. The sclera was punctured with a fine needle to expedite entry of fixative. After washing in buffer, the retina was dissected from the choroid and cut into small squares. These were post-fixed in 1% osmium tetroxide, dehydrated in ethanol and propylene oxide, and embedded in epon for thin sectioning and subsequent electron microscopy.

Results

Retinal basement membranes were obtained. After homogenization, lysis in distilled water, and DNAse and detergent extraction, the resulting basement membrane preparations were relatively free of contaminating cellular debris. Collagen-like fibrils were observed in the matrix of the basement membrane structure. Other fiber-like components having a delicate, net-like appearance could also be observed. These were probably derived from the associated pericytes, which have numerous fibrous elements in the cytoplasm. After enzyme digestion, only fibrils with typical collagen periodicity were present. PAGE of the supernatant from RBM digests demonstrated 4 major bands and several minor components (Fig.1).

Rabbits immunized with RBM (or RBM-S) developed antibodies that bound to heterologous (bovine, rat) retinal vessels, to glomerular and tubular basement membranes, to capillary and epithelial cell basement membranes in intestinal villi, to isolated glomeruli and RBM preparations, and to EHS-mouse sarcoma tissue. Control rabbit sera were consistently negative for these tissues.

Immunized BN and Lewis rats developed antibodies that reacted with bovine retina, and bovine, rabbit, and human glomerular and tubular basement membranes. Serum

ologous (rabbit) antisera had deposits of rabbit IgG along the endothelial cell membranes of some, but not all, retinal vessels (Fig. 2A). The large vessels were more likely to demonstrate rabbit IgG by immunofluorescence than the smaller retinal vessels. Retinas from control rats injected with normal rabbit IgG did not show any retinal vessel immunofluorescence (Fig.2B).

These animals also had linear deposits of rabbit IgG along the glomerular basement membranes, but not on the renal tubular basement membranes. This was evident shortly after injection (24 hours) and could still be detected after 30 days. Rat IgG deposits were observed at least 14 days after intravenous injection of heterologous serum, and their distribution correlated with the linear deposits of rabbit IgG along the glomerular basement membranes. We could not demonstrate unequivocally either rat IgG or C3 deposits in the retinal vasculature.

Retinal vessels became stenotic in passively immunized rats after 30 days (Fig. 3). Adjacent retinal neural tissue appeared atrophic and vacuolated, and with swollen mitochondria. Platelets were rarely observed in normal rat retinal vessels. They were often present in retinal vessels in the passively immunized experimental animals (Fig. 4). Platelets were firmly attached to the vessel endothelium (Fig. 5), and some were degranulated. Platelet adherence could be observed within 24 hours after serum transfer and was maximal at 7 days. Other inflammatory cells (lymphocytes and polymorphonuclear cells) were not found in affected vessels or adjacent tissues.

Discussion

Retinal basement membrane antigens have been solubilized by a number of biochemical procedures followed by digestion with elastase. Ultrastructural examination has confirmed the purity of the RBM preparations, which were consistently free of adhering cellular components and debris. The amorphous structure of the purified basement membranes is consistent with that observed in untreated retinas.

PAGE of solubilized RBM demonstrated

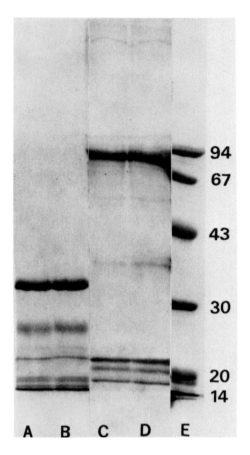

FIG. 1. PAGE of RBM-S obtained after 24 hour digestion with elastase. Lanes A,B: elastase prior to incubation at a concentration four times greater than actually used in digestions. Lanes C,D: subunit proteins of RBM-S. Lane E: molecular weight markers (94,000; 67,000; 43,000; 30,000; 20,000; and 14,000 daltons).

from these animals were also able to bind to the collagen matrix in the EHS mouse sarcoma tissue. IgG from immunized rats, which bound to tissue substrates from other species, was demonstrated to be IgG_1 and IgG_{2a}.

Immunized rats and rabbits did not show any IgG bound to autologous retina or kidney, nor did their serum IgG bind to cryostat sections of allogeneic tissues. IgG antibodies that were reactive with heterologous tissue did not activate guinea pig complement *in vitro*, as determined by immunofluorescence microscopy.

Rats injected intravenously with heter-

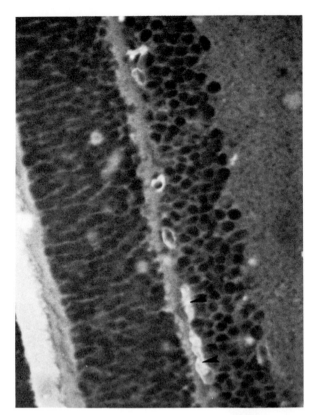

FIGS. 2A and B. Passive transfer of immune rabbit sera in Lewis rats. *(A)* Localization of rabbit IgG in retinal vessels (arrows). *(B)* Control rat retina injected with normal rabbit serum.

FIG. 2B

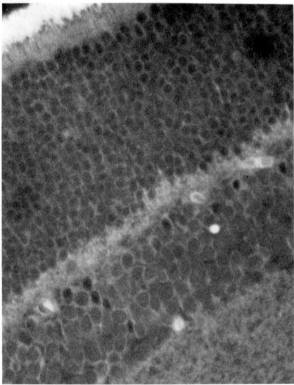

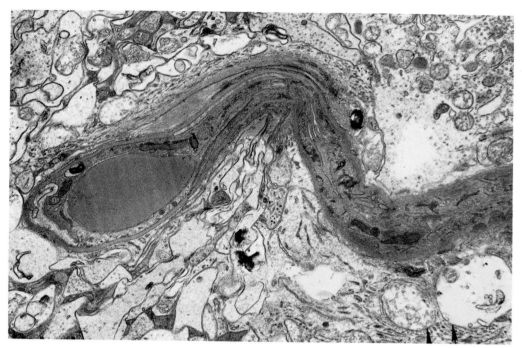

FIG. 3. Retina from rat passively immunized with rabbit-anti RBM-S and fixed for electron microscopy after 30 days. Note collapsed vessel, typically observed in rats from the experimental group, and abnormal mitochondria (swollen) in adjacent tissue (arrow) ($\times$ 10,500).

the presence of four major subunit proteins. The molecular weight and electrophoretic mobility of these are similar to the major proteins reported from solubilized human glomerular basement membranes.[6] Trace protein contaminants may also be present, derived from the elastase used in digestion.

Rabbits immunized against RBM and RBM-S produce antisera with similar specificities. Basement membrane antigenic determinants are apparently not altered by the solubilization procedure. These antibodies bind to a variety of tissues from different species and have affinity for the basement membranes present in vascular tissue. Probably they are directed toward collagen and laminin; elevated titers of these antibodies (1:1000) for collagen (Type IV) and laminin (ELISA assays) were demonstrated. Antibody also bound to the EHS-sarcoma (Type IV collagen).

Rats (and rabbits) were not able to recognize shared antigenic determinants on bovine retinal basement membranes. This tolerance for self proteins has been noted

in other experimental autoimmune disease models. Immune sera reacted strongly with heterologous tissues including bovine retinal vessels. The rat antibodies that bound to these tissues were immunoglobulins of the IgG_1 and IgG_{2a} subclass. These immunoglobulins did not activate complement, and this may reflect differences between the amount of IgG fixed and the amount required for detection. Alternatively, these antibodies may not activate complement; the failure of certain Ig subclasses to interact with complement *in vivo*, even though they are expected to, has been observed in human glomerulo-nephropathies,[8] and is believed to be related to the amount and affinity of the tissue-fixing antibody present.[9]

A thin linear deposit of rabbit IgG and platelet aggregates along the endothelial membrane was detected in some retinal vessels in passively immunized rats. The absence of immunoglobulin in the majority of retinal vessels indicates a possible selectivity, the cause of which is uncertain. In some forms of clinical retinal vascular dis-

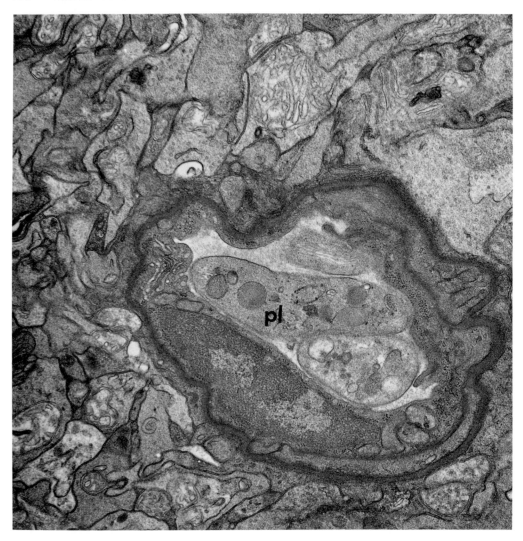

FIG. 4. Rat injected (I.V.) seven days earlier with rabbit anti-RBM-S (IgG). Note platelets (pl) in lumen which were readily observed in the majority of sectioned vessels in the experimental (passively immunized) group, but not in the control animals (× 12,500).

ease only certain retinal vessels are affected. Explanations for this observation are conjectural.

Common antigens may be shared among the kidney, retina, and uvea.[10]After primary inflammation in any of these tissues, immunological cross reactivity may result in the formation of immune complexes distant from the original insult. In patients with Goodpasture's syndrome, there are linear IgG deposits in choroidal vessels and in Bruch's membrane.[11] It is possible that distantly formed immune complexes preferentially deposit in vessels already compromised by prior inflammation.

Acknowledgments

This work was supported in part by a grant from That Man May See and by NIH Grants EY01441, EY01759, and EY03675. Dr. Char is a recipient of NIH Career Research Development Award (KO4-EY00117).

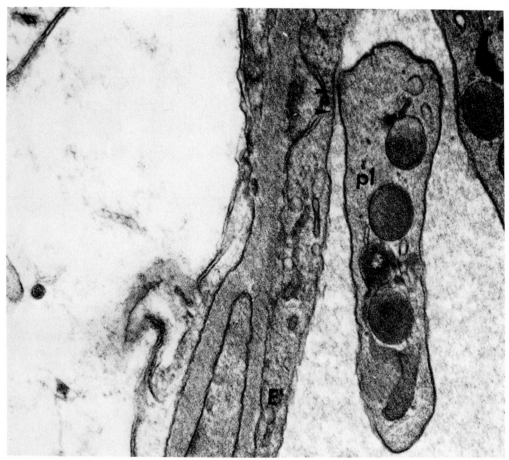

FIG. 5. Higher magnification of platelets (pl) adhering to endothelium (E) in retinal vessel from rat passively immunized with rabbit anti-RBM-S. Arrows indicate regions of close association between platelets and endothelial cell membranes (× 50,000).

References

1. Char DH et al.: *Amer J Ophthalmol* **87**:678, 1979.
2. Andrews BS et al.: *Clin Exp Immunol* **29**:23, 1977.
3. Carlson EC et al.: *J Ultrastruct Res* **62**:26, 1978.
4. Stuffers-Heiman M et al.: *J Immunol Res* **32**:93, 1980.
5. Timpl R et al.: *Eur J Biochem* **84**:43, 1978.
6. Mahieu P and Winand RJ: *Eur J Biochem* **12**:410, 1970.
7. Williams MR and Moore MJ: *J Exp Med* **138**:775, 1973.
8. Wilson CB and Dixon FH: *Kid Int* **3**:74, 1973.
9. Bolton WK et al.: *Clin Exp Immunol* **33**:463,1978.
10. Roberts DSC: *Br J Ophthalmol* **41**:338, 1957.
11. Jampol LM *Am J Ophthalmol* **79**:452, 1975.

chapter 33

Alpha 1 Antitrypsin Phenotypes and HLA Antigens in Anterior Uveitis

Denis Wakefield, Samuel N. Breit, Peggy Clark, and Ronald Penny

Considerable interest has been generated in recent years in the study of the immunogenetics of inflammatory eye disease. The relationship between several HLA antigens and anterior uveitis is now well established. This is dramatically demonstrated in the strong association of HLA B27 with anterior uveitis[1,2] and of HLA B5 and HLA DRw5 with Behçet's syndrome[3] and chronic anterior uveitis of juvenile chronic pauciarticular arthritis[4] respectively. However, the association of alpha 1 antitrypsin-deficient phenotypes and uveitis remains controversial.

The aim of this study was to investigate the relationship between alpha 1 antitrypsin phenotypes and HLA antigens and the nature and severity of inflammatory eye disease.

Patients and Methods

We studied 90 patients with anterior uveitis referred to a Uveitis Research Clinic at Sydney Eye Hospital over a 12-month period. Patients with anterior uveitis were divided clinically into four groups: acute (AAU), chronic (CAU), bilateral (BAU) (simultaneous involvement of both eyes at presentation), or recurrent disease (RAU).

Department of Immunology, St. Vincent's Hospital and the Department of Medicine, University of New South Wales; Uveitis Research Clinic, Sydney Eye Hospital, Sydney, Australia.

The alpha 1 antitrypsin phenotype of all patients was determined using isoelectric focusing on polyacrylamide gel. All patients were typed for the HLA B27 antigen. Additionally, the first 50 patients studied were also typed for HLA A, B, and DR locus determinants by the N.S.W. Red Cross Blood Transfusion Service. All patients were examined by two independent ophthalmologists and by an internist. Where relevant, additional investigations were undertaken to determine the etiology of the uveitis. A population of 339 blood donors had previously been studied to determine the normal distribution of alpha 1 antitrypsin phenotypes and HLA antigens.

Results

Table I summarizes the etiology of anterior uveitis. The HLA B27 antigen was most frequently associated with males suffering from ankylosing spondylitis or Reiter's syndrome. Table II summarizes the incidence of different alpha 1 antitrypsin phenotypes. Compared with controls, deficient phenotypes of alpha 1 antitrypsin (MS or MZ) occurred in those patients with more severe forms of AU, that is chronic (4/10 patients; $p < 0.001$), bilateral (3/5 patients; $p < 0.001$) and recurrent disease (8/39 patients; $p < 0.05$). Table III outlines the HLA B27 data and relative risk values. In HLA B27 positive patients the relative risk of developing severe or recurrent AU was

TABLE I
Etiology of Anterior Uveitis

Disease	No. of Patients	Acute	Bilateral	Recurrent	Chronic
Idiopathic	58	26	3	19	10
Ankylosing spondylitis	16	6		10	
Reiter's syndrome	4			4	
Behçet's syndrome	4	3		1	
Herpes simplex	3	1		2	
Ulcerative colitis	2		1	1	
Sarcoidosis	2			2	
Juvenile chronic arthritis	1		1		

TABLE II
Alpha 1 Antitrypsin Phenotypes in Anterior Uveitis

Anterior uveitis	Phenotype %			
	MM	MS	MZ	MS/MZ
Acute (36)	100^a	0	0	0^a
Chronic (10)	$60^{a,b}$	$40^{a,b}$	0	$40^{a,b}$
Bilateral (5)	40^b	$40^{a,b}$	20	$60^{a,b}$
Recurrent (39)	80^b	$15^{a,b}$	5	$20^{a,b}$

[a] $p < 0.001 - 0.05$ compared with controls.
[b] $p < 0.05$ compared with acute anterior uveitis.

2.1 and 3.7 respectively, compared with patients who were alpha 1 antitrypsin deficient and/or HLA B27 positive in whom the relative risks were 6 and 7.9 respectively (Table III). Four of six patients with the MS phenotype were HLA DR5 positive. There was no increased incidence of this deficient phenotype in 10 healthy controls who were known to have the HLA DR5 antigen.

Discussion

The results of the present study indicate a significantly increased incidence of alpha 1 antitrypsin-deficient phenotypes and the HLA B27 antigen in patients with anterior uveitis, especially in those with more severe forms of the disease. Comparison of patients experiencing a single uncomplicated attack of anterior uveitis with those with a more severe form of disease (chronic, bilateral, or recurrent AU) shows that only the latter group have deficient phenotypes (Table II). Alpha 1 antitrypsin deficiency thus predisposes to more severe and recurrent AU.

Two previous reports on the association of deficient alpha 1 antitrypsin phenotypes and uveitis have revealed conflicting results. Brewerton et al[5] initially described an increased association of the MZ deficient phenotype in 25% of their 80 patients with anterior uveitis, compared with an incidence of 3% in the control population. Brown et

TABLE III
Distribution of Genetic Risk Factors in Anterior Uveitis

	No. of Patients	% Alpha 1 AT deficient phenotypes	% HLA B27 +	% Alpha 1 AT deficiency and/or HLA B27 +
Acute uveitis	36	0	22	22
Severe uveitis	54	$28(3.2)^a$	$37(2.1)^a$	63(6.0)
Recurrent	39	20(2.0)	51(3.7)	69(7.9)
Chronic	10	40(5.4)	0	40(8.0)
Bilateral	5	60(12.1)	0	6(8.0)

[a] Number in parentheses indicates relative risk compared with acute uveitis or in the case of alpha 1 antitrypsin deficient phenotypes, compared with controls.

al[6] were unable to substantiate this in 133 patients. The results of the present study may serve to explain these discrepancies. If patients are categorized in terms of the nature and severity of their disease, although there is only a slightly increased but significant incidence of deficient phenotypes (MZ) in anterior uveitis, yet examination of the more severe subtypes of anterior uveitis reveals a much higher association (Table II). There is no increased incidence of alpha 1 antitrypsin deficient phenotypes in the HLA B27 subgroup of anterior uveitis patients. Both HLA B27 and alpha 1 antitrypsin deficient phenotypes predispose to recurrent disease but appear to act as independent variables. Together, these independent genetic factors are present in over 60% of patients with severe anterior uveitis and represent the most significant predisposing and prognostic factors so far detected. The mechanism by which such genetic factors influence disease is not known. Anterior uveitis is believed to be an immunologically mediated disease,[7,8] and it would appear not unreasonable to assume that the HLA antigen and alpha 1 antitrypsin phenotype influence immunological aspects of this disease.

Alpha 1 antitrypsin is the major serum inhibitor of proteases, has an ubiquitous intra- and extravascular distribution, and acts as an acute phase reactant. Work in our department has shown that it has a significant influence on the regulation of several components of the immune and inflammatory response[8,9] (summarized in Table IV). Deficient phenotypes of alpha 1 antitrypsin have a reduced protease inhibitor, and acute phase responses. Hence they manifest reduced inactivation of enzymes and mediators of the inflammatory response. This may

TABLE IV
Influence of Alpha 1 Antitrypsin Deficiency on the Immune Response

	Increased in $\alpha 1 AT$ deficiency	Suppressed by purified $\alpha 1 AT$
Complement levels (C3, C5, B, PH_{50})	+	−
Delayed hypersensitivity	+	−
Lymphocyte proliferation	+	+
MN and PMN phagocytosis	+	−
Chemokinesis of MN and PMN	−	+
Antibody dependent cell cytotoxicity	−	+
NK cell activity	−	+

predispose such patients to more prolonged and severe damage at inflammatory foci, as revealed in the present study in anterior uveitis patients.

Acknowledgments

Dr. D. Wakefield and Dr. S.N. Breit are recipients of National Health and Medical Research Grants.

References

1. Brewerton DA et al.: *Lancet* **1**:994, 1973.
2. Wakefield D et al.: *Human Immunol* **7**:89, 1983.
3. Lehner T and Barnes CG (Eds.): *Behçet's Syndrome: Clinical and Immunological Features.* Academic Press, New York, 1979.
4. Glass D et al.: *J Clin Ivest* **66**:426, 1980.
5. Brewerton DA et al.: *Lancet* **1**:1103, 1978.
6. Brown WT et al.: *Lancet* **1**:646, 1979.
7. Wakefield D et al.: *Med J Aust* **1**:229, 1982.
8. Wakefield D et al.: *Aust J Ophthalmol* **11**:15, 1983.
9. Wakefield D et al.: In *Immunogenetics in Rheumatology.* Excerpta Medica, Amsterdam, 1982.

chapter 34

The Significance of HLA-B27 Determination in Patients with Acute Anterior Uveitis

A. Linssen,[a] **A.J. Dekker-Saeys,**[b] **A. Kijlstra,**[a] **B.J. Christiaans,**[c]
P.T.V.M. de Jong,[e] **M.R. Dandrieu,**[d] **G.S. Baarsma,**[e]
and T.E.W. Feltkamp[a]

The early diagnosis of ankylosing spondylitis (AS) is difficult. After the onset of the complaints it often takes ten years before the diagnosis is made.[1] Acute anterior uveitis (AAU) is the first associated sign of the disease in only 1%–10% of AS patients.[2,3]

Both AS and AAU are diseases associated with the histocompatability antigen HLA-B27, the relative risks being respectively 87.4 and 10.4.[4] In the present study we tried to reveal the significance of HLA-B27 determinations in patients with AAU for an early diagnosis of AS.

Patients and Methods

In three separate university outpatient clinics, 103 patients were diagnosed as AAU. All patients were x-rayed for sacroileitis in an anteroposterior view of the pelvis at a craniocaudal angle of 25°. The x-rays were examined "blindly," i.e., irrespective of the other results of the study. All patients also underwent a rheumatological examination. The rheumatologist was not aware of the other findings of the study, i.e., the HLA-B27 type.

The diagnosis of AS was made according to the New York criteria.[5] In short these are as follows:

1. Radiological criteria
   a. sacroileitis, grade 3 or 4, bilateral.
   b. sacroileitis, grade 2, bilateral, or grade 3, unilateral, or grade 4, unilateral.
2. Physical criteria
   a. limitation of motion of the lumbar spine in all three planes.
   b. low back pain.
   c. limitation of chest expansion (< 2.5 cm).

The diagnosis of AS is made if:
—1a plus 2a or 2b or 2c, or
—1b plus 2a, or
—1b plus 2b plus 2c are encountered.

Since we were also interested in the number of patients who had not fulfilled these criteria, but nevertheless had some symptoms of AS, we suggested criteria for possible AS or, as designated by Agarwal[6] and Mladenovic,[7] pre-AS. The proposed criteria for pre-AS were as follows:

1. Low back pain and/or peripheral arthritis and/or arthralgia and/or heel pain.

[a]Dept. of Opthalmo-Immunology The Netherlands Ophthalmic Research Institute, Amsterdam; [b]St. Elisabeth's Gasthuis, Haarlem; [c]Dept. of Ophthalmology, University of Amsterdam; [d]Dept. of Ophthalmology, Free University, Amsterdam; [e]Dept. of Ophthalmology, Erasmus University, Rotterdam, The Netherlands.

2. Limitation of lumbar spine movement (Schober $\leq$ 20 cm) and /or chest expansion ($\leq$ 5 cm).
3. The description of a relative of the patient affected with AS.
4. Sacroileitis on x-ray, grade 2, uni- or bilateral.

If two or more of these criteria were met, the patient was considered to be pre-AS.

Results

Of 103 patients with AAU, 49 (48%) were positive for HLA-B27. The diagnosis of AS was made in 28% of the AAU patients. Among the HLA-B27 positive AAU patients, the New York criteria for AS were met in 55%; while among the HLA-B27 negative AAU patients the criteria were met in only 4%.

We were especially interested in the 22 AAU patients who were HLA-B27 postiive but who had not fulfilled the New York criteria. On checking these patients for the pre-AS criteria, it was found that 17 could be diagnosed as such. From the results given in Table I, one can see that pre-AS was only diagnosed in seven of the 52 HLA-B27 negative AAU patients without AS. This difference is statistically significant ($p <$ 0.001).

Taking the diagnosis of AS and pre-AS together, 90% of the HLA-B27 positive AAU patients fulfilled these criteria in contrast to only 17% of the HLA-B27 negative AAU patients (Table II).

Discussion

In the present study AS was found in 28% of 103 AAU patients. This frequency is somewhat higher than the frequency found by other authors of 15–20%.[6–8] Only a few of the AAU patients had been previously

diagnosed as AS. In most cases the diagnosis was a new one.

The term pre-AS, used in this paper, suggests a future development of AS. The only argument for this is the significantly (Student t test) lower mean age of the HLA-B27 positive AAU patients with pre-AS than the mean age of the HLA-B27 positive AAU patients with AS. A great number of our pre-AS patients will possibly never develop the full clinical picture of AS. Nevertheless, they belong to the poorly defined group of HLA-B27 associated diseases. In a previous study[9] we included HLA-B27 positivity as one of the pre-AS criteria. This, of course, hampered the demonstration of an association between pre-AS and HLA-B27 since this had been fulfilled *per se*. The present (extended) figures do not show a large difference.

Of the HLA-B27 positive AAU patients 55% had AS and 90% belonged to the category of AS or pre-AS. HLA-B27 is therefore a marker of importance for this "spondylitic diathesis."

One may question whether it is permissible to designate (by HLA-B27 typing) a patient with an eye disease as being prone to develop spondylitis in some form. The patient visits the ophthalmologist, not the rheumatologist, for his complaints. We are of the opinion that HLA-B27 typing facilitates the early diagnosis of AS. Thereby adequate therapy with phenylbutazone and/or physiotherapy can be instituted in time. This may lead to the prevention of the development of a serious disability.

Of even greater importance may be the psychological effect of taking the vague low back pains of AAU patients more seriously than those of other patients with such complaints. Not only will the medical profession take such patients more seriously; the general population may become alerted to the significance of back pain in patients with

TABLE I
Number of Pre-AS Criteria Met by HLA-B27 Positive and Negative Patients with AAU but without AS

Number of criteria	0	1	2	3	4	Total
HLA-B27 pos.	2	3	9	7	1	22
HLA-B27 neg.	23	22	7	0	0	52
Total	25	25	16	7	1	74

TABLE II
Association between AS, Pre-AS and HLA-B27 in 103 Patients with AAU

	AS+	Pre-AS	AS−	Total
HLA-B27 pos.	27 (55%)	17 (35%)	5 (10%)	49 (100%)
HLA-B27 neg.	2 (4%)	7 (13%)	45 (83%)	54 (100%)
Total	29 (28%)	24 (23%)	50 (49%)	103 (100%)

ocular inflammation. Of course, in some cases knowledge of the symptoms may cause unwarranted anxiety for the future. The clinician has to decide in individual cases how to handle this problem.

References

1. Hill HFN et al.: *Ann Rheum Dis* **35**:267–270, 1976.
2. Wright V and Moll JMH: *Seronegative Polyarthritis*. North Holland Publishing, Amsterdam, 1976, p. 67.
3. Käss E.: *Acta Rheum Scand* **14**:197–209, 1968.
4. Svejgaard A et al.: In *Immunology 80* M. Fongereau and J. Dausset (Eds.). Academic Press, London, p. 530, 1980.
5. Bennet PH and Burch TA: *Bull Rheum Dis* **17**:453–458, 1967.
6. Perkins ES: *Trans Ophthalmol Soc UK* **96**:105–107, 1976.
7. Stanworth A and Sharp J: *Ann Rheum Dis* **15**:140–150, 1956.
8. Lenoch F et al.: *Ann Rheum Dis* **18**:45–48, 1959.
9. Linssen A et al.: *Immunogenetics in Rheumatology*. Exerpta Medica, Amsterdam, 1982, p. 191.

chapter 35

Free Radicals and Phacoanaphylaxis

George E. Marak, Jr., Narsing A. Rao, Joan M. Scott, Ricardo Duque, and Peter A. Ward

Immune complexes are formed when a host responds to the presence of microbial or other recognized foreign materials. If the amount of complex formed exceeds the amount that can be handled or cleared by the reticuloendothelial system, an inflammatory reaction resulting in tissue injury may develop. It should be noted that a variety of immune complexes are formed during an immune response, but only certain types of immune complexes are pathogenic. Those characteristics that imply pathogenicity for immune complexes are not well understood. The features of immune complexes that have been traditionally associated with disease may relate either to the ability of complexes to bring about release or generation of factors resulting in the concentration and localization of the complexes and neutrophils (PMN), or they may relate to other properties of the complexes such as complement fixation. It has generally been assumed that once PMN are attracted to the site where complexes have been deposited, tissue damage usually follows. It has been presumed that the tissue injury is related to the proteolytic enzymes of the PMN.

The belief that tissue injury associated with immune complex disease results from the destructive effects of proteolytic enzymes released by PMN has at least two features that are inconsistent with experimental observations. The first inconsistency is that the presence or absence of an inflammatory reaction is often not associated with levels of immune complexes in serum or in other body fluids. The second problem is that there is little direct evidence that lysosomal enzymes of the PMN's are involved in the early stages of tissue destruction in immune complex disease.

Discrepancies between presence or absence of complement-fixing antibody or of serum levels of immune complexes and inflammation in tissues are frequently reported. Rahi's studies in rabbits and our own observations on strain differences in immunoresponsiveness to lens proteins are two examples of discrepancies between serologic responses and tissue inflammation in phacoanaphylactic endophthalmitis.[6–8,10] Immunoreactant levels measure a polyclonal response, so that the levels of the phlogogenic components may not be accurately reflected in determinations of total immune complexes or of complement-fixing antibodies. One recent observation suggests that immune complexes that activate complement and stimulate enzyme release from PMN are different from those complexes that activate PMN to produce toxic oxygen products.[12]

While there is no doubt that the lysosomal enzymes of PMN play a role in tissue alteration, there is little evidence that PMN proteases play a significant role in the initial tissue injury developed in immune complex induced reactions. It has been demonstrated that antiproteases provide little protection

---

Department of Ophthalmology, Georgetown University Hospital, Washington, D.C.; and Department of Pathology, University of Michigan, Ann Arbor, Michigan.

against immune complex lung injury.[12] Human leukocyte elastase induces little acute lung damage.[11] Finally, mice deficient in leukocytic neutral immune protease activity are still able to manifest immune complex-induced tissue injury.[3]

On the other hand, toxic oxygen products from PMN are well known to play an important role in the destruction of both bacteria and tumor cells.[1,13] For example, in chronic granulomatous disease in which neutrophils are deficient in the production of oxygen metabolities, *in vitro* killing of bacteria is profoundly depressed as are the cytotoxic effects of PMN on antibody-coated tumor cells.[2] Recently evidence has been presented that oxygen-derived free radicals and other products such as O_2^-, H_2O_2, the hydroxyl radical, and possibly single oxygen may be implicated in the early phases of tissue damage in immune complex disease.[4,9,12,13] H_2O_2 and O_2^- have been shown to be important in the tissue damage observed in both dermal immune complex vasculitis and immune complex disease in the lungs.[3,4]

Discussion of the various mechanisms of oxygen-derived free radical production by PMN, the relative importance of the various enzyme systems involved, and the relative importance of the various oxygen radicals in tissue damage are beyond the scope of this presentation. However, hydrogen peroxide is the most stable of the products of oxygen metabolism. This presentation will concentrate on the role of hydrogen peroxide in tissue inflammation in experimental phacoanaphylactic endophthalmitis.

The purpose of this report is to evaluate the relative roles of PMN proteases and oxidizing agents in the production of tissue damage in a well-defined ocular immune complex disease, experimental phacoanaphylactic endophthalmitis.

Female Mai, W.F. rats were sensitized to lens protein according to a protocol which produces a massive intraocular Arthus reaction that develops in 100% of animals within 48 hours of lens injury.[5] One group of animals received 43,000 units (in 0.1 ml saline) of soybean trypsin inhibitor (Sigma T-9003), which inhibits the principal protease of the rat PMN lysosome. The other group received 146,000 units (in 0.1 ml saline) of bovine catalase (Sigma C-100) injected into the anterior chamber at the time of lens injury. The animals were killed 24 hours after lens injury; the eyes were removed and fixed in 10% buffered formaldehyde. Paraffin-embedded sections were stained with H & E. Inactivation of the proteolytic enzymes had no effect on the inflammation. The inflammation in the catalase-treated eyes was significantly reduced when compared with the lens injured controls.

Our observations confirm prior work in other systems that oxygen products play an important role in the immune complex inflammation of phacoanaphylactic endophthalmitis, but proteolytic enzymes are less important in early tissue inflammation.

References

1. Babior BM: *Engl J Med* **298**:659, 1978.
2. Clark RA and Klebanoff SY: *Immunol* **119**:1413 1977.
3. Johnson KJ et al.: *J Immunol* **112**:1807, 1979.
4. Johnson KJ and Ward, PA: *J Immunol* **126**:2365–2369, 1981.
5. Marak GE et al.: *Ophthalmic Res* **9**:162–170, 1977.
6. Marak GE et al.: *Ophthalmic Res* **13**:320, 1981.
7. Marak GE et al.: *Ophthalmic Res* **14**:241, 1982.
8. Misra RN et al.: *Br J Ophthalmol* **61**:285–296, 1977.
9. Petrone WF *Proc Natl Acad Sci USA* **77**:1159, 1981.
10. Rahi AHS et al.: *Br J Ophthalmol* **61**:164, 1977.
11. Senior RM et al.: *Ann Rev Respir* **116**:469, 1977.
12. Weiss SJ and Ward PA *J Immunol* **129**: 309–13, 1982.
13. Weiss SJ and LoBuglio AF: *Lab Invest* **47**:5, 1982.

chapter 36

Serologically Active Alloantigens Determine the Immunologic Fate of Allogeneic Neoplasms Placed in the Anterior Chamber of the Eye

J. Wayne Streilein and Jerry Y. Niederkorn

The observation that the anterior chamber of the eye is an immunologically privileged site is many decades old. The capacity of this unique site within the eye to accept—for prolonged and even indefinite intervals—solid tissue grafts from allogeneic and even xenogeneic sources was originally ascribed to an anatomic idiosyncrasy.[1] Unlike most other body sites, no lymphatic drainage pathway from the anterior chamber to regional lymph nodes could be demonstrated. It was presumed that in this circumstance alien antigenic tissues placed within this site never reached the systemic immune system and could never, therefore, initiate a destructive immunologic response. The inadequacy of this hypothesis has been realized over the past decade as it has become increasingly apparent that active regulatory immune processes operate to maintain immune privilege within the anterior chamber.[2-4] In our laboratory, we have developed a model system to study these processes in mice, a system in which histoincompatible tumor cells are inoculated into an anterior chamber of adult mice.[5] For example, when P815 mastocytoma cells are inoculated into the anterior chamber of al-logeneic BALB/c mice, the tumor cells establish a progressively growing neoplasm that will kill the host eventually by direct extension into the cranial vault. We have demonstrated that in these animals development of specific cell-mediated immunity directed at the alloantigens on the P815 cells is impaired: animals with intracameral P815 tumors failed to reject DBA/2 skin allografts placed orthotopically.[9] These same animals, however, were able to make circulating antibodies directed at antigens on the tumor cells.[7] Thus, alloantigenic tumor cells placed into the anterior chamber of the murine eye evoke a deviant immune response characterized by: (a) progressive intraocular tumor growth, (b) impaired systemic cell-mediated immunity, and (c) intact humoral immunity. We have employed the term anterior chamber-associated immune deviation (ACAID) to designate this unusual spectrum of immune responsiveness. Recently we have been concerned with understanding the factors that permit the three expressions of the ACAID phenomenon to exist simultaneously in the same animal. One approach, used in the experiments to be reported, employed tumors and hosts of varying degrees of histoincompatibility. Each combination was examined for the type of local intraocular tumor growth as well as the character of immune responses recipients made to the ocular tumor.

Departments of Cell Biology, Internal Medicine, and Ophthalmology, University of Texas Health Science Center at Dallas, Dallas, Texas.

TABLE I
Immunogenetic Disparity between Tumor and Host

		Tumor	Origin	Host
Type I.	Minor H loci, H-2 Identity	P815	DBA/2	BALB/c
Type II.	H-2K^{bm} mutation (coisogenic)	B16F10	C57BL/6	Hzl (Kbml)
Type III.	One class I H-2 locus ($\pm$ minor H loci)	P815	DBA/2	A/J
Type IV.	Entire H-2 complex ($\pm$ minor H loci)	P815	DBA/2	C57BL/6

Experimental Approach

Four degrees of immunogenetic disparity between tumors and hosts were selected for these studies. A description of each type is presented in Table I along with examples of tumors and hosts for each. Type I comprises tumors that differ from their recipients at multiple minor histocompatibility loci but are syngeneic at the H-2 chromosomal region. Type II disparities resulted from mutations that have occurred in the gene that encodes for the K region molecule of the H-2^b haplotype. Thus, these strains of mice are co-isogenic, differing genetically and antigenically only at the site of the mutation.[8] Type III disparities include tumors that differ from recipients at one class I H-2 region (either K or D); some of these tumors also differ from their host at minor histocompatibility loci. Tumors in the Type IV disparate category differ from recipients across the entire H-2 chromosomal segment; some also differ at multiple minor H loci. At least four different tumor–host combinations were tested in each type of disparity.

These four types of immunogenetic disparities were selected because of critical differences in their immunogenic properties, which are listed in Table II. In every in-

TABLE II
Immunogenic Properties of Diverse Alloantigens

Skin graft rejection (days)[a]		Primary MLR and CML	Humoral antibody
Type I.	12–25	No	Weak (if at all)
Type II.	12–28	Yes	None
Type III.	10–18	Yes	Strong
Type IV.	9–14	Yes	Strong

[a] Range of first set graft rejections.

stance, the disparity results in the induction of skin allograft immunity; however, the ranges of survival time are somewhat different for each type, the most vigorous rejection being induced by Type IV disparities. Other differences include the fact that Type I disparities are unable to induce mixed lymphocyte reactions (MLR) and cell-mediated lympholysis (CML) reactions *in vitro* without prior *in vivo* priming of the responding lymphocytes. By contrast, Types II, III, and IV readily give rise to primary MLR and CML *in vitro*. Type III and IV disparities regularly elicit vigorous humoral antibody responses; by contrast, the H-2K^b mutant disparities are serologically silent.[8] Type I disparities, i.e., those resulting from minor H-loci differences alone, are relatively inefficient in inducing humoral antibodies and frequently induce no antibody at all following conventional immunizing procedures.

Patterns of Growth of Allogeneic Tumors after Intracameral Inoculation

Tumor cell suspensions were prepared from the P815 line of mastocytoma cells (originated from DBA/2 mice with H-2^d haplotype) and the B16F10 melanoma (originated from C57BL/6 mice with H-2^b). Tumor cells (5×10^5) were inoculated intracamerally into appropriate recipient mice as described previously.[5] Growth of the tumors *in situ* was evaluated daily thereafter using gross inspection as well as biomicroscopy. Degree of tumor growth was gauged according to the proportion (percent) of the anterior chamber that appeared to be occupied by tumor at the time of observation.[5] If inflammation accompanied the growth of the tumor, this was described.

At the termination of the experiment, the anatomical integrity of the eye was assessed. Three distinctly different patterns of intraocular tumor growth were observed. Pattern A was characterized by progressive intraocular growth without evidence of host resistance, that is, little or no evidence of inflammation in or surrounding the tumor in the anterior chamber. Animals exhibiting this pattern of progressive tumor growth ultimately died as a consequence of direct local invasion of the orbit and cranial vault.

Pattern B was characterized by transient intraocular tumor growth during which a significant portion of the anterior chamber became filled with tumor. However, commencing at approximately 8–10 days after inoculation, an intense inflammatory reaction developed in and around the tumor attended by impressive neovascularization of limbus and cornea. During the next several days this inflammatory reaction gradually subsided, leaving in its wake a totally destroyed globe. In these animals, all evidence of tumor within the eye disappeared. In fact the mice were cured of their tumor, although they had become blind in the tumor-bearing eye.

In growth pattern C, transient growth similar initially to that described in pattern B was observed. However, commencing at approximately 8–10 days, a mild inflammatory reaction was observed in some but not all of the tumor-bearing eyes. From this time on, the tumor mass within the eye became progressively smaller and finally disappeared. In the vast majority of instances, the eye, now divested of the tumor, appeared to be anatomically intact. Thus, these animals were not only rid of their intraocular tumor, but they retained the anatomical integrity of their eyes.[9]

We next matched the patterns of intraocular tumor growth with the four types of tumor–host immunogenetic disparity. These data are summarized in Table III. It can be observed that animals that were cured of their intraocular tumor and retained an anatomically intact eye differed from their tumor across the entire *H-2* complex. Animals that were able to rid their eyes of tumors at the cost of destruction of the tumor-bearing eye differed from their tumor at only one Class I *H-2* locus. Progressive tumor growth that proved to be fatal to the host was characteristic of both Types I and II immunogenetic disparities, a surprising finding since the latter represents a strong alloantigenic difference dictated by a mutation at a Class I *H-2* gene. Under normal circumstances, mice disparate at *K* region mutations reject reciprocal skin and tumor grafts readily.[8]

Immunogenic Properties of Diverse Intraocular Tumors

Panels of intraocular tumor-bearing animals representing each type of immunogenetic disparity were also tested for their capacity (a) to reject skin grafts syngeneic with the ocular tumor and (b) to produce humoral antibodies directed at alloantigens on the tumor. These data are also included in Table III. The capacity to accept an orthotopic skin graft syngeneic with the tumor, one of

TABLE III
Immunogenic Properties of Diverse Intraocular Tumors

Immunogenetic disparity		Patterns of IC tumor growth	Induce orthotopic skin graft rejection	Induce humoral antibodies
Type I.	Minor *H* loci, *H-2* identity	A Progressive, host killed	No	Low titer
Type II.	*H-2K^{bm}* mutation	A Progressive, host killed	Yes	None
Type III.	One class I *H-2* locus	B Transient, eye and tumor destroyed	Yes	Moderate titer
Type IV.	Entire *H-2* complex	C Transient, eye preserved, tumor destroyed	Yes	High titer

the hallmarks of the ACAID phenomenon, was observed only in Type I disparity. Despite the fact that progressive intraocular tumors were characteristic of animals receiving tumor cells representing an *H-2* mutational disparity, these mice were able to reject skin grafts bearing these same alloantigenic specificities; clearly, this aspect of ACAID had not been induced. As anticipated, animals disparate from their tumor by Types II and IV disparities were fully able to reject skin grafts syngeneic with the tumor.[9]

When serum was harvested from animals bearing intraocular tumors and was analyzed for antibodies directed at antigens on the tumor, animals of Type III and IV disparities displayed moderate to high titers of antibodies. These results have been reported in detail elsewhere.[10] By contrast, animals differing from the intracameral tumor only at minor *H* loci inconsistently developed low titers of anit-tumor antibodies. Since antigens of the *H-2K^b* mutants are serologically silent, it was not surprising that animals bearing intraocular tumors of this type failed to form antibodies against them.

Discussion and Conclusion

Based on these data, several conclusions seem appropriate:

1. The ability of an alloantigenic tumor to grow progressively in the anterior chamber of the eye depends in part upon the degree of immunogenetic disparity between the host and the tumor. In general, antigens that are serologically weak or silent—such as minor histocompatibility antigens and *H-2* Class I mutational differences—correlate positively with progressive intraocular tumor growth that eventually kills the host.

2. Minor histocompatibility antigens on intraocular tumors impair the induction of systemic cell-mediated immunity, whereas *H-2* antigenic differences regularly induce systemic cell-mediated immunity. *H-2* mutant antigens resemble conventional *H-2* antigens in this regard.

3. Taken together, these two conclusions suggest that the capacity of an alloantigenic tumor to induce systemic cell-mediated immunity is insufficient in and of itself to cause immune destruction of the intraocular tumor. The *H-2K^b* mutants are a case in point: these tumors grow progressively in the eye without local evidence of host defense, yet the animals are systemically immune to the mutant antigen since they reject skin allografts syngeneic with the tumor.

4. These data are consistent with the hypothesis that antibodies alone (or perhaps in consort with T-lymphocytes) are sufficient to account for the failure of allogeneic tumors to grow progressively in the anterior chamber of the eye. Once again, the *H-2K^b* mutants are critical to this conclusion. These mutational differences are serologically silent, although they induce vigorous cell-mediated immunity. Thus, we presume that it is the capacity of conventional *H-2* alloantigens to induce *both* antibody and T-cell mediated immunity that leads to the eventual destruction of the intraocular tumors bearing these strong transplantation antigens.

One final point seems important to emphasize. Even when *H-2* disparity between tumor and host is involved, the degree of this disparity matters a great deal in the ultimate resolution of the local tumor. When disparity across the entire *H-2* complex is involved, the tumor-destructive host response is of sufficiently tempered intensity and specificity that the tumor alone is destroyed, but the anatomical integrity of the eye is preserved. By contrast, when only a single *H-2* Class I disparity exists between tumor and host, the tumor-destructive response is of great intensity but is imprecise. As a consequence, destruction of the tumor leads to innocent bystander destruction of the ocular structures in which it is enmeshed. It is possible, although our data do not address this tissue directly, that the exquisite specificity of antibody accounts for this important difference between these two degrees of *H-2* disparate tumor–host combinations. It is reasonable to suppose that cell-mediated immunity, especially that mediated by Lyt 1^+ $2,3^-$ delayed hypersensitivity cells, necessarily evokes an inflammatory reaction comprised of largely nonspecific effects. By contrast, when antibody is the dominant effector modality, nonspecificity becomes less of a factor. With a view toward developing therapeutic strategies, immunotherapy devised for intraocular tumors might be optimal if it produced predominantly humoral rather than cellular effector modalities.

Acknowledgment

This work was supported by U.S.P.H.S. grants EY03319 and CA30276.

References

1. Barker CF and Billingham, RE: *Adv Immunol* **25**:1-54, 1977.
2. Franklin R and Prendergast RA: *J Immunol* **104**:463-469, 1970.
3. Kaplan HJ and Streilein JW: *J Immunol* **118**:809–814, 1977.
4. Subba Rao DSV and Grogan JB: *Transplantation* **24**:377–381, 1977.
5. Niederkorn JY et al.: *Invest Ophthalmol Vis Sci* **20**:355–363, 1981.
6. Streilein JW et al.: *J Exp Med* **152**:1121–1125, 1980.
7. Niederkorn JY and Streilein JW: *J Immunol* **128**:2470–2474, 1982.
8. Klein J: *Adv Immunol* **26**:55–146, 1978.
9. Niederkorn JY et al.: *Immunogen* **13**:27–236, 1981.
10. Niederkorn JY and Streilein JW: *Transplantation* **33**:573–577, 1982.

chapter 37

The Immune Response to Lens Antigens

Hans Otto Sandberg and Otto Closs

About half the population have naturally occurring antibodies against alpha crystallin.[1,2] In the majority the concentration is low but in about 10% of individuals the concentration is definitely higher.[2]

It has been demonstrated that lens damage can lead to increased concentration of lens antibodies in the serum.[3] It is uncertain how this increase in antibodies is related to lens-induced uveitis. The present paper is an attempt to elucidate (1) the relationship between naturally occurring antibodies and the serum immunoresponse to lens injury, and (2) the relationship between the immunoresponse in serum and the occurrence of uveitis. This may be of particular interest since extracapsular cataract extractions are becoming increasingly frequent in connection with intraocular lens implantation procedures.

Materials and Methods

The anti-alpha crystallin activity was assayed by a previously described RIA.[2,4,5–7] The activity was measured by counting the radioactivity bound and expressed as percent of the total amount of radioactivity added.

The anti-alpha crystallin activity of selected sera was titrated in a dilution sequence from 1:10 to 1:160. The titer was expressed as the serum dilution that bound 20% of the added label and was calculated by interpolation.

Human sera were collected from eleven patients after extracapsular cataract extraction. Five of these were above 60 years of age. Of the six patients below 50 years of age, two had a chronic uveitis with precipitates and four had a perforating injury to the eye that also perforated the lens. Sera were also collected from two patients with perforating injury to the eye that did not perforate the lens, and from five patients who underwent uncomplicated intracapsular cataract extraction. Sera were collected serially from each patient. The first sample was taken on the day of injury or cataract extraction. Sera were later collected after the first and second weeks and at 1, 3, 6, and 12 months. Sera from the group of patients having intracapsular cataract extraction were only collected for a four week period.

Statistical analysis was done with a Student t-test for small samples.[8]

Results

Figure 1*A* shows that of the 11 patients who underwent extracapsular cataract extraction, nine had an antigen-binding capacity of less than 15% at the time of operation, and two bound between 30 and 40%. Four of the nine showed an insignificant increase in anti-alpha crystallin activity after the operation; the other five showed an average increase in alpha crystallin binding capacity from 11 to 48%. Of the patients with high initial anti-alpha crystallin activity, one showed only a slight increase (from 33 to 39%) and one showed an increase from 37 to 55%.

The Eye Department, The Central Hospital, Fredrikstad, Norway.

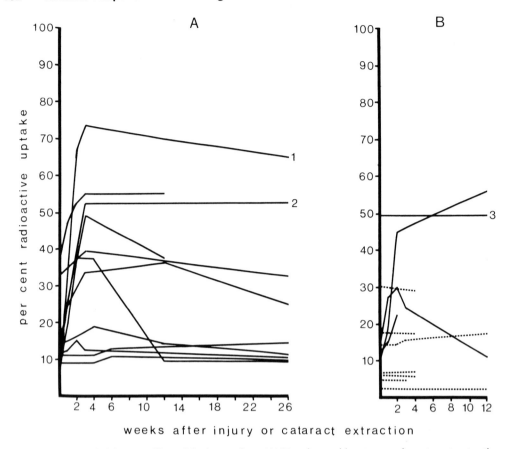

weeks after injury or cataract extraction

FIG. 1. The anti-alpha crystallin activity in sera from *(A)* 11 patients with extracapsular cataract extraction and *(B)* four patients with traumatic injury to the lens (continuous line), five patients with intracapsular cataract extraction, and two patients with perforating injury to the eye without damage to the lens (dotted line). The first serum sample was in all cases taken on the day of operation or injury. The sera indicated by numbers were titrated (see Fig. 2).

Two patients had a complicated cataract with uveitis and keratitic precipitates. In one, a substantial cataract remained in the eye after the first operation. One month later evacuation of the cataract was done and finally, after six months, a discission. During this period the uveitis remained largely unchanged. The second case with uveitis had secondary glaucoma. This case showed the greatest increase in anti-alpha crystallin activity (from 15 to 73% binding). Even though this activity remained high all signs of uveitis, including precipitates and glaucoma, had disappeared approximately 1 year after the extracapsular extraction. Thus, of the two cases with signs of uveitis at the time of operation, one improved while the other was

unaffected. In the former there was a great increase in antialpha crystallin activity, but only a slight increase occurred in the latter. In none of the cases in Figure 1A was any uveal reaction observed which appeared to be more than the ordinary post-operative reaction. That is, no correlation could be found between the occurrence of uveitis and any increase in anti-alpha crystallin activity.

In Figure 1B are shown four patients who suffered a perforating injury to the eye including the lens. Three of these had low anti-alpha crystallin activity on the day of injury, and all three showed a clear increase in activity during the first 3 weeks. In the patient who showed the greatest increase in activity, it increased until the 12th week. Evac-

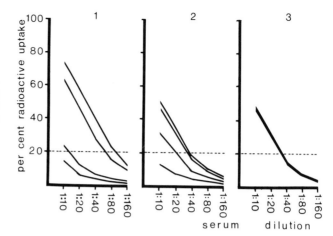

FIG. 2. Titration of the anti-alpha crystallin activity in the three sera indicated in Figure 1. The titer is expressed as the serum dilution that bound 20% of the added labeled antigen.

uation of the cataract had to be done 5 days after injury because of lens swelling. After this the eye quickly became quiescent with no signs of uveitis. In the second of these patients the anti-alpha crystallin activity decreased after the third week. The third patient failed to comply with the serum collection schedule after the 21st day, at which time he had a distinct increase in serum anti-alpha crystallin activity and the injured eye was without signs of uveitis. The fourth patient with lens injury had the highest initial anti-alpha crystallin activity of all patients included in the study. In this case no increase in activity could be demonstrated and no signs of uveitis developed.

Figure 1B also shows the anti-alpha crystallin activity in serum from five patients who underwent uncomplicated intracapsular cataract extraction. None of the patients showed any increase in anti-alpha crystallin activity. Also included in Figure 1B are two cases with perforating injury to the eye without lens damage. They were followed for several months and neither showed any increase in anti-alpha crystallin activity.

Of the 15 patients who were exposed to lens antigens, five were above 60 and 10 below 50 years of age. Statistical analysis could not demonstrate any significant differences in the immune response in the two groups.

Of the same 15 patients, 12 had an initial antigen binding capacity of 15% or less, while three bound more than 30% of the la-

belled antigen. The response in the latter group was less than the response in the former at a significance level of 92%(t-test).

The observations show that liberation of lens antigens caused an increase in anti-alpha crystallin antibodies in serum in a high proportion of the cases studied. In patients in whom the anti-alpha crystallin activity increased, it rose rapidly and reached its peak in three weeks after surgery (except in one patient in whom it continued to increase).

Figure 2 shows the results of the titration of three selected sera. These sera are identified by Arabic numerals in Figures 1A and 1B. In none of the sera could a plateau be demonstrated. The maximum capacity to bind the labelled alpha crystallin was far less in all the patient sera than in our hyperimmune anti-alpha crystallin sera. This indicates that only small amounts of anti-alpha crystallin were present. In some patients the titer increased from less than 10 to more than 80 while in others it remained stable. The shape of the titration curves does not suggest that the increase in anti-alpha crystallin activity is caused by new antibody populations. The fall of the titration curves indicates fairly good avidity in the antibody population.

Discussion

Several reports have described increase of lens antibodies in serum following liberation of lens antigens in the eye.[3, 9] These antibodies have probably been directed mainly, or exclusively, against alpha crystallin.[2] The present paper supports this finding. An increase in anti-alpha crystallin activity occurred in 11 of the 15 patients with free lens material in the anterior chamber. The immune response did not seem to be related to the age of the patient.

Approximately 50% of the normal population have serum anti-alpha crystallin antibodies. It has been suggested that a high antilens activity at the time of injury would be likely to lead to a more pronounced increase in antibody activity.[9] Our results do not confirm this because the sera with low initial activity showed a significantly stronger immune response than those with a high initial activity.

A correlation has been suggested between an increase in serum anti-alpha crystallin and the occurrence of lens-induced uveitis.[3] The present results do not confirm this, as none of the patients who demonstrated rather pronounced increases in antibody activity developed any accompanying uveitis.

It has previously been suggested that, as a result of a steady leakage of crystallins from the lens, man has developed a low-zone tolerance to these substances and that in about half the population this tolerance has been broken for alpha crystallin.[10,11]

These individuals produce small amounts of anti-alpha antibodies in response to this normal leakage of crystallins from the lens. The present results suggest that when larger quantities of lens crystallins reach the circulation, this general production of antibodies is increased. Thus the increase in anti-alpha crystallin that occurs because of increased lens protein liberation is the result of a general immune response and not of a local response in the eye.

References

1. Hackett E and Thompson A: *Lancet*, 2:663–66, 1964.
2. Sandberg HO and Closs O: *Scand J Immunol* **10**:549–54, 1979.
3. Wirostko E and Spalter HF: *Arch Ophthalmol* **78**:1–7, 1967.
4. Sandberg HO and Closs O: *Exp Eye Res* **27**:457–69, 1978.
5. Rosa V et al.: *Biochem Biophys Acta* **86**:519–26, 1964.
6. Sandberg HO and Closs O: *Exp Eye Res* **27**:701–12, 1978.
7. Arvidsson S et al.: *Acta Pathol Scand* (B) **79**:399–405, 1971.
8. Spiegel MR: *Theory and Problems of Statistics*, Shaum Publishing Company, New York, 1961.
9. Nissen SH et al.: *Brit J Ophthalmol* **65**:63–66, 1981.
10. Marak GE Jr. et al.: In *Immunology and Immunopathology of the Eye*, AM Silverstein and GR O'Connor (Eds.). Masson Publishing USA Inc., New York, 1979, pp. 135–37.
11. Sandberg HO: *Antigens of the Crystalline Lens*. Thesis. Fredrikstad, 1980.

chapter 38

Lens as Inducer and Target in Uveitis

Comparison of Different Types of Naturally Occurring and Experimentally Induced Phacogenic Reactions

**H. Konrad Müller-Hermelink,[a] Ellen Kraus-Mackiw,[b]
Wilfried Daus,[b] and Herbert Adams[c]**

Various types of morphological changes previously observed in human lenses following injury [1] can be reproduced experimentally in inbred rats (Strains CAP and LEW) following discission alone. These have been classified as Types 1–6. Progressive lens-induced inflammation occurred only rarely in rats (15 of 300) and only rarely in humans (11 of 96 eyes enucleated 1936–1981.) Despite the use of inbred rats and the production of extensive lens damage, progressive inflammation occurred in only 5% of the rats. Therefore, factors other than lens damage may be important in the production of lens-induced uveitis in both rats and man.

Intraocular infections. Lens discission, combined with a single injection of 2×10^8 *Staphylococcus aureus* into the anterior chamber of Lewis rats produced an inflammation (in 15 of 16 eyes) consisting of infiltrations of macrophages, granulocytes, lymphocytes, and occasionally giant cells (Type 2 reaction.) These infiltrates were seen in the vicinity of the exposed lens material, in the lens capsule, and in a concentric pattern characteristic of "phacoanaphylactic endophthalmitis". Bacteria could sometimes be seen in stained sections up to 10 days, but not consistently thereafter.

Presensitization. Presensitization with a suspension of 20 mg of cow lens material (including capsule), administered subcutaneously and by foot-pad injection along with complete Freund's adjuvant at intervals ranging from 50 days to 10 days prior to discission, produced massive accumulations of giant cells near the lens capsule.

Skin tests. Only two of 48 rats skin tested with 6 mg of syngeneic lens material 70, 100, or 140 days following simple lens discissions showed positive skin-test reactions. The latter, on biopsy, consisted of severe lymphocytic infiltrations with some giant cells. The eyes of these two animals showed Type 3 reactions (epithelioid cell granulomas seen mainly in the area of the lens capsule).

These results make advisable long-term studies with presensitization as well as long-term studies with presensitization combined with intraocular infection. In addition, the connection between skin-test positivity and progressive inflammatory reactions in the lens should be investigated.

Acknowlegment

The authors thank Siglinde Lehr, Gunhild Liszy, and Elisabeth Vorreuther for their valuable technical assistance.

Reference

1. Müller-Hermelink HK: In *Uveitis, Pathophysiology and Therapy*. Kraus-Mackiw E and O'Connor GR (Eds.). Thieme-Stratton, New York, 1983, pp. 152–197.

The Department of Pathology, University of Kiel[a], and the University Eye Hospitals of Heidelberg[b] and Freiburg,[c] West Germany.

chapter 39

Anti-receptor Antibodies

A Tool for Analyzing Hormone-receptor Actions

George L. King

The biological actions of peptide hormones are initiated by the binding of hormones to a specific surface receptor on cellular plasma membrane.[1-4] However, subsequent events after the formation of hormone-receptor complex leading to the onset of the hormone's biological activities are generally unknown. Recently, our laboratories and others have used various immunological techniques to study the structure of receptors for the hormone insulin and its mechanism of actions at the cellular level.[1,3,5-15] In this review are summarized some of the available data on the mechanism of insulin action derived from studies using specific autoantibodies to the insulin receptor purified from the sera of patients with insulin resistance.[5-15] In addition, included are some of the studies my laboratory has recently done using these specific anti-insulin receptor antibodies in order to understand the role of insulin in the development of vascular complications of diabetes mellitus. We have investigated the effect of insulin in the metabolism and growth of vascular cells that are involved in the development of vascularopathies, in particular the retinopathy of diabetes mellitus.[16]

Clinical Syndrome of Patients with Autoantibodies to the Insulin Receptor

The anti-insulin receptor antibodies used for our studies were partially purified from sera of patients with insulin resistance and acanthosis nigricans. A summary of the clinical features of these patients is presented in Table I. In general, these patients had no particular age distribution, but there was a female predominance. In addition, most of the patients were non-Caucasian: of 17 patients, eleven were black individuals, three were Caucasian, two were Japanese, and one was a Mexican-American.[1,5,6] Clinically, the patients presented with symptoms of diabetes such as polyuria, polydipsia, polyphagia.[1,5,6] However, five of the patients either initially had or eventually suffered from symptoms of hypoglycemia.[17] Some of the patients presented with associate signs of other autoimmune disorders. On physical examination, almost all of the patients had a skin abnormality acanthosis nigricans, which involves the neck, axilla, trunk, and face, with skin tags and velvety lesions. Other associated findings indicating the autoimmune nature of the disease included arthralgia, alopecia, vitiligo, splenomegaly, and Raynaud's phenomenon.[1,5,6] Laboratory findings in these patients were striking for abnormalities in glucose metabolism and autoimmune parameters. Foremost, these patients had severe degrees of

Department of Medicine, Brigham and Women's Hospital, Harvard Medical School, Joslin Diabetes Center, Boston, Massachusetts.

TABLE I

Clinical Findings in Patients with Insulin Resistance, Acanthosis Nigricans, and Anti-Insulin Receptor Antibodies (Type B)

1. Age predominance: all ages over 20
2. Sex: More female
3. Race: non-Caucasian
4. Symptoms: polyuria, polydipsia, polyphagia, ketoacidosis, insulin resistance, acanthosis nigricans, hypoglycemia
5. Associated autoimmune diseases: systemic lupus erythematosus, glomerulonephritis, ataxia telangiectasis, Sjögren syndrome, Hashimoto's thyroiditis, hemolytic anemia
6. Physical findings: vitiligo, Raynaud's phenomenon, polyarthritis, acanthosis nigricans, alopecia, hirsutism, secondary amenorrhea, splenomegaly
7. Laboratory findings: elevated sedimentation rate, positive LE cell preparation, Coomb's positive, decreased complement level, antithyroglobulin antibody, anti-DNA antibody, antinuclear antibody, proteinuria, cryoglobulins, hypergammaglobulin, anti-insulin antibody, and *all* have anti-insulin receptor antibody

glucose intolerance especially during stress; however, ketoacidosis occurred only very infrequently, usually in periods of infection, and it was mild. Basal insulin measurements showed that the fasting insulin levels in these patients' sera were greatly elevated, ranging from 85 to 950 µU/ml as compared to undetectable to 10 µU/ml in the normal population. When tested, these patients had an exaggerated insulin response when stimulated to release by glucose, tolbutamide, and leucine. Their insulin resistance was shown most dramatically when they were challenged by exogenous insulin. Doses ranging up to 150,000 units per day were given intravenously without normalization of their blood glucose or any apparent effect.[18]

In addition to severe insulin resistance, most of these patients had laboratory features of autoimmune diseases, such as elevation of erythrocyte sedimentation rate, leukopenia, gamma globulin, and anti-DNA antibodies, and also decreased complement levels. Other patients had positive Coombs

tests, cryoglobulins, positive LE cell preparation, and anti-thyroid antibodies.[1,5,6,18]

The common denominator in all these patients, which makes the definitive diagnosis, is the biochemical demonstration of an anti-insulin receptor antibody immunoglobulin in their sera that can inhibit the binding of insulin to its receptor on the cell membrane.[5-8] The existence of a serum inhibitor in the sera of these patients was initially shown by Flier et al.[5] in the following manner. First, the insulin receptors on each patient's circulating red blood cells and monocytes were measured using ^{125}I-insulin as a marker and found to be much lower than the normal range, suggesting that either the patient's serum is preventing the binding of ^{125}I-insulin.[12] However, the sera from these patients, unlike sera from normals, were found to have a potent inhibitory effect on the binding of insulin to normal cells grown in tissue culture. Therefore, the data strongly indicate that the insulin resistance in these patients was due to an abnormal factor in their sera.[5-9,12]

In addition, the insulin receptors on skin fibroblasts from these patients were grown in tissue culture and subsequently shown to be normal (Fig. 1). The inhibitory biochem-

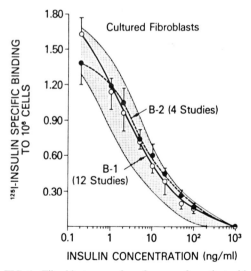

FIG. 1. Fibroblast grown from forearm of a patient with insulin resistance Type B. Insulin receptor binding is normal.

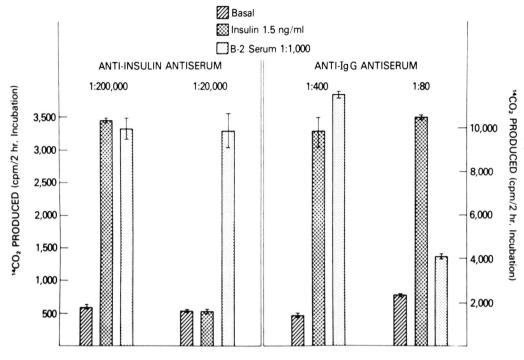

FIG. 2. Precipitation of anti-receptor antibody activities by antibodies to human IgGs.

TABLE II
Insulin-like Effects of Anti-insulin Receptor Antibody

Biological effects	Cell types studied
Stimulation of transport processes	
–2-Deoxyglucose	Adipocytes, 3T3-L1 cells, muscle, human fibroblasts
–Amino acid (A1B)	Adipocytes
Stimulation of enzymatic processes	
–Insulin receptor phosphorylation	3T3-L1 cells
–Cytoplasmic protein phosphorylation	Adipocytes, 3T3-L1 cells
–Glycogen synthesis	Adipocytes, hepatocytes, muscles
–Tyrosine amino transferase	Hepatoma cells
–Pyruvate dehydrogenase	Adipocytes
–Acetyl CoA carbohylase	Adipocytes
–Lipoprotein lipase	3T3-L1 cells
Stimulation of glucose metabolism	
–Glucose incorporation into glycogen	Adipocytes, 3T3-L1 cells, muscle
–Glucose incorporation into lipids	Adipocytes
–Glucose oxidation to CO_2	Adipocytes
Stimulation of macromolecular synthesis	
–Leucine incorporation into protein	Adipocytes, human fibroblasts
–Uridine incorporation into RNA	Hepatoma cells
–Thymidine incorporation into DNA	Hepatoma cells, melanoma cells, retinal pericytes
Miscellaneous effects	
–Inhibition of lipolysis	Adipocytes
–Stimulation of glycosaminoglycan secretion	Chondrosarcoma cells

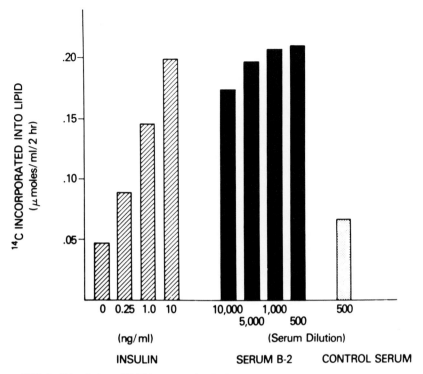

FIG. 3. Stimulation of lipid incorporation by insulin and anti-receptor antibodies from the patients.

ical activities from the sera were analyzed and shown to be of the immunoglobulin family of proteins.[5,8–10] Most of these auto-antibodies were IgG's since the inhibitory activities can be precipitated by antibodies raised against human IgG's (Fig. 2) and bind to Protein-A, which specifically binds to IgG's only.[12] However IgM has only been identified in one patient. The anti-receptor antibodies are of polyclonal nature. Using antilambda or antikappa antibodies to light chains of IgG's, only part of the IgG can be removed, whereas precipitation with both antikappa and antilambda antisera are needed in order to remove the blocking activities from the serum of a single patient.

Even more interestingly, when the auto-antibodies were assayed in a biologicial system for measuring insulin's biological effects, anti-receptor antibodies were shown to be able to mimic insulin's biological effects on many types of cells (Table II). These include acute metabolic effects of in-sulin, such as the stimulation of glucose transport and oxidation in addition to some chronic effects of insulin, such as the stimulation of protein synthesis and the activation of enzyme systems (Fig. 3).

These data clearly showed that (1) the receptor of a hormone-receptor complex contains most of the information required for biological activities and (2) the hormone initiates a cascade of effects because the biological effects of the insulin can be reproduced by the occupation of the insulin receptor with two dissimilar proteins such as insulin and anti-receptor antibodies. When (Fab)$_2$ and Fc fragments were made from the patients' IgG by digestion with pepsin or papain, the biological activity and the blocking effects of the IgG were found to be with the (Fab)$_2$ fragments.[12] When IgG was digested by papain and reduced by cystein, the resultant monovalent (Fab) did not retain any insulin-like binding activity although it retained the ability to inhibit the

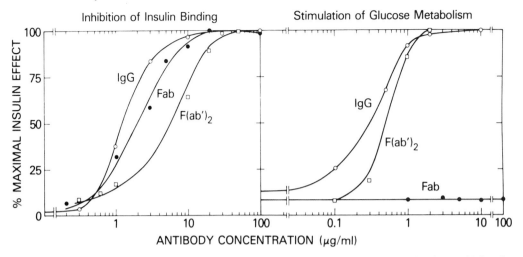

FIG. 4. Insulin-like effect of bivalent IgG's (Fab)₂ and monovalent (Fab) from the sera of patients with insulin resistance Type B. The role of valence in the action of antibodies to the insulin receptor.

binding of insulin to the receptor (Fig. 4).[19] When a second antibody, anti-Fab antibody was added (Fig. 4), the biological activity of the monovalent Fab fragment from the patients' IgG could be restored.[19] These data showed that the simple occupation of the insulin receptor is not enough for the transmission of the biological signal. The aggregation of the hormone-receptor complex induced by the second antibody is important to the biological effects of antibody. Even when insulin is occupying the receptor, the aggregation of the receptor may be important.[19] Studies have shown that the addition of low concentration of anti-insulin antibody with insulin and adipocytes would actually enhance the biological potency of insulin.[19] Recent studies have demonstrated that the aggregation of occupied peptide receptors may be a general phenomenon and could be important in the expression of biological actions of the hormone.[20,21] Therefore, the anti-receptor antibodies have demonstrated another important concept of hormone action: the role of aggregation of receptors.

The monovalent fragments (Fab) of the anti-receptor antibody, having the properties of being able to block insulin binding without inducing biological activity, are a good competitive antagonist of insulin at the receptor. This property of the Fab fragments was used by us to investigate the growth-promoting actions of insulin, which are the stimulation of DNA synthesis and cellular proliferation.[13] The growth-promoting action of insulin, unlike its acute metabolic effects, requires a long period of preincubation time (measured in hours) and pharmacological concentrations of insulin. These remarkable differences in requirements between the two types of biological activities have always puzzled investigators of insulin action. We have showed that when the insulin receptor is blocked using the monovalent (Fab) of the anti-receptor antibody, insulin's growth-promoting effects are not affected. In contrast, insulin's dose-response curve for metabolic effects (as in the stimulation of glucose oxidation in rat adipocytes) is shifted to the right, indicating competitive inhibition. From these data we concluded that insulin can mediate its pleotypic effects through different sets of receptors. This finding also explains the inability of the divalent anti-insulin receptor antibody to stimulate DNA synthesis, although it readily can mimic insulin's metabolic effects. We showed that the antireceptor antibody does not bind to the growth-effect receptor of insulin in sufficient quantity.[3]

Studies of Insulin Actions on Vascular Cells

Recently, my laboratory has become interested in the vascular complications of diabetes. Although vascular and neural tissues are not considered traditionally the insulin-sensitive organs, complications of diabetes develop mainly in these tissues. We decided to see whether insulin has a role in the metabolism and growth of vascular cells from capillaries and large arteries, which are prominently involved in the microangiography and macroangiopathy of diabetes mellitus.[16]

Pure cultures of endothelial cells from bovine retinal capillary and aorta were established. Vascular support cells (pericytes from retinal capillaries and smooth muscle cells from the aorta) were grown in pure culture at the same time. Insulin receptors and the effects of insulin on all four types of vascular cells were studied. Using ^{125}I-insulin, we found very specific and high affinity insulin receptors on all four types of cells. Scatchard analysis of these insulin receptors studied showed curvilinear lines similar to previously reported insulin receptors in other cell types. From the X-intercept of the Scatchard curves, receptor concentration on these cells can be approximated. The endothelial cells and retinal pericytes had approximately $2-4 \times 10^4$ receptors per cell, whereas the aortic smooth muscle cells had only 5×10^3 receptors per cell. The affinities of binding for these insulin receptors are difficult to determine accurately with the Scatchard analysis due to the curvilinear nature of the lines. However, the affinity of binding can be approximated by the concentration of unlabeled insulin required to cause a 50% displacement of ^{125}I-insulin binding on the competition studies. Since unlabeled insulin can displace ^{125}I-insulin at 1×10^{-9}M range, all four types of vascular cells appeared to have very high affinity binding to their receptors.

Next, we studied the effect of insulin on the ability of the cells to incorporate ^{14}C-glucose into glycogen, measuring a metabolic effect of insulin. Unlike the omnipresence of insulin receptors, insulin was effective in retinal capillary endothelial cells and pericytes, and aortic smooth muscle cells, but insulin had no effect on aortic endothelial cells even at very high concentrations. In retinal endothelial cells and pericytes, insulin was quite potent and was able to achieve a maximal stimulation at physiological concentrations. For aortic smooth cells, insulin had an effect at physiologic range (< 10 ng/ml); however, the maximal effect was not attained until 100 ng/ml.

In addition to metabolic effects, we studied the growth-promoting effects of insulin as measured by the stimulation of DNA synthesis in these four types of vascular cells. Like its effect on glucose metabolism, insulin was able to increase DNA synthesis in all cells except aortic endothelial cells. However, among the responsive cells, difference in insulin potency can be detected. Retinal capillary endothelial cells and aortic smooth muscle responded to insulin starting at 20 ng/ml and did not reach a maximal effect until 10 μg/ml of insulin. In contrast, the pericyte was able to respond at 1–2 ng/ml and maximal effect was reached at 1 μg/ml. The results on the retinal pericytes showed it to be extremely sensitive to insulin for growth-promoting effects; this finding is very interesting and unusual. As stated previously, in most culture cells insulin has growth-promoting effects only at pharmacological concentrations, 10–100-fold higher than that used in the pericytes. This requirement of a very high concentration of insulin for growth responses was shown to be due to the fact that insulin has to bind to any growth factor receptor in order to mediate the chronic effects. The uniqueness of the pericytes for the growth effects of insulin was confirmed when we demonstrated that the specific anti-insulin receptor antibody from the patients was able to stimulate DNA synthesis in pericytes, whereas the monovalent Fab fragment from patients' sera and IgG's from normal sera were ineffective. The data from these studies have shown that endothelial cells from retinal capillaries and large blood vessels respond differently to insulin. This differential response to insulin could be one factor responsible for the differential pathology observed in micro- and macroangiopathy. In addition, the extreme responsiveness of pericytes to insulin for both metabolic and growth effects suggests that insulin may play a role in retinal pericytes *in vivo*. Extrapolating further, it is easy to postulate that if pericytes require insulin for metabolism, then the relative deficiency of insulin in early diabetes may contribute to the dysfunctioning or even the death of pericytes, as is observed in early diabetic retinopathy.

In summary, presented in this brief review are the various ways that we have used a specific anti-receptor antibody isolated from patients' serum to elucidate important steps in the mechanism of hormone and insulin action. Now we are using the same antibodies to study retinal capillary cells, which may ultimately lead us to understand the pathological process for the development of diabetic retinopathy.

References

1. Kahn CR et al.: *Recent Prog Horm Res* **37**:477–537, 1981.

2. Roth J et al.: *Recent Prog Horm Res,* **31**:95–139, 1975.
3. Kahn, CR: *Trends Biochem Sci* **263**:4–7, 1979.
4. Czech, MP: *Annu Rev Biochem,* **46**:359–384, 1977.
5. Flier JS et al.: *Science* **190**:63–68, 1975.
6. Kahn CR et al.: *N Engl J Med* **294**:739–745, 1976.
7. Bar RS et al.: *Diabetologia,* **18**:209–216, 1980.
8. Flier JS et al.: *J Clin Invest,* **60**:784–794, 1977.
9. Kahn CR et al.: *J Clin Invest* **60**:1094–1106, 1977.
10. Kasuga M et al.: *J Clin Endocrinol Metab* **47**:66–77, 1978.
11. VanObberghen E et al.: *Nature* **280**:500–501, 1979.
12. Flier JS et al.: *J Clin Invest* **58**:1442–1449, 1976.
13. King GL et al.: *J Clin Invest* **66**:130–140, 1980.
14. Foley TP et al.: *J Biol Chem* **257**:663–669, 1982.
15. Lawrence JC et al.: *Mol Cell Biochem* **22**:153–158, 1978.
16. King GL et al.: *J Clin Invest* **71**:914–919, 1983.
17. Taylor SI et al.: *N Engl J Med* **307** (23):1422–1428, 1982.
18. Kahn CR and Harrison LC In *Insulin Receptor Antibodies in Carbohydrate Metabolism and Its Disorders, Vol. IV* Randle PJ et al. (Eds.). Academic Press, London, pp. 279–330, 1980.
19. Kahn CR et al.: *Proc Natl Acad Sci* **75**:4209–4213, 1978.
20. Schechter Y et al.: *Proc Natl Acad Sci USA* **76**:2720–2724, 1976.
21. Schlessinger J et al.: *Nature* **286**:729–730, 1980.

chapter 40

The Role of Prostaglandin D$_2$ in Allergic Ocular Disease

**Mark B. Abelson[a,b] Nalini A. Madiwale,[b]
and Judith H. Weston[b]**

Prostaglandins are products of the cyclo-oxygenase pathway of arachidonic acid metabolism.[1] The ocular effects of exogenously applied prostaglandins have been well characterized. PGE$_1$ and PGE$_2$ cause increased intraocular pressure after systemic,[2] intraventricular,[3] intraocular,[4] and topical administration.[5] Associated with this effect are a breakdown of the blood–aqueous barrier, increased aqueous humor protein, and miosis.[2,4,6]

Prostaglandin D$_2$ is synthesized primarily by mast cells and to a lesser extent by platelets.[7] The central role of the mast cell in allergic ocular disease is well known. Since PGD$_2$ is the main prostaglandin produced by the mast cell, we believe that an investigation into the role of PDG$_2$ in ocular inflammation will further our understanding of allergic ocular disease. We present the effects of topical administration of prostaglandin D$_2$ to the eyes of guinea pigs and humans.

Materials and Methods

Preparation

Prostaglandin D$_2$ was prepared in phosphate-buffered saline to concentrations of 100, 500, 750, and 1000 μg/ml and adjusted

to pH 7.0 to 7.4 with 0.1N sodium hydroxide. Tonicity of the solutions was within the physiologic range. Phosphate-buffered saline was used as a control.

Experimental procedure (guinea pig)

Five strain 13 guinea pigs weighing approximately 600 g each were treated topically with 20 μl of prostaglandin D$_2$ (1000 μg/ml) in one eye and 20 μl of phosphate-buffered saline in the fellow eye in a double-masked fashion. Slit-lamp examinations were performed at baseline and every 15 minutes for 1 hour after treatment for evidence of conjunctival redness, chemosis, tearing, and mucous discharge. Blepharospasm and scratching around the eyes with the hind paw were assumed to denote itching. Injection intensity was graded on a scale of 0 to 3+ (0 = none, 1+ = mild redness, 2+ = moderate redness, and 3+ = severe redness).[8] External ocular findings were documented photographically.

One hour after treatment, conjunctival scrapings were obtained from the superonasal quadrant, and biopsy specimens were taken from the superotemporal aspect of both eyes of each guinea pig. Both procedures were performed under topical proparacaine hydrochloride (0.5%).

Experimental procedure (human)

The right eyes of two normal, healthy volunteers were treated with 20 μl of prostaglandin D$_2$ in concentrations of 100 and

[a]From the Department of Ophthalmology, Harvard Medical School, and the [b]Immunology Unit, Eye Research Institute of Retina Foundation, Boston, Massachusetts.

TABLE I
Histological Response of Guinea Pigs to Topically Applied PGD$_2$ (1000 μg/ml)

Guinea pig no.	No. of eosinophils per mm^2		% of Blood vessels with eosinophil margination	
	PGD$_2$-treated eye	Control eye	PGD$_2$-treated eye	Control eye
1	42.5	2.5	53.5	24.5
2	18.0	9.0	58.5	0
3	11.0	0	0	0
4	171.0	29.0	58.0	7.0
5	152.0	4.5	45.0	0

750 μg/ml (subject 1) and concentrations of 500, 750, and 1000 μg/ml (subject 2). The left eyes received 20 μl of phosphate-buffered saline. A rest interval of at least 2 months was required for repeat participation.

The subjects were examined clinically at baseline and every 15 minutes for 1 hour after treatment for evidence of conjunctival redness, chemosis, tearing, and mucous discharge. Injection intensity was scored as described above. Subjective evaluation of itching was graded on a scale of 0 to 3+ (0 = none, 1+ = mild itching, 2+ = moderate itching, and 3+ = severe itching that required scratching). Two ophthalmologists performed independent slit-lamp examinations at 30 and 60 minutes after treatment.

One hour after treatment, conjunctival scrapings of the inferonasal quadrant of both eyes, and biopsy specimens of the inferotemporal conjunctiva of the eyes treated with prostaglandin D$_2$ (750 μg/ml) were obtained under topical proparacaine hydrochloride (0.5%) anaesthesia.

Cytologic method

The conjunctival scrapings were heat fixed, stained by Luna's method,[9] and screened for the presence of eosinophils.

Histologic method

Biopsy specimens were fixed in 10% neutral-buffered formalin, dehydrated in alcohol, and embedded in methyl methacrylate. Tissues were then sectioned at 4 μm and stained by Luna's method.[9]

Counting technique

All biopsy specimens were evaluated in a double-masked fashion for (1) number of eosinophils per mm,[2] (2) number of eosinophil granule clusters per mm,[2] and (3) percentage of blood vessels with eosinophil margination. Planimetry was used to measure the area of each section.

Results

Guinea pig response

The clinical response to topically applied PGD$_2$ (1000 μg/ml) included 1+ conjunctival redness in all five animals, conjunctival chemosis in four, and tearing in three; none had mucous discharge or itching. Eyes treated with phosphate-buffered saline remained white and quiet. None of the conjunctival scrapings contained eosinophils. However, histologic evaluation disclosed a significantly greater number of eosinophils

TABLE II
Clinical Response of Humans to Topically Applied PGD$_2$

Dosage (μg/ml)	Itching	Conjunctival redness	Chemosis	Tearing	Mucous discharge
100	Absent	1+	Absent	Absent	Absent
500	Absent	2+	Present	Absent	Absent
750[a]	Absent	2+	Present	Absent	Absent
1000	Absent	3+	Present	Absent	Present

[a] Two eyes received the 750 μg/ml dosage; one eye received each of the remaining dosages.

per mm^2 in eyes treated with PGD$_2$ (1000 µg/ml) than in control eyes ($p = 0.008$, paired t-test) (Table I). The percentage of blood vessels with eosinophil margination was also significantly greater in the prostaglandin D$_2$-treated eyes ($p = 0.024$, paired t-test) (Table I). There was no significant difference in the number of eosinophil granule clusters per mm.^{2}

Human response

The clinical response to topical PGD$_2$ (Table II) included a dose-dependent dilation of scleral, episcleral, limbal, and to a lesser extent, conjunctival vessels which began at five minutes, peaked at 20 minutes, and resolved within 90 minutes. Stringy mucous discharge, similar to the type found in patients with vernal conjunctivitis, was noted at only the 1000 µg/ml dose. The subjects described a gritty feeling and burning sensation in the PGD$_2$-treated eye; neither subject experienced itching. No elevation of intraocular pressure was detected at one hour after treatment. Control eyes remained white and quiet.

Eosinophils were recovered in the conjunctival scraping at only the highest dose (1000 µg/ml). The conjunctival biopsy of subject #1 contained 38 (average) eosinophil granule clusters per mm^2 and no intact eosinophils; the conjunctival biopsy of subject #2 contained 30 (average) eosinophil granule clusters per mm^2 and 12 (average) intact eosinophils per mm^2. Control biopsies were not done because the absence of eosinophils in normal human conjunctiva has been well-documented.[10] None of the human biopsy sections had blood vessels with eosinophil margination.

Discussion

Topical administration of prostaglandin D$_2$ to the eyes of guinea pigs and humans causes vasodilation, increased vascular permeability, and eosinophil chemotaxis. The previous *in vitro* finding that PGD$_2$ is strongly chemokinetic for eosinophils supports our work.[11]

Eosinophils have long been considered the hallmark of allergic ocular disease. Al-though the presence of eosinophils in conjunctival scrapings can corroborate the diagnosis of ocular allergy, we have recently shown that eosinophils present in deep and superficial conjunctival tissues may not be recovered in scrapings, and their absence should not preclude the diagnosis of allergic ocular disease.[12] In fact, only 63% of the patients with vernal conjunctivitis had eosinophils in their conjunctival scrapings.[12]

The stringy mucous discharge produced by topical application of PDG$_2$ (1000 µg/ml) resembled the mucus found in allergic ocular conditions. Based on our finding that prostaglandin D$_2$ causes conjunctival redness, chemosis, stringy mucous discharge, and eosinophil chemotaxis, we believe that a better understanding of PDG$_2$, the mast cell's most prevalent prostaglandin, may further our ability to treat allergic ocular disease. Although prostaglandin D$_2$ is inactivated by PGD dehydrogenase,[13] to our knowledge, there are no drugs available that selectively block the effects of prostaglandin D$_2$. Nonsteroidal anti-inflammatory agents, such as aspirin or piroxicam, block the synthesis of PGD$_2$ by acetylating the active site of the enzyme cyclo-oxygenase.[14,15] Inhibitors of arachidonic acid metabolism will surely play a role in the future treatment of allergic ocular disease. Preliminary results indicate that systemic aspirin may be a useful adjuvant in the management of patients with vernal conjunctivitis.

Acknowledgment

This work was supported in part by CooperVision Pharmaceuticals Inc. and the Massachusetts Lions Eye Research Fund, Inc.

References

1. Weksler BB: *NESA Proceedings* 2:56-61, 1981.
2. Beitch BR and Eakins KE: *Br J Pharmacol* 37:158–167, 1969.
3. Krupin T et al.: *Am J Ophthalmol* 81:346–350, 1976.
4. Waitzman MB and King CD: *Am J Physiol* 212:329–334, 1967.
5. Kass MA et al.: *Invest Ophthalmol* 11:1022–1027, 1972.
6. Whitelocke RAF and Eakins KE: *Arch Ophthalmol* 89:495–499, 1973.

7. Oelz O et al.: *Prostaglandins* **13**:225–234, 1977.
8. Abelson MB et al.: *Ann Ophthalmol* **13**:1225, 1981.
9. Luna LG (Ed.): *Manual of Histologic Staining Methods of the Armed Forces Institute of Pathology.* McGraw-Hill, New York, 1968, pp. 111–112.
10. Allansmith MR et al.: *Am J Ophthalmol* **86**:250–259, 1978.
11. Goetzl EJ et al.: In *Advances in Inflammation Research, Vol. 1,* Weismann G et al. (Eds.). Raven Press, New York, 1979, pp. 157-167.
12. Abelson MB et al.: *Arch Ophthalmol* **101**:555–556, 1983.
13. Watanabe K et al.: *J Biol Chem* **255**:1779–1782, 1980.
14. Rome LH et al.: *Prostaglandins* **11**:23–30, 1976.
15. Roth GJ et al.: *Proc Natl Acad Sci USA* **72**:3073–3076, 1975.

chapter 41

Biological and Biochemical Properties of a Corneal Epithelial Cell-Derived Thymocyte-Activating Factor (CETAF)

G. Grabner, T.A. Luger,[a] G. Smolin, and J.O. Oh

A variety of hormone-like factors produced by lymphoid as well as by nonlymphoid cells regulate the interactions of cells participating in inflammatory reactions. The macrophage, a cell with a pivotal role in immunological reactions, has been demonstrated to release a nonspecific, immunoregulatory helper factor, Interleukin 1 (IL 1, LAF).[1,2] Similarly keratinocytes, the major constituents of epidermal cells, can produce a cytokine, epidermal cell-derived thymocyte-activating factor (ETAF) that is biologically and biochemically indistinguishable from IL 1.[3,4] We briefly report our data on an IL 1-like factor secreted by corneal epithelial cells under *in vitro* conditions, named corneal epithelial cell-derived thymocyte-activating factor (CETAF).

Generation and Stimulation of CETAF

Primary rabbit corneal epithelial cell cultures as well as a rabbit corneal cell line (SIRC from the American Type Culture Collection) release a soluble mediator that is able to enhance the proliferative capacity of lectin-stimulated C3H/HeJ mouse thymocytes.[5] This indirect thymocyte-assay is a quantitative microassay that is widely used to measure IL 1 and ETAF activity.[3] The maximal CETAF activity was recovered from SIRC cultures seeded at a cell density of 1×10^5 cells per milliliter. Serum-free conditions as well as the addition of a variety of stimulants, which are known to increase the release of IL 1 by macrophages and of ETAF by murine keratinocytes (such as lipopolysaccaride, hydroxyurea, silica, phorbol myristate acetate, mitomycin C), to the culture medium significantly augmented CETAF production by SIRC cells.[6,7,8]

Disruption of a monolayer with the use of a rubber policeman was found to result in a significant increase of the factor production when supernatants were compared to those of untreated plates.[5] In general, increased levels of CETAF activity were associated with an increased number of cells that enter the G1/S interphase either by blocking the cell cycle or by shifting the cells into G1. Rabbit CETAF, similar to rabbit IL 1, exhibited two peaks with molecular weight ranges from 95,000 to 55,000 (= CETAF I) and 30,000 to 15,000 (= CETAF II) daltons when partially purified by gel filtration.[5,9]

The Francis I. Proctor Foundation for Research in Ophthalmology, UCSF, San Francisco, California and [a]The National Institute of Dental Research, Laboratory of Microbiology and Immunology, NIH, Bethesda, Maryland.

Dr. Günther Grabner was on leave from the Second Department of Ophthalmology, University of Vienna, Austria, and Dr. Thomas A. Luger was on leave from the Second Department of Dermatology, University of Vienna, Austria.

Biological Properties of CETAF

CETAF (like IL 1 and ETAF) in conjunction with suboptimal doses of lectins activated murine C3H/HeJ thymocytes to proliferate. In contrast, CETAF by itself stimulated thymocytes only to a minimal degree.[5] One of the most important functions of IL 1 is to promote T-cell production of lymphokines such as lymphocyte-derived chemotactic factor (LDCF) and Interleukin 2 (IL 2).[10,11] In a similar way CETAF enhanced the production of IL 2 by murine spleen cells and human peripheral blood lymphocytes (unpublished observations) that were concomitantly stimulated with optimal doses of concanavalin A and phorbol myristate acetate.[8] Neither SIRC supernatants, nor partially purified CETAF had a direct stimulatory effect on the growth of the IL 2-dependent murine T-cell line CT 6, indicating that CETAF does not possess any IL 2 activity.[8] This finding may have important implications since IL 2, besides its T-cell growth promoting capacity, has been demonstrated to generate cytotoxic T-cells, to enhance the antibody production by B-cells and to induce the γ-interferon production by T-lymphocytes.[12–14] Thus through the production of CETAF, corneal epithelial cells, which do not belong to the immune system, may be endowed with the capacity to enhance local immunological reactions in the cornea and lymphoid tissues of the conjunctiva.

Previous data indicate that corneal epithelium may produce a mediator which is chemotactic for polymorphonuclear granulocytes (PMN).[15,16] When tested in a modified Boyden-chamber assay crude SIRC cell supernatants, as well as partially purified CETAF, were significantly chemotactic for rabbit PMNs.[8] This again parallels activities of IL 1 and ETAF.[17] However, high concentrations of the different cytokines led to an aggregation of PMNs in vitro, which might be due to the release of specific granule contents of PMNs.[18] When rabbit macrophages were used as target cells, CETAF was chemotactically not active in a wide range of concentrations (both in its crude and partially purified forms), but again led to an aggregation of macrophages on the

membranes at high concentrations (unpublished observations).

Macrophage-derived mediators have effects on different cell types participating in acute or chronic inflammations. Recently it was shown that IL 1 and ETAF directly stimulate the in vitro growth of human fibroblasts.[7,19] Different preparations of CETAF, when tested for their effects on fibroblasts, significantly enhanced their proliferation in the absence of other mitogenic signals.[8] In addition, this epithelial cytokine is possibly closely related, if not identical, to a regulator of connective tissue collagenase production derived from epithelia of different origin[20] and, therefore, may have an in vivo effect in corneal ulceration and wound healing through the activation of stromal fibroblasts.

A variety of closely related if not identical monokines, such as serum amyloid A inducer (SAA inducer), endogenous pyrogen (EP), and IL 1, that are released by macrophages, have been observed to stimulate SAA production by hepatocytes following an intraperitoneal injection into C3H/HeJ mice.[21] Like ETAF, crude SIRC cell culture supernatants as well as partially purified CETAF were found to increase the serum levels of this late acute-phase reactant.[22,23] It is unknown, however, what SAA species are induced by CETAF and whether there is a relationship between a local production of CETAF in vivo and primary or secondary amyloidosis afflicting the cornea and conjunctiva.

The Possible Role of CETAF in Corneal Disease

Our findings indicate that CETAF, derived from a primary rabbit corneal epithelial cell culture as well as from a transformed corneal cell line (SIRC), acts on a variety of target cells (Fig. 1). Therefore, it seems warranted to study patients suffering from healing problems of the corneal epithelium, as observed in nontraumatic recurrent corneal erosions (in epithelial dystrophies), following different insults (such as lye burns, infections, or trauma) or in severely dry eyes (as in pemphigoid or the Stevens-Johnson syndrome). The injured, and possibly con-

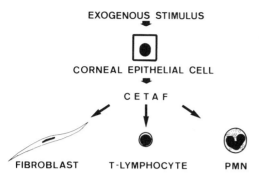

EXOGENOUS STIMULUS

CORNEAL EPITHELIAL CELL

C E T A F

FIBROBLAST   T-LYMPHOCYTE   PMN

FIG. 1. Pathway of CETAF formation and effects.

tinuously stimulated, corneal epithelial cells may release an increased amount of CE-TAF. This sets the stage for stromal keratitis and scarring or even severe ulceration, eventually leading to corneal perforation through the stimulation of collagenase release by stromal cells.[24] Furthermore, CE-TAF may play a role in amplifying cellular and humoral immune responses during the course of corneal or conjunctival diseases caused by infectious agents, such as viruses, chlamydia, or bacteria. We would like to caution, however, that all our hypotheses regarding the possible *in vivo* role of CETAF are based on *in vitro* experimental work and that there is as yet no clear evidence for an *in vivo* role of CETAF in the pathogenesis of corneal diseases.

In summary, like macrophages and epidermal cells, rabbit corneal epithelial cells have been shown to produce a nonspecific immunoregulatory mediator, corneal epithelial cell-derived thymocyte-activating factor (CETAF). CETAF is produced *in vitro* by primary rabbit corneal epithelial cells and by a transformed rabbit corneal cell line (SIRC). Biochemically CETAF is similar to macrophage-derived rabbit IL 1 in that they both elute within a similar molecular weight range of gel filtration exhibiting two peaks of activity from 95,000 to 55,000 (CETAF I) and 30,000 to 15,000 (CETAF II) daltons.

The biological properties of CETAF so far investigated are identical to those of IL 1 or epidermal cell-derived ETAF. CETAF enhances the proliferative capacity of thymocytes in the presence of suboptimal doses of lectins, induces a further increase of IL 2 production by T-cells, and is chemotactic

for PMNs. In addition, CETAF affects non-lymphoid target cells: It is directly mitogenic for fibroblasts and induces SAA production by hepatocytes. Therefore corneal epithelial cells, through the production of CETAF, may modulate inflammatory reactions in the cornea and would influence corneal healing, thereby playing an important role in the pathogenesis of corneal diseases.

Acknowledgments

This work was supported in part by the Max Kade Foundation, New York.

The authors thank Marcela A. Kopal and Petros Minasi for their help with the chemotaxis assays and Joost J. Oppenheim for his critical comments.

References

1. Gery I et al.: *J Exp Med* **136:**128–142, 1972.
2. Oppenheim JJ et al.: In *Role of macrophages in Selfdefense Mechanisms,* Ishida (Ed.). Elsevier, North Holland, 1982.
3. Luger TA et al.: *J Immunol* **127:**1493–1498, 1981.
4. Luger TA and Oppenheim JJ: In *Advances in Inflammation Research, Vol. 5,* Weissman G (Ed.). Raven Press, New York, 1983.
5. Grabner G et al.: *Invest Ophthalmol Vis Sci* **23:**757–763, 1982.
6. Oppenheim JJ et al.: In *Biology of the Lymphokines,* Cohen S. et al. (Eds.). Academic Press, New York, 1979, pp. 291–323.
7. Luger TA et al.: *J Immunol* **128:**2147–2152, 1982.
8. Grabner G et al.: *Invest Ophthalmol Vis Sci,* **24:**589–595, 1983.
9. Simon PL and Willoughby W: In *Lymphokines, Vol. 6,* Mizel S (Ed.). Academic Press, New York, 1982, p. 47–63.
10. Mizel SB: In *The Lymphokines: Biochemistry and Biological Activity,* Hadden JW and Steward WE II (Eds.). Humana Press, New Jersey, 1981, p. 347–365.
11. Farrar JJ et al.: *J Immunol* **125:**793–798, 1980.
12. Farrar JJ et al.: *J Immunol* **121:**1353–1360, 1978.
13. Farrar WL et al.: *J Immunol* **126:**1120–1125, 1981.
14. Gillis S and Smith KA: *Nature* **268:**154–158, 1977.
15. Weimar V: *J Exp Med* **105:**141–152, 1957.
16. Kenyon KR et al.: *Invest Ophthalmol Vis Sci* **18:**570–587, 1979.
17. Luger TA et al.: *J Invest Derm* **81:**187–193, 1983.
18. Klempner MS et al.: *J Clin Invest* **61:**1330–1336, 1978.
19. Schmidt JA et al.: *J Immunol* **128:**2177–2182, 1982.
20. Johnson-Wint B and Gross J: *J Cell Biol* **91:**158, 1981.
21. Sztein MB et al.: *Cell Immunol* **63:**164–176, 1981.
22. Sztein MB et al.: *J Immunol* **129:**87–90, 1982.
23. Luger TA et al.: *Ophthalmic Res* **15:**121–125, 1983.
24. Johnson-Muller B and Gross J: *Proc Natl Acad Sci USA* **75:**4417–4421, 1978.

chapter 42

Studies on Interferon Systems in Endogenous Uveitis

Shigeaki Ohno, Satoshi Kotake, Fujiko Kato, Hidehiko Matsuda, Nobuhiro Fujii, and Tomonori Minagawa

Recent studies on the interferon (IFN) system have revealed that IFN is classified into three groups, α, β, and γ.[1] IFN has nonantiviral functions, such as cellular modulations, in addition to classical antiviral function and has a close association with the immune response of the host.

The chronicity and recurrences that are common to endogenous uveitis are highly suggestive of an underlying immunological process. We have been studying the IFN system in endogenous uveitis.[2-4] The purpose of this study is to investigate the level, type, and the cellular source of IFN in patients with endogenous uveitis.

Materials and Methods

A total of 358 serum samples were collected from 101 patients with endogenous uveitis at each visit. Sixty of them had the diagnosis of Behçet's disease, 17 had Vogt–Koyanagi–Harada's disease (VKH), and 24 had ocular sarcoidosis. All patients were seen at the Uveitis Survey Clinic of Hokkaido University Hospital. IFN assay was performed by the semimicro, dye-binding assay method[5] based on quantitation of inhibition of cytopathic effects. Cultures of human amnion cells (FL) were exposed to serially two-fold-diluted serum samples,

Departments of Ophthalmology and Microbiology, Hokkaido University School of Medicine, Sapporo, Hokkaido, Japan.

challenged with vesicular stomatitis virus, and stained with gentian violet. The bound dye was then eluted and measured colorimetrically. A laboratory standard that had been calculated from NIH standard (G-023-901-527) was inoculated in each microplate. IFN titer of samples was calibrated with the standard and expressed as IU per ml. Neutralization of IFN was performed by anti-IFN-α, anti-IFN-β, and anti-IFN-γ antisera. These antisera were diluted to neutralize 100 IU/ml of standard IFN-α and IFN-β. IFN samples were diluted serially two-fold and each dilution was mixed with the equal volume of the antisera. After incubation at 37°C for 30 minutes, the residual IFN activity was assayed. The IFN samples were also dialyzed against 0.2 M KCl-HCl buffer (pH 2.0) for 48 hours at 4°C and further dialyzed against RPMI 1640 medium to adjust to pH 7.4. The antiviral activity was then measured by the same method.

Results and Discussion

Serum IFN activity of the controls was generally negative; only 26 of 79 controls (33%) showed positive IFN titers. The average titer was 5.9 ± 1.1 IU/ml (mean $\pm$ standard error) in this group. Table I shows that the patients with endogenous uveitis had significantly increased serum IFN activity. The average titer of IFN was 38.5 ± 3.4 IU/ml in the patients with Behçet's disease ($p < 0.0001$), 48.8 ± 4.2 IU/ml in the patients with VKH ($p < 0.0001$), and

TABLE I

Serum IFN Activity in Patients with Endogenous Uveitis

	IFN titers (IU/ml)
Behçet's disease (n = 233)	38.5 ± 3.4*
Vogt–Koyanagi–Harada's disease (n = 94)	48.8 ± 4.2*
Sarcoidosis (n = 49)	47.9 ± 6.2*
Controls (n = 79)	5.9 ± 1.1*

*p < 0.0001

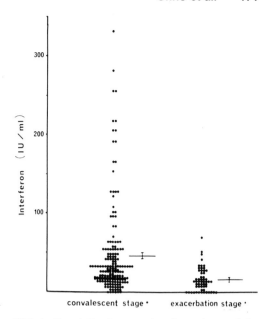

FIG. 1. Correlation between interferon titers and the stage of ocular disease in Behçet's disease.

47.9±6.2 IU/ml in the patients with sarcoidosis (p<0.0001), respectively. Characterization of serum IFN showed that all serum samples tested were acid labile, and dialysis at pH 2.0 completely destroyed the antiviral activity both in Behçet's disease and VKH, as was the case with treatment at 56°C for 30 minutes. IFN activity showed no change with treatment of either anti-IFN-α or anti-IFN-β antisera, but it was completely inactivated by treatment with anti-IFN-γ antisera in both diseases. Therefore, it was concluded that the detected IFN in Behçet's disease and VKH was primarily IFN-γ. In sarcoidosis, however, IFN activity was partially inactivated by treatment of serum at pH 2.0, as was the case with treatment of serum with anti-IFN antisera. It was considered that the detected IFN is a mixture of IFN-α and IFN-γ in sarcoidosis, and IFN-α may play a major role in this disease, similar to systemic lupus erythematosus.[6]

Correlation between IFN titers and the duration of ocular disease was studied in Behçet's disease. No significant difference was observed between them, and the increased IFN activity had no direct correlation with the duration of ocular lesions. In VKH, the kinetics of serum IFN showed that significantly increased IFN activity was detected in 1 to 2 months after the onset compared with that within 2 weeks (1 month: p<0.005; 2 months: p<0.0006).

Correlation between IFN titers and the stage of ocular disease was also studied. In Behçet's disease, the mean IFN titer was 45.6±4.3 IU/ml in the convalescent stage and 15.8±2.1 IU/ml in the exacerbation stage (Fig. 1). The convalescent stage showed significantly increased IFN titers as compared with both the exacerbation stage (p<0.0001) and the controls (p<0.0001). In VKH, the posterior involvement group (Harada type) showed slightly higher IFN titers (51.9±5.1 IU/ml) than the anterior involvement group (Vogt–Koyanagi type) (38.3±6.6 IU/ml). However, the difference was not statistically significant (p>0.1).

Investigation of the cellular source of IFN detected in sera of the patients with endogenous uveitis was then performed. Peripheral blood mononuclear leukocytes (PBML) obtained from the patients were cultured in RPMI 1640 medium with 10% fetal calf serum at 37°C in a CO_2 incubator. In was shown that the PBML preparations obtained from the patients with Behçet's disease produced IFN spontaneously in culture fluids without any stimulation.[4] The average IFN titers in cultures of the patients' PBML were significantly higher than those of the control cultures (p<0.0001). The IFN activity was significantly higher in cultures of patients in the ocular convalescent stage than in the exacerbation stage (p<0.0001), as was the case with serum interferon activity. Fractionation of PBML showed that only T-lymphocytes of patients produced IFN-γ spontaneously in the cultures. Our preliminary studies showed that the PBML of patients with VKH also produced IFN-γ spontaneously *in vitro* without any stimulation, but none was found in cultures from patients with sarcoidosis.

It seems reasonable to think that IFN-γ detected in the sera of patients with Behçet's disease and VKH is produced by their own T-lymphocytes *in vivo*. In contrast, no IFN activity was found in PBML cultures of patients with sarcoidosis, and the cellular source of serum IFN in this disease remains to be determined. The exact correlation between serum IFN-α and IFN-γ also has to be investigated in this disease.

The mechanism of IFN production in endogenous uveitis is obscure. Immune complexes may be one of the causative IFN inducers, since they may stimulate the host's PBML to produce IFN. It may also be possible that the T-lymphocytes of the patients have been altered to produce IFN spontaneously by some exogenous agents. However, it seems difficult at the present time to understand the biological role of detected IFN in patients with endogenous uveitis, because it is unknown whether the increased IFN is the cause or the result of these diseases. The production of IFN during the host's immunological response to these diseases may protect the host from contracting viral disease. Another possibility is that IFN activates the cytotoxic T-lymphocytes, natural killer cells, or macrophages, and causes tissue damage. It has been reported that IFN can induce lesions in several different organs in mice and rats.[7]

The IFN system seems to play an important role in the immunopathophysiology of endogenous uveitis, and further extensive studies are needed to elucidate the mechanism of production and the function of IFN in endogenous uveitis.

Acknowledgments

This investigation was supported in part by grants from the Ministry of Education, Science, and Culture and from the Ministry of Health and Welfare, Japan.

References

1. Committee on Interferon Nomenclature. *Nature* **286**:110, 1980.
2. Ohno S: *Trans Ophthal Soc U K* **101**:335–341, 1981.
3. Ohno S et al.: *Infect Immun* **36**:202–208, 1982.
4. Ohno S et al.: *Ophthalmologica* **185**:187–192, 1982.
5. Armstrong JA: *Appl Microbiol* **21**:723–725, 1971.
6. Friedman RM et al.: *Arthritis Rheum* **25**:802–803, 1982.
7. Gresser I: *Regulatory Functions of Interferons.* New York Academy of Sciences, New York, 1980.

chapter 43

Regeneration of Mast Cells after Ocular Anaphylaxis

Mathea R. Allansmith,[a,b] **Robert S. Baird,**[b] **Antonio S. Henriquez,**[b] **and Kurt J. Bloch**[c]

Injection of antigen into ocular tissues of immunized animals causes swelling, enhanced vascular permeability, and a gain in weight of the tissues.[1] Presumably these changes are from materials released from mast cells undergoing anaphylaxis-induced granulation. Therefore, it is important to evaluate the number and changes of mast cells in ocular tissues.

Material and Methods

Seventy-five Sprague-Dawley rats were divided into nine groups. The first group was not immunized nor were the tissues injected. Groups 2–5 were immunized and injected as previously described.[1] Groups 6–9 were nonimmunized antigen-injected controls. Place and amount of injection were as previously described.[1] After appropriate time had elapsed after induction of anaphylaxis, animals were exsanguinated from cervical vessels. Tissues were fixed in glutaraldehyde, processed,[2] and embedded in Epon for 1 micron Giemsa-stained sections. Mast cell recognition and criteria for degranulation were as previously described.[3] In addition to the counting of mast cells, the granulated cells were observed for (1) irregularity of cell outline, (2) localized irregularity of the cell surface, (3) sparseness of granules, (4) variability of granule size, and (5) variation of color intensity of the granules. A scoring system was made to favor low scores for granulated mast cells from normal animals and high scores for granulated cells of animals recovering from anaphylaxis. The scoring system was made on pilot observations of tissues.

Results

The mean number of mast cells for tissues from normal, from antigen-injected controls, and from animals undergoing anaphylaxis was about 3000 cells/mm³. No significant difference in the number of mast cells was observed among the groups ($p>0.05$). The percentage of mast cells showing evidence of degranulation had a median of 0 at time zero, 75% at 0.5 hour, 85% at 1 hour, 70% at 6 hours, and 30% at 24 hours (Fig. 1). These were significantly different from injected controls at the 0.5, 1, and 6 hour periods. At 24 hours there was no significant difference between the number of degranulated mast cells of the injected controls and those animals that had undergone anaphylaxis. Although a degranulated mast cell was declared when the cell had 4 or more pink granules in the cytoplasm which was otherwise filled with dark purple granules, the granule changes were far more extensive in tissues undergoing anaphylaxis at 0.5 and 1 hour periods as compared with other time

[a]Department of Ophthalmology, Harvard Medical School, Boston; [b]Department of Cornea Research, Eye Research Institute of Retina Foundation, Boston; and [c]Allergy Units, Medical Services, Massachussetts General Hospital, Boston, Massachusetts.

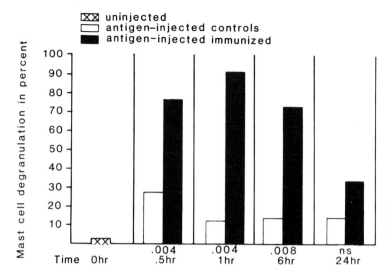

FIG. 1. Percentage of degranulated mast cells in lid tips of uninjected, antigen-injected control, and antigen-injected immunized rats at various intervals prior to histologic examination. The criterion for degranulation was the presence of four or more bright red-pink granules in the cytoplasm of a mast cell. Note the enormous increase in degranulated mast cells at the 0.5 hour time and the return to normal number of degranulated cells by 24 hours.

periods, normals, and controls. By 6 hours after anaphylaxis, all mast cells had resumed their oval contour, even though they still had changes in the granules. At 24 hours all cells were oval, and the cell contour resembled that of the uninjected controls. However, by the five criteria stated above, it could be seen that the granulated mast cells at 24 hours after anaphylaxis had more irregularity of the cell outline, had localized outpouching of the cell surface, had fewer granules in the cytoplasm, had greater variability of granule size, and had greater variation of color intensity of these granules ($p = 0.05$ for all groups) (Fig. 2.) Therefore, mast cells that had presumably undergone

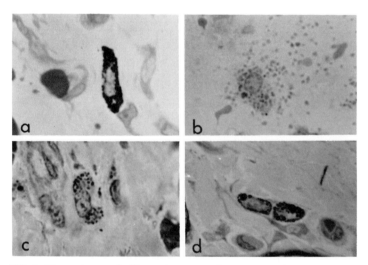

FIG. 2. Photomicrographs illustrating mast cell appearances in 1 micron sections from lid tips of rats prior to anaphylactic challenge and at various times after challenge (1000 × original magnification); *(a)* granulated mast cell in an uninjected rat, *(b)* extensively degranulated cells with exocytosis of granules in tissue of an immunized rat 0.5 hour after antigen injection, *(c)* moderately degranulated mast cell 6 hours after anaphylaxis, *(d)* granulated mast cell 24 hours after induction of anaphylaxis. (Granules of various sizes are present.)

anaphylaxis had regranulated but remained morphologically distinguishable from granulated mast cells of normal animals.

Discussion

The number of mast cells in the lid tip of rats undergoing ocular anaphylaxis remained unchanged in the course of local anaphylaxis. The number of degranulated cells reached 70–100% at 30–60 minutes after injection of antigen in immunized animals. Severe local anaphylaxis had been induced in these animals. Yet at 24 hours the number of degranulated mast cells was not greater than was observed in the tissue of control injected or uninjected animals. Thus considerable recovery of the mast cell population to its preanaphylaxis level of granulation had occurred. This is the first demonstration of rapid (less than 24 hours) mast cell recovery in an *in vivo* model of anaphylaxis.

Others had proposed, on the basis of observations of an *in vitro* model, that mast cells begin to recover almost immediately after undergoing anaphylaxis. Ginsburg et al.[4] triggered degranulation which affected more than 90% of cultured mast cells. The thoroughly degranulated mast cells in culture reverted to pregranulation appearance after overnight incubation.[5] Whether the reconstituted mast cells observed in our experiments are available for a second response to antigen administered 24 hours after the first challenge remains to be determined. This information would be helpful in assessing the continued contribution of the IgE antibody mast cell mediator system to chronic injury of the eye tissues in humans.

Acknowledgments

This work was supported by grants EY-02099 and T32-AM-07258, from the National Institutes of Health, and by grants from the Massachusetts Chapter of the National Arthritis Foundation and the HOR Foundation.

References

1. Allansmith MR et al.: *Immunology* **44**:623–627, 1981.
2. Allansmith MR et al.: *Am J Ophthalmol* **86**:250–259, 1978.
3. Allansmith MR et al.: *Invest Ophthalmol Vis Sci* **19**:1521–1524, 1980.
4. Ginsburg H and Lagunoff D: *J Cell Biol* **35**:685–697, 1967.
5. Ginsburg H et al.: *Immunology* **35**:485–502, 1978.

chapter 44

Genetic Regulation of Natural Resistance of the Mouse Cornea to *Pseudomonas aeruginosa* Infection

Possible Allelic Role

R. S. Berk, L. D. Hazlett,[a] and K. W. Beisel

During the past several years we have examined the genetic regulation of natural resistance of the murine cornea to challenge with *Pseudomonas aeruginosa* ATCC strain 19660. [1–3] We have established that young adult mice (4–6 weeks old) Swiss-Webster, DBA/1J, DBA/2J (D2), BALB.B and BALB.K strains are naturally resistant (R) to intracorneal infection since the infection remains localized and spontaneously heals within several weeks in virtually all of the mice. This natural resistance is not present in 5–10-day-old mouse pups, but progressively develops as the animals enter young adulthood (21–30 days).[4] On the other hand, the C57BL/6J (B6), BALB/cJ, NZB, C3H/HeJ and A/J strains are classed as susceptible (S) since the infection results in failure to restore corneal clarity. A third type of response has also been noted in some strains of mice in which a certain percentage of mice recover corneal clarity while the rest fail to do so. We will refer to this "mixed" response as intermediate susceptibility (IS) as opposed to the "all or none" effect seen in other inbred strains.

Recent studies suggest that at least two autosomal complementing genes control natural resistance to intracorneal infection by *P. aeruginosa* and that they both reside outside of the *H-2* complex.[2] We have provisionally designated the B6 resistance gene as *PsCRI* and the BALB/cJ resistance gene as *PsCR2* because complementation studies between these two strains result in resistant F_1 hybrids that carry both resistance genes.[2,3] However, current studies with C3H/HeJ suggest that the genetic control of natural resistance is more complex than expected. There appear to be two linked genes that express a dominant susceptibility effect (PsCS) (manuscript for this is in preparation). The purpose of this paper is to examine further the interaction between all these genes and their alleles.

Materials and Methods

Stock cultures of *P. aeruginosa,* strain 19660 ATCC, were used in these studies as previously described.[2] The culture was hemolytic, proteolytic, and lecithinolytic, and it produced exotoxin A.

Inbred congenic lines, all F_1, F_2, and backcross matings, were bred in our animal facility or purchased from Jackson Laboratories (Bar Harbor, ME). In all experimental studies involving either F_1, F_2, or backcross animals, both male and female progeny were used in approximately equal

---

Departments of Immunology/Microbiology and Anatomy,[a] Wayne State University School of Medicine, Detroit, Michigan.

numbers. Mice (18–22 g) were infected at 5–6 weeks of age and monitored over a six-week period as previously described.[2] All experimental data represent the results of two or more independently performed experiments.

Results

Previous studies have shown that inbred mouse strains either spontaneously recover or lose their sight when intracorneally challenged with *P. aeruginosa*.[1] Certain mouse sublines of the A background express an intermediate type response that is manifested by the ability of several mice within each strain to recover corneal clarity while the remainder do not. To determine whether this intermediate type of response is under genetic control, we mated the parent A/J strain with several inbred prototypes previously classified in our ocular system as either susceptible (S) or resistant (R).[1] The results are shown in Table I. The F_1 hybrids resulting from matings between A/J (S) and D2 (R), C3H (S) and BALB/c (S) yielded intermediate responses similar to that seen with the sublines of A. Genetic complementation was observed in the F_1 of (A/J x C3H) and (A/J x BALB/c). The gene(s) controlling D2 resistance was masked by the intermediate phenotype, contributed by A/J. In order for the intermediate response of (A/J x D2)F_1 to occur, the D2 gene(s) must complement with the A/J gene(s), and the intermediate response is dominant over the expression of the D2 resistance gene. When A/J was mated with B6, all of the infected F_1 progeny

(45/45) failed to recover corneal clarity. These data indicated that either B6 is unable to complement with A/J to produce the intermediate response as did the BALB/c and C3H strains, or B6 carries a dominant susceptibility gene.

In order to test for the presence of susceptibility genes in B6, we mated the susceptible B6 with the resistant D2 strain. All of the resultant B6D2F$_1$ hybrid mice were susceptible, as evidenced by the fact that 35/35 mice failed to regain corneal clarity. When the B6D2F$_1$ was backcrossed to the parent D2 strain, 95/106 of these progeny were susceptible. These data suggest that multiple susceptibility genes may be present in B6. Since BALB/c has previously been shown to complement with B6 to yield 100% resistant progeny, one possible hypothesis is that BALB/c can override the B6 susceptibility gene(s). If this were true, then one could expect two possible results from examination of the (B6D2)F$_1$ x BALB/c progeny. The first would be that all the offspring are resistant. This is possible if the resistant D2 gene is an allele of *PsCR1*. However, if the D2 gene is not *PsCR1*, then 75% of the offspring should be resistant. Neither was the case. Instead, 52% (26/50) of the animals were resistant. The *PsCR2* gene of BALB/c does not exert an influence on the B6 susceptibility gene(s) as proposed, but only produces resistance in combination with the *PsCr1* allele of B6.

We have previously noted that BALB.B and BALB.K strains are resistant to ocular challenge with *P. aeruginosa*.[2] Since genetic complementation between BALB/c and B6 mice resulted in resistant progeny, we tested if this were also the case with BALB.B and BALB.K. Progeny from BALB.B x B6 matings yielded 24/26 susceptible mice, but the (BALB.K x B6)F$_1$ mice yielded 27/27 susceptible, indicating expression of the susceptibility phenotype and no complementation. These studies were extended to examine the segregation pattern of the F_2 progeny. The results showed that 33/97 (34%) of the (BALB.B x B6)F$_2$ animals and 28/29 (31.4%) (BALB.K x B6)F$_2$ animals were resistant. These data suggest a multigenic and/or multiallelic control of both susceptibility and resistance to corneal in-

TABLE I
Response of F_1 Hybrids to Intracorneal Challenge with *P. aeruginosa*[a]

Crosses[b]	Response	
	Resistant	Susceptible
A/J × D2	7/43	36/43
A/J × C3H	15/45	30/45
A/J × BALB/c	8/27	19/27
A/J × B6	0/45	45/45

[a] All mice received a topical application of 10^8 cfu of *Pseudomonas aeruginosa*.
[b] ♀ × ♂.

jury by *P. aeruginosa* infection. These results indicate that the genetic control of natural resistance to infection is much more complex than had been previously anticipated.[1-3]

Discussion

A number of infectious agents to which mice exhibit some form of natural resistance are under polygenic control. These are herpes simplex virus-1,[5] *Toxoplasma gondii*[6] *Salmonella typhimurium*,[7,8] and *Pseudomonas aeruginosa*.[2,3] The proposed genetic control of natural resistance to *P. aeruginosa* corneal eye infection is influenced by the *PsCR1*, *PsCR2*, and two linked *PsCS* genes. The alleles of B6 *PsCR1* and BALB/c, *PsCR2* interact to produce a resistant phenotype.[3] Another subline BALB/cKh and its congenic, BALB.K and BALB.B, are resistant to Pseudomonas corneal damage, unlike BALB/c. The difference observed between the BALB/c sublines is under unigenic control.[2] Since the F_1 hybrids between either BALB.B or BALB.K and B6 are susceptible, the resistance allele of BALB.B and BALB.K is unable to complement with B6 and also is masked by the B6 susceptibility genes. It is, therefore, most likely that this resistance gene is an allelic form of *PsCR2*. So far we have identified three alleles of *PsCR2* with differing functional abilities. These are the alleles of B6, BALB/cJ, and BALB/cKh. The existence of multi-allelic effects have undoubtedly increased the complexity of this model.

The three inheritable phenotypes, resulting from Pseudomonas infection, are resistance (R), intermediate susceptibility (IS), and susceptibility (S). Examination of all the possible hybrid combinations of the five "phenotype" strains (A/HeJ, BALB/c, C3H, B6, and D2) has suggested a hierarchy of dominance balance in the response to *P. aeruginosa* corneal infection. The genes controlling the resulting phenotype have the following dominance patterns in the F_1 hybrids where B6 (S) > A/HeJ (IS) > C3H (S) > D2 (R) > BALB/c (S). The exception is that the progeny of the two susceptible strains, BALB/c and B6, are resistant to eye damage. Segregation analyses of backcross and F_2 animals had suggested the existence of four genes (*PsCr1*, *PsCR2*, and two linked *PsCS*) whose interactions determine the resulting susceptibility phenotype. Thus, the data presented herein suggest the presence of complex intra- and intergenic interactions of the alleles that ultimately determine the outcome of the corneal response to *P. aeruginosa*. In order to understand completely the intricacies of these interactions, the identification of the *PsCR1*, *PsCR2*, and *PsCS* loci must first be determined. We are presently examining several recombinant inbred lines to accomplish this task.

Acknowledgments

This study was supported by Public Health Service grants EY-01935-05 and EY-02986-04 from the National Eye Institute, National Institutes of Health.

References

1. Berk RS et al.: *Infect Immun* 26:1221–1223, 1979.
2. Berk RS *Infect Immun* 34:1–5, 1981.
3. Berk RS et al.: *Rev Infect Dis* (in press, 1983).
4. Hazlett LD et al.: *Infect Immun* 20:25–29, 1978.
5. Lopez C *Nature* 258:152, 1975.
6. Williams DM et al.: *Infect Immun* 19:416–420, 1978.
7. Plant J and Glynn AA: *Clin Exptl Immunol* 37:1–6, 1979.
8. O'Brien AD et al.: *J Immunol* 124:20–24, 1980.
9. O'Brien AD et al.: *J Immunol* 123:720–724, 1979.

chapter 45

The Murine Ocular Response to *P. aeruginosa*

Immunological Studies

L.D. Hazlett and R.S. Berk[a]

Mice (Swiss-Webster and DBA/2) ocularly challenged with *P. aeruginosa* have been termed resistant if they are able to restore corneal clarity after bacterial challenge. Other strains of mice (BALB/c and C57B1/6) mice are termed susceptible, since the infection results in corneal perforation and often in eye shrinkage.[1,2] Several genetic loci govern susceptibility or resistance to Pseudomonas eye infection,[2,3] but the mechanism of the process remains incompletely defined.

Previous studies have shown that cyclophosphamide depression of circulating PMNs in resistant mice followed by corneal challenge with *P. aeruginosa* results in a loss of ocular resistance and death.[4] Cleavage products of complement components, particularly C3 and C5, are significant in neutrophil chemotaxis, immune adherence, and metabolic activation. In this regard, this report reviews the role of these complement components in the regulation of susceptibility and resistance to Pseudomonas ocular infection and also describes the potential role of cellular and humoral immunity as protective factors.

Materials and Methods

Mice. DBA/2 (adult and retired breeders) and C57BL/6 mice were obtained from Cumberland View Farms (Clinton, TN), or

Departments of Anatomy and [a]Immunology/Microbiology, Wayne State University, Detroit, Michigan.

from Jackson Laboratories (Bar Harbor, ME). Swiss-Webster mice were obtained from Harlan Industries (Indianapolis, IN). Female normocomplementemic B10.D2n/Sn and congenic C5 deficient B10.D2o/Sn mice were purchased from Jackson Laboratories (Bar Harbor, ME).

Verification of C5 deficiency. Previous studies have established that DBA/2, B10.D2o/Sn and Swiss-Webster mice were genetically deficient in the fifth component (C5) of complement while B10.D2n/Sn mice were normocomplementemic.[5,6]

Cobra venom factor treatment. Partially purified cobra venom factor (CoVF) was administered intraperitoneally (ip) at a dose of 200 units/kg (Cordis Laboratories, Miami, FL) as described before.[5] Control mice received an equivalent volume of the appropriate diluent. Plasma C3 levels were monitored on Ouchterlony double diffusion gels as reported.[5]

Bacterial challenge. Cell suspensions of *P. aeruginosa* strain 19660 were prepared and mice were inoculated intracorneally as previously described.[1]

Antibody titers and classes. ELISA analysis was used to measure both antibody level and class.[7] The antigen used was heat-inactivated *P. aeruginosa* strain 19660. Rabbit antiserum to mouse IgG, IgM, and IgA was purchased from Bionetecs Laboratory Products, Kensington, MD.

Spleen cell transfer. DBA/2 mice were sacrificed, and the spleen was surgically removed and placed in RPMI 1640 medium

(GIBCO, Grand Island, NY) or balanced salt solution (BSS). Spleen cells were prepared and 1.8×10^8 CFU were injected via the tail vein into DBA/2 mice. Mice were challenged with *P. aeruginosa* (19660) 24 or 96 hours later.

Flagella. Susceptible C57BL/6 mice were immunized by intradermal (id) inoculation of 1 g of purified flagella prepared from *P. aeruginosa* strain 1244 (courtesy of T. Montie).[8] The mice were ocularly challenged 2 weeks later with 1.0×10^8 CFU of strain 1244.

Results

Course of ocular infection in C3 depleted mice

The course of ocular infection with Pseudomonas has been examined in DBA/2 mice depleted of C3 following treatment with CoVF.[5] In brief, these experiments determined that depletion of C3 during the course of Pseudomonas eye infection altered the response of resistant DBA/2 mice. At 6-weeks after infection, 10 of 15 normal mice exhibited only slight corneal opacity. In contrast, 19 of 25 C3-depleted mice had perforated corneas and eye shrinkage. Six of 25 C3-depleted mice regained corneal clarity. A single inoculation of CoVF given 24 hours prior to Pseudomonas corneal challenge similarly altered the response of DBA/2 mice.

In both experiments, a small but consistent proportion of normal DBA/2 mice did not fully recover from Pseudomonas corneal infection. To be certain that these mice were capable of clearing such infections, mice from the experiment described above were challenged with Pseudomonas on the contralateral, unchallenged eye following recovery of normal C3 levels. More than 90% of mice challenged in this manner regained corneal clarity in 4 weeks. To show the correlation between C3 depletion and altered response to *P. aeruginosa* eye infection in DBA/2 mice, these same mice were again depleted of C3 and rechallenged on the recovered, contralateral eye. The results again indicated that depletion of C3 rendered previously resistant DBA/2 mice susceptible to Pseudomonas corneal infection with subsequent corneal perforation and eye shrinkage.

The restored resistance of contralaterally rechallenged DBA/2 mice whose C3 levels were normal and whose antibody titer to *P. aeruginosa* was 1:2000, suggested that some degree of ocular immunity had developed as a result of the first corneal infection. To explore this possibility, susceptible C57BL/6 mice were unilaterally infected with *P. aeruginosa* strain 19660. As expected, all of the mice (35/35) were blind within 4–5 weeks. At 6 weeks after the first challenge, the contralateral eye of these mice was infected. Of these, 20/35 (57%) were able to regain corneal clarity. In contrast, control C57BL/6 mice not infected unilaterally and similarly challenged with Pseudomonas on the contralateral eye, failed to regain corneal clarity.

Effects of circulating antibody

An attempt was made to determine if circulating antibody to *P. aeruginosa* was formed following corneal infection. Five weeks after ocular challenge, 14-week-old C57BL/6 mice were exsanguinated and the sera pooled. ELISA analysis, using an alkaline phosphatase assay, showed serum titers ranging from 1:1500 to 1:2000. Using rabbit antimouse antiserum, net titers above background controls of 1:1600 for IgG and 1:100 for IgM were obtained. No significant serum IgA was detectable.

Protection using flagella as an immunogen

The above studies indicated that corneal immunity of the contralateral eye is due in part to production of specific anti-Pseudomonas antibody. Rather than using whole organisms for immunization, flagella purified from *P. aeruginosa* strain 1244 were used to immunize C57BL/6 mice. Animals were given a single (ip) 1 g dose and challenged 2 weeks later. Protection varied from experiment to experiment, but a substantial number (up to 56%) of the mice were able to regain a clear cornea. Multiple injections of flagella weekly for 2–3 weeks prior to infection, did not substantially increase protection. These studies are yet to be repeated with flagella purified from strain 19660.

TABLE I

Effect of Spleen Cell Transfer from 8-Week-Old Resistant DBA/2 to 17-Month-Old Susceptible Aged DBA/2 Mice

Mouse strain	No. of mice	Spleen cells/ RPMI inoculum[a]	Pseudomonas challenge	Corneal response (4 weeks later)[b]				
				Clear	Slight	Full opacity	Phthisis	Death
DBA/2	8	RPMI	1.8×10^8 CFU				7	1
DBA/2	9	RPMI	1.8×10^8 CFU				8	1
DBA/2	12	1.2×10^8 CFU	1.8×10^8 CFU	3	8	1	0	0
DBA/2	9	1.2×10^8 CFU	1.8×10^8 CFU	6			3	0

[a] Total volume = 0.5 ml.

[b] Number of mice showing the indicated response.

Spleen cell transfer

Spleen cells were transferred via tail vein injection from resistant 8-week-old donor DBA/2 into recipient 17-month-old (aged) DBA/2 female retired breeders that are susceptible to ocular *P. aeruginosa* challenge (Hazlett and Berk, unpublished data). Aged mice were challenged with Pseudomonas 96 hours after cell transfer. Table I shows that 52% of the recipient mice recovered from bacterial challenge. These data were significant with $X^2 = 24.9$, $p < 0.001$. The reciprocal of this experiment was performed also. Spleen cells from susceptible 17-month-old DBA/2 females were transferred into resistant 8-week-old DBA/2 animals. Table II shows that 90% of the mice were susceptible to Pseudomonas challenge, whereas 100% of the similarly challenged controls recovered clear or nearly clear corneas. These data were significant with $X^2 = 16.4$, $p < 0.001$.

Discussion

The work described reflects our endeavors to explore further the role of humoral and cellular immunity as regards resistance or susceptibility to Pseudomonas eye infection. The murine response to ocular challenge has been shown to be dependent upon genetic background,[2,3] animal age,[9-12] and immune status.[4,5,13] Studies have shown that C3 complement has an essential role in regulating the murine response to ocular bacterial infection, whereas C5 does not appear to be critical.[5]

C3 depletion of resistant DBA/2 mice diminished the ability of these mice to regain corneal clarity following Pseudomonas challenge. After recovery of C3 levels, if the contralateral eye of these mice was challenged, all of them recovered corneal clarity. If C3 was depleted and the contralateral recovered eye again challenged, this depletion in C3 also resulted in susceptibility, despite an antibody titer of 1:2000. This suggested that the role of anti-Pseudomonas antibody in resistance be examined. When susceptible C57B1/6 mice were ocularly challenged with Pseudomonas, none of the mice regained corneal clarity. At 4 weeks, more than 50% of these same mice challenged on the contralateral eye regained corneal clarity. ELISA analysis showed serum antibody ti-

TABLE II

Effect of Transfer of Spleen Cells from 17-Month-Old Susceptible Aged DBA/2 to Resistant 8-Week-Old DBA/2 Mice

Mouse	No. of mice	Spleen cells/BSS[a] inoculum	Pseudomonas challenge	Corneal response (4 weeks later)[b]			
				Clear	Slight	Full opacity	Phthisis
DBA/2	10	BSS	1.8×10^8 CFU	8	2	0	0
DBA/2	10	1.3×10^8 CFU	1.8×10^8 CFU	1	0	0	9

[a] Total volume = 0.5 ml.

[b] Number of mice showing the indicated response.

ters of at least 1:1500, predominantly of the IgG class. Others also have shown that immune serum offers some protection to ip[14,15] and ocular Pseudomonas challenge.[16]

P. aeruginosa also inhibits cellular immunity as evidenced by prolonged survival of skin homografts in humans and laboratory animals,[17,18] suppression of tuberculin skin reactions in guinea pigs,[17] and depression of contact sensitivity to oxazolone in mice.[19,20] The mechanism remains unclear, although various reports have implicated suppression of effector T-cells, activation of B-suppressor cells, and development of adherent, macrophage-like suppressor cells in spleen and peritoneal cavity.[21]

Spleen cell transfer techniques were used to explore the role of cellular immunity in age-related susceptibility to ocular challenge. Spleen cells from resistant adult DBA/2 mice conferred some protection when transferred to 17-month-old susceptible aged mice, and, conversely, aged spleen cells from susceptible 17-month-old DBA/2 mice transferred to resistant adults significantly reduced their resistance to bacterial infection. Although the number of mice used in these studies was small, the data appear promising and suggest that further work in this area will aid our understanding of age-related increased bacterial susceptibility.

Acknowledgments

This study was supported by Public Health Service grants EY02986,EY01935 and in part by Core Vision grant EY04068 from the National Eye Institute.

The authors are indebted to Drs. R. Cleveland, M. Leon, C. Hinman and T.C. Montie for their advice and assistance.

References

1. Hazlett LD et al.: *Ophthalmic Res* **8**:311–318, 1976.
2. Berk RS et al.: *Infect Immun* **26**:1221–1223, 1979.
3. Berk RS Beisel K et al.: *Infect Immun* **34**:1–5, 1981.
4. Hazlett LD, Rosen D et al.: *Invest Ophthalmol* **16**:649–652. 1977.
5. Cleveland RP et al.: *Invest Ophthalmol Vis Sci* **24**:237–242.
6. Nilsson UR and Müller-Eberhard HJ *J Exp Med* **125**:1–16, 1976.
7. Voller A et al.: *The Enzyme-Linked Immuno-Sorbent Assay*. Flowline Publications, Guernsey, Europe.
8. Montie TC et al.: *Infect Immun* **35**:281–88, 1982.
9. Hazlett LD et al.: *Infect Immun* **20**:25–29, 1978.
10. Berk RS et al.: *Infect Immun* **33**:90–94, 1981.
11. Hazlett LD et al.: *Invest Ophthalmol Vis Sci* **19**:694–697, 1980.
12. Hazlett LD et al.: In: *Proceedings IV International Symposium on the Structure of the Eye* Joe G. Hollyfield (Ed.). Elsevier North Holland, New York, 1982, pp. 279–296.
13. Hazlett LD and Berk RS *Infect Immun* **22**:926–933, 1978.
14. Fisher W: *J Infect Dis* **136**:181–185, 1977.
15. Sokalaska M and Maresz-Babczyszyn: *Arch Immunol Ther Exp* **29**:643–647, 1981.
16. Gerke JR and Nelson JS: *Invest Ophthalmol Vis Sci* **16**:76–80, 1977.
17. Floersheim GL et al.: *Clin Exp Immunol* **9**:241–247, 1971.
18. Stone HH et al.: *Surg Gynecol Obstet* **124**:1067–1070, 1967.
19. Colizzi V et al.: *Med Microbiol Immunol* **167**:181–188, 1979.
20. Colizzi VC et al.: *Infect Immun* **21**:354–59, 1978.
21. Petit JC et al.: *Infect Immun* **35**:900–908, 1982.

chapter 46

Aspects of the Immunology of Lymphoproliferative Pseudotumors of the Orbit

Alec Garner,[a] Amjad H.S. Rahi,[a] and John E. Wright[b]

It is well-known that swellings in the orbit attributable to proliferation and accumulation of lymphoid cells cover a spectrum that extends from unequivocal chronic inflammation to frank malignant neoplasia.[2,8] Due largely to the introduction of immunological methods of investigation, the intervening "gray area" of clinically and histologically indeterminate lesions is gradually being whittled down. In this presentation we describe some of the interim findings in a prospective study of all patients presenting to the Orbital Clinic of Moorfields Eye Hospital with proptosis due to presumed inflammatory pseudotumor.

Methods

Thus far 19 patients have been investigated. In each case biopsy was performed, and the tissue was divided into three parts for routine light microscopy, for transmission electron microscopy, and for the preparation of monolayer cell impressions. The latter involved pressing the unfixed tissue against a dry glass slide with immediate air drying. The slides were examined by epifluorescent microscopy for the differential distribution of kappa and lambda immunoglobulin light chains.

[a]Department of Pathology, Institute of Ophthalmology, University of London, and [b]Moorfields Eye Hospital, London.

At the time of biopsy, blood was collected for immunological studies. Tests of T-cell function involved lymphocyte transformation in response to phytohemagglutinin and tuberculin stimulation and assessment of their prevalence in the circulating blood using conventional T-cell rosetting and immunofluorescence techniques. Serological testing included the determination of IgG, IgA, and IgM levels, and the detection of autoantibodies to nuclear and mitochondrial antigens, smooth muscle, gastric mucosa, and reticulin.

Discussion

The immunological findings in individual cases are provided in Table I.

There is no definite indication in the results thus far obtained of any correlation between the histological categorization and the immunological data. Nevertheless, it is noteworthy that seven of the 19 cases had peripheral blood T-cell counts significantly below the lower end of the normal range of 65–75%. In one patient, treated with systemic corticosteroids on the basis of a histological diagnosis of lymphoid hyperplasia and in whom the immunological tests were repeated, it was found that the proportion of T-cells in the blood increased over a 6-month period from 58% to 72%. This coincided with regression of the proptosis. Despite the reduced proportion of T-cells,

TABLE I

Immunological Findings in Patients with Orbital Pseudotumors

Case No.	Circulating T-cells		Serum immunoglobulins (mg/dl)			Auto-antibodies	Tissue Ig	Histopathology
	% of total lymphocytes	Trans. Index	IgG	IgA	IgM			
1	68		1240	232	110	—	Poly	Chronic inflammation
2	67		1670	322	63	—	Poly	Lymphoid hyperplasia
3	58		860	95	456	SMA	Mono	Lymphoid hyperplasia
4	54	76 PHA	822	118	706	SMA	Poly	Chronic inflammation
5	44	5 PHA	1130	110	178	—	Mono	Lymphoid hyperplasia
6	82	9 PHA	1430	250	63	—	Poly	Lymphoid hyperplasia
7	53	3 PPD	1230	413	122	SMA	Poly	Chronic inflammation with fibrosis
8	73	21 PHA 10 PPD	1700	319	112	—	Poly	Vasculitis
9	57	7 PHA	970	82	161	ANA	Poly	Chronic inflammation
10	63	11 PHA	1560	411	84	—	Poly	Chronic inflammation
11	60		803	279	65	ANA	Mono	Lymphoid hyperplasia
12	65	38 PHA	908	225	159	—	Mono	Lymphoid hyperplasia
13	60	6 PHA	781	154	64	—	Poly	Chronic inflammation
14	71	22 PHA	828	181	74	ANA	Mono	Lymphoid hyperplasia
15	62	49 PHA	985	499	109	SMA	Mono	Lymphoid hyperplasia
16	62	35 PHA	1450	185	233	AR	Poly	Lymphoid hyperplasia
17	53		471	86	66	—	Poly	Lymphoid hyperplasia
18	60		1390	482	264	—	Poly	Chronic inflammation (granulomatous)
19	45		1030	184	376	—	Poly	Lymphoid hyperplasia

Chronic inflammation denotes infiltration with lymphocytes and other leucocyte types with or without fibrosis. Lymphoid hyperplasia indicates an overwhelming predominance of lymphocytes. Mono = antibody confined to a single type of light chain. Poly = antibody of more than one sort. PHA = phytohemagglutinin. PPD = purified protein derivative (tuberculin). SMA = smooth muscle antibody. ANA = antinuclear antibody. AR = antireticulin antibody.

the normal lymphocyte transformation index in all patients so tested suggests that there was no defect of function. Whether the finding that four of the seven patients with low T-cell counts also had circulating auto-antibodies (compared with four of the 12 with normal counts) is significant remains to be seen. If it is, the possibility that the T-cell reduction relates to the suppressor cell fraction will need to be considered.[4] Another possible indicator of diminished suppressor T-cell activity is the finding that of 13 patients with normal IgM levels 12 had normal T-cell counts, whereas four of eight patients with elevated IgM levels (above 200

mg/dl) had reduced T-cell numbers. The contrasting and restricted association of low IgA levels with diminished proportions of circulating T-cells (all five patients with low IgA titres had low T-cell counts) is also intriguing. The finding of autoantibodies in three of the five cases is, however, in line with other experiences with selective IgA deficiency.[3] It is reasoned that the lack of IgA involves an enhanced predisposition to assault by both infective and noninfective antigens resulting in an increased incidence of infection, autoantibody formation, and neoplasia. In an analysis of over 4000 patients suffering from a variety of ocular disorders, Addison and Rahi[1] found that IgA levels below 60 mg/dl occurred in 15%. This supports the concept of increased susceptibility to inflammatory disease and to neoplasia (there was a 17% incidence of reduced IgA in patients with malignant uveal melanoma) in individuals with low levels of circulating IgA. IgA deficiency, particularly in minor degree, is a common finding, and it will be interesting to see whether such persons have an increased tendency to develop lymphoproliferative and inflammatory lesions of the orbit.

The potential for the usefulness of monoclonal antibody detection in differentiating between neoplastic and inflammatory disorders has been confirmed by others.[5-7,9] In the present study, six patients showed a monoclonal antibody pattern in the lymphoid cells of their orbital lesions, but there was no discernible correlation with circulating antibody findings. The histology of these cases revealed a predominance of lymphocyte proliferation in each of them with a relative absence of other types of leukocytes. On the other hand, 11 of the 13 patients showing a polyclonal antibody tissue response had clearly inflammatory lesions and only two showed an indeterminate lymphoid hyperplasia. The inference is that

lesions producing just one type of antibody stem from a single cell and are neoplastic. Such lesions may remain relatively innocuous for many years,[6] but they retain a potential for ultimate dissemination. A polyclonal antibody response represents an essentially inflammatory reaction.

In conclusion, it would be foolish to read too much into these preliminary results, but already there are prospects of improved diagnostic precision and understanding of the indeterminate group of lymphoproliferative disorders so shamefully labeled as pseudotumors.

References

1. Addison DJ and Rahi AHS: *Trans Ophthalmol Soc UK* **101**:9–11, 1981.
2. Chavis RM et al.: *Arch Ophthalmol* **96**:1817–1822, 1978.
3. Doniach D and Bottazo GF: In *Clinical Immunology Update 1981* Franklin EC (Ed.). Elsevier, North Holland, Amsterdam, p. 95.
4. Elson CJ: *Immunology Today* **1**:42–45, 1982.
5. Hautzer NW and Nikolai V: *IRCS Med Sci: The Eye* **10**:229, 1982.
6. Jakobiec FA et al.: *Arch Ophthalmol* **100**:84–98, 1982.
7. Knowles DM and Jakobiec FA: *Human Pathol* **13**:148–162, 1982.
8. Knowles DM et al.: *Am J Ophthalmol* **87**:603–619, 1979.
9. Saraga P et al.: *Human Pathol* **12**: 713–723, 1981.

chapter 47

A Mechanism for Immunoglobulin-A Plasma Cell Lodging in Lacrimal Gland

Rudolph M. Franklin and Dennis W. McGee

The majority of immunoglobulin in tears is of the immunoglobulin-A (IgA) isotype, which is produced mainly by plasma cells of the lacrimal gland (LG).[1,2] The mechanism responsible for the appearance of predominantly IgA plasma cells in the LG is unknown but may involve a glandular influence on circulating B-cells already committed to the IgA isotype.[3] The selective lodging of these committed B-lymphocytes eventually leads to differentiation of the cells into IgA secreting plasma cells commonly observed in LG as well as other secretory tissues.[4]

Previous *in vivo* studies suggested a glandular substance capable of interacting with B-cells and leading to an increase in IgA plasma cells in the vicinity of glandular tissue.[5,6] In the present experiments, the effects of LG epithelial cells on lymphocyte populations were examined in 4-day cultures in order to observe more directly epithelial cell influences on B-cell differentiation. We observed direct suppression of B-cell differentiation on certain B-cell populations, but noted an increase in IgA secreting cells from Peyer's patch. These experiments demonstrate that LG epithelial cells can have a direct influence on lymphocytes, which could explain the large proportion of IgA plasma cells in the LG.

Department of Ophthalmology, Louisiana State University Eye Center, LSU Medical Center School of Medicine, New Orleans, Louisiana.

Materials and Methods

Animals. Female BALB/c mice weighing 16–20 g were used in all experiments (Dominion Labs., Dublin, VA). The mice were killed by cervical dislocation, and various tissues were obtained immediately.

Isolation of cells from lymphatic tissues. Spleen (SP), Peyer's patch (PP), and mesenteric lymph node (MLN) were removed from the mice and placed in Hanks' balanced salt solution (HBSS) supplemented with 5% fetal calf serum. The tissues were gently teased apart and centrifuged through a Ficoll-Hypaque gradient.[7] The resulting lymphocyte band was washed once in buffer and then treated with an erythrocyte-lysing buffer[7] before washing a second time. Cell counts and viability were obtained using the trypan blue exclusion method and counting with a hemacytometer.

Isolation of mouse lacrimal gland epithelial cells. Single cell suspensions of the exorbital gland were prepared by a modification of a previously described method[8] in which the glands were sequentially treated with 0.188% collagenase and 0.15% hyaluronidase, then 2.0% pronase. Afterwards, the cells were washed once in buffer and resuspended in culture medium.

Co-culture conditions. The culture medium consisted of RPMI 1640 with 200 units/ml penicillin, 200 µg/ml streptomycin, 2mM L-glutamine, 48 µg/ml gentamicin, 5×10^{-5} M 2-mercaptoethanol, and 20% fetal calf serum. Lymphoid cells and epithelial cells were diluted in culture medium and combined to yield 5 x 10^5 lymphocytes/ml and 4.17 x 10^4 large granular epithelial cells/ml (approximately 2 x 10^5 total LG cells/ml) in a 1 ml well of a Linbro 24-well plastic tissue culture plate. Generally, six such wells were prepared for each assay. All cultures were driven with lipopolysaccharide (LPS) at a final dose of 52.5 µg/ml, a dose determined in earlier studies to give a maximal response for PP, SP, and MLN. The cultures were then incubated at 37°C in a 5% CO_2-95% air, humid atmosphere. After 4 days, wells containing cells stimulated under the same conditions were combined, washed twice in HBSS, and finally suspended in HBSS for the plaque assay.

Protein A plaque assay for immunoglobulin production. Staphylococcal Protein A (SpA, Pharmacia, Piscataway, NJ) was coupled to sheep red blood cells (SRC) by combining 10% washed SRC, 1 mg/ml (in 0.9% saline) SpA, and 1:100 dilution of 1% $CrCl_3$-$6H_2O$ (in 0.9% saline) at a ratio of 1.0:0.15:0.15 (vol:vol:vol), then incubating in a 37°C water bath for 60 minutes. After washing twice in 0.9% saline, the coupled cells were suspended to 30% in HBSS and a 50 µl aliquot was added to 50 µl of the developing antibody (rabbit antimouse IgA, IgG, or IgM diluted to 1:8 in HBSS, Bionetics Labs, Kensington, MD). Next, 100 µl of the cell suspension and 50 µl of guinea pig complement (diluted 1:3 in HBSS, Gibco, Grand Island, NY) were added, and the mixture was transferred to 0.5 ml of 0.5% agarose (in HBSS). A 200 µl aliquot was then placed onto a 15 mm x 100 mm plastic culture dish and a 22 mm x 30 mm cover slip was placed on it. Three aliquots were placed on each culture dish. After incubation for a minimum of 6 hours at 37°C in a 5% CO_2–95% air, humid atmosphere, the number of plaques formed was counted.

For each group of six wells, sufficient cells were available for counting plaques of the three major immunoglobulin isotypes. The number of plaques was adjusted to 10^6 lymphocytes after 4 days in culture. The number of plaques of each isotype was compared between cultures containing epithelial cells and cultures without epithelial cells. The percent change between cultures with and without epithelial cells represents the amount of stimulation or suppression obtained in each co-culture.

Depletion of T-cells in the lymphoid tissues. In some experiments, SP and PP were depleted of T-cells prior to co-culture by treatment with monoclonal anti-Thy 1.2 serum (New England Nuclear, Boston, MD) and guinea pig complement. The method of T-cell depletion used was similar to a published method.[9] Cell numbers for co-cultures were adjusted in these T-cell depleted populations by considering the depletion of T-cells and, therefore, the resulting increase in B-cells.

Results

Lacrimal gland cell dissociation. Each animal yielded 2.0 to 2.5 x 10^6 viable cells. The heterogenous population of LG cells showed a viability of 75–85% following the enzyme treatment. Wright-Giemsa stained smears displayed numerous cell types. The two predominant cell types showed a relatively high ratio of cytoplasm to nucleus. The larger of these two cell types had a granular appearing cytoplasm, and represented 21% of the cells. The other large cell type, representing 46% of the cells, displayed a relatively clear cytoplasm. The remaining 33% of the cells were a mixture of lymphocytes and unidentified cells (Fig. 1).

Co-culture of lacrimal gland epithelial cells with lymphoid cells. After 4 days of culture, at a ratio of one LG epithelial cell to 12 lymphocytes, or controls consisting of no epithelial cells, the recovery of cells for all cultures ranged between 110 and 249% as compared with starting cell numbers. In most instances, the percent recovery was slightly lower for cultures containing LG epithelial cells compared with controls. The viability of recovered cells was approximately 70%.

In the presence of LG epithelial cells, SP lymphocytes showed a significant suppression for the three major isotypes. MLN lymphocytes demonstrated a similar degree of suppression of the three isotypes. PP lymphocytes demonstrated suppression for

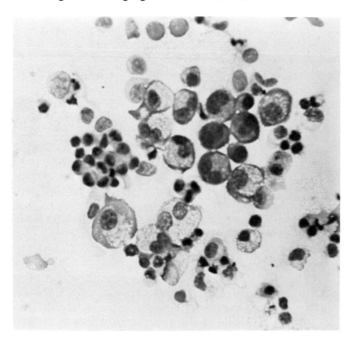

FIG. 1. Cytocentrifuge preparation of dissociated normal mouse lacrimal gland. Two larger cell types, one with bilobed nuclei, are epithelial cells. Smaller mononuclear cells are a mixture of plasma cells, lymphocytes, and macrophages. (Wright-Giemsa stain × 512.)

IgM and IgG with an increase in the number of plaques relative to control for IgA (Table I).

The dose-response relationship of LG epithelial cells on SP cells showed a lessening of suppression at lower relative concentrations of epithelial cells to lymphocytes. At a ratio of one epithelial cell to 40 lymphocytes, no difference was observed between controls and cells stimulated for IgA and IgM. IgG responses were still slightly suppressed. PP lymphocytes at the same ratios showed a decreasing level of suppression for IgM and IgG, but even at a ratio of 1:40 significant suppression still existed. IgA responses of SP lymphocytes showed no in-

crease at the 1:24 dilution and only a slight increase at 1:40. In general, the 1:12 ratio of epithelial cells to lymphocytes showed the highest degree of suppression for IgG and IgM in both SP and PP, and showed the maximum IgA suppression in SP but IgA stimulation in PP.

T-cell depletion experiments. A series of experiments was performed with SP and PP lymphocyte populations depleted of T-cells. After one treatment with anti-Thy 1.2 + C, SP showed a 28% and PP a 16% depletion of lymphocytes. When the remaining cells were co-cultured with LG epithelial cells at a ratio of one epithelial cell to 12 lymphocytes, SP lymphocytes showed a similar de-

TABLE I

Four-Day LPS-Driven Cultures of Lymphocytes with Lacrimal Gland Epithelial Cells

Lymphocyte source	Isotype–plaque assay		
	IgA	IgG	IgM
Spleen	−60[a](6.2)[b]	−55(4.7)	−45(4.4)
Mesenteric lymph nodes	−48(9.5)	−42(9.8)	−45(11)
Peyer's patch	+35(8.5)	−30(9.2)	−50(7.9)

[a] Percentage change between control (no epithelial cells) and experimental cultures.

[b] S.E.M.

TABLE II
Four-Day Cultures of T-Cell-Depleted Lymphoid Tissues and Lacrimal Gland Epithelial Cells

Lymphocyte source	Percent change		
	IgA	IgG	IgM
Spleen	−57(11.3)[a]	−51(11.4)	−47(5.1)
Peyer's patch	0	−27	−29

[a] S.E.M.

gree of suppression of all three isotypes. PP showed a suppression of IgG and IgM with no change from control levels for IgA (Table II).

Discussion

The phenomenon of selective lodging of IgA-committed B-cells to mucosal tissues has been observed repeatedly but is not yet explained.[4,10] Although mucosally applied antigen may play a role in influencing IgA cells to lodge in certain tissues, many experiments have demonstrated lodging without antigen.[5,11,12] Other experiments have examined the role of high endothelial venules[13] and of secretory component,[14] but these approaches have not provided an ex-

planation for the predominance of IgA plasma cells in glandular tissues. Previous *in vivo* experiments suggested a glandular factor capable of influencing mixed B-cell populations to express a higher proportion of IgA-producing cells.[5,6] The present experiments examined the ability of LG cells to affect directly lymphocytes from SP, PP, and MLN. It was observed that glandular cells suppressed IgG and IgM production directly, but IgA responses were increased using PP lymphocytes in the presence of T-cells. In SP and MLN, IgA was directly suppressed by LG cells.

It has been suggested that T-cells might participate in regulating IgA expression within mucosal tissues.[15] A high level of T-cell helper activity specific for IgA was

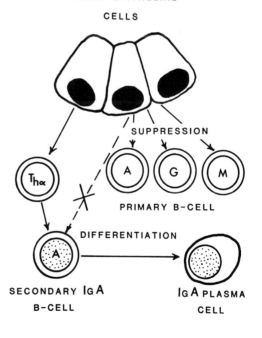

FIG. 2. Hypothesis for selective lodging of IgA B-cells in lacrimal gland (see text).

found in PP but not in SP, using an LPS-driven culture of various mixes of T- and B-cells. The same investigators also found that T-suppressor cells existed in PP for IgG and IgM and in SP for all three isotypes. In the experiments reported here, it was shown that the increase in IgA responses in PP were related to T-cells. Since no attempts were made to mix PP T-cells and SP B-cells, it is uncertain if differences also exist for B-cells.

A difference for B-cells between PP and SP is suggested, however, because SP IgA responses are dramatically depressed in the absence of T-cells, compared with PP responses (Table II). Tseng[16] has demonstrated an increased proportion of secondary IgA B-cells in PP compared with other tissues. It is possible that the high proportion of secondary IgA B-cells responds to the T-cell help from PP, and the primary B-cells are more easily suppressed by LG epithelial cells.

Based on the above considerations, a sequence of events can be proposed to explain the selective lodging of IgA B-cells in the LG. Primary B-cells of all isotypes, randomly passing through the LG, are directly prevented by epithelial cell factor(s) from differentiating into plasma cells. Specific IgA helper T-cells, activated by LG epithelial cell factor(s), stimulate secondary IgA cells to differentiate into plasma cells (Fig. 2).

Experiments reported here demonstrate the first example of epithelial cell effects on lymphocyte differentiation. Certainly the details of the proposed mechanism for the preponderance of IgA plasma cells in LG need clarification, but the ability of LG epithelial cells to influence B-cell responses is established.

Acknowledgment

Supported in part by NEI grant R01 EY 03028.

References

1. Franklin RM et al.: *J Immunol* **110**:984-992, 1973.
2. Franklin RM et al.: *Invest Ophthalmol Vis Sci* **18**:1093-1096, 1979.
3. Weisz-Carrington P et al.: *J Immunol* **123**:1705-1708, 1979.
4. Lamm ME: *Adv Immunol* **22**:223-290, 1976.
5. Franklin RM et al.: *J Exp Med* **148**:1705-1710, 1978.
6. Franklin RM: In *Proceedings, Immunology of the Eye; Workshop III,* Suran A et al. (Eds.). Special Supplement Immunology Abstracts, Information Retrieval, Inc., Washington, DC, 1981, pp 181-183.
7. Mishell BB and Shiigi SM: *Selected Methods in Cellular Immunology*. W.H. Freeman and Co., San Francisco, 1980.
8. Wienhes GJ and Prop FJA: In *Tissue Culture: Methods and Applications*. Kruse PF and Patterson MK (Eds.). Academic Press, New York, 1973.
9. Mongini PKA et al.: *J Exp Med* **153**:1-12, 1979.
10. Craig SW and Cebra JJ: *J Exp Med* **134**:188-200, 1971.
11. Pierce NF and Cray WC Jr: *J Immunol* **127**:2461-2464, 1981.
12. Parrott DMV and Ferguson A: *Immunology* **26**:571-588, 1974.
13. Husband AJ: *J Immunol* **128**:1355-1359, 1982.
14. McWilliams M et al.: *J Immunol* **115**:54-58, 1975.
15. Elson CO et al.: *J Exp Med* **149**:632-643, 1979.
16. Tseng J: *J Immunol* **128**:2719-2725, 1982.

chapter 48

Complement and Lactoferrin in Normal Tears

A. Kijlstra, R. Veerhuis, and S.H.M. Jeurissen

The complement system is one of the main effector mechanisms of the humoral immune response.[1] Whether complement is present in the tear film is not clear. Earlier investigators were not able to detect C1 and found undetectable or low concentrations of C3 and C4 in normal human tears.[2,3] In contrast to these findings, Yamamoto and Allansmith[4] reported a number of experiments showing that human tears contained an intact classical and alternative complement system. In light of the controversial findings mentioned above, it was decided to reinvestigate the question as to whether tears contain complement activity.

The study reported here shows that human tears do not contain hemolytic classical complement activity but instead contain a potent complement inhibitor, which was identified as tear lactoferrin.

Materials and Methods

Tear samples were collected from staff members of our institute after stimulation with tear gas or pressurized air flow. Complement activity in tears was measured as described earlier.[5,6] The effect of tears or lactoferrin on serum complement activity was performed as reported elsewhere.[5-7] Lactoferrin was isolated from human tears by DEAE-52 anion exchange chromatography.[6]

Department of Ophthalmo-Immunology, The Netherlands Ophthalmic Research Institute, Amsterdam, The Netherlands.

Results

To investigate whether human tears contain hemolytic complement activity, a classical CH50 assay was performed. Despite small modifications to increase the sensitivity of the assay, no hemolytic activity was detected in tears, whereas an 800-fold dilution of human serum still contained detectable complement activity.

A reason for these negative findings could be the presence of complement inhibitors in tears. This was subsequently investigated by mixing tears with serum and measuring the effect of the tears on the complement mediated lysis of antibody-coated red cells. A representative experiment is shown in Table I. Tears completely inhibited serum (0.5%) complement activity up to a 20-fold tear dilution. Tears diluted 1/160 still caused approximately 50% complement inhibition. Thus these findings showed that although tears could possibly contain complement, this could not be detected due to the presence of strong complement inhibitors.

The observation that the inhibitory effect could be detected at high tear dilutions facilitated its isolation. Purification of the complement inhibitor resulted in a fraction containing anticomplementary activity, showing one band on SDS-polyacrylamide gel electrophoresis, and this was identified as lactoferrin by immunochemical means.[6] Figure 1 shows a dose response curve of the anticomplementary effect of tear lactoferrin and human tears. Between 10 and 20 μg of lactoferrin in this assay causes a 50% inhi-

TABLE I

The Effect of Tears on Serum Hemolytic Complement Activity[a]

Tear dilution	Residual serum complement activity (units)
1/10	0.04
1/20	0.08
1/40	0.15
1/80	0.30
1/160	0.61
0	1.16

[a] 100 μl tear dilution + 100 μl human serum (0.5%) + 100 μl EAs. Complement activity was measured after 60 minutes incubation at 37° C.

bition of the complement activity. Analysis of anticomplement activity in human tears by sucrose gradient ultracentrifugation showed that lactoferrin was not the only complement inhibitor present in tears.[5] Besides the large molecular weight anticomplementary activity that is always found in the pellet of the sucrose gradient,[5] some tear samples contain a peak of anticomplement

activity, which is somewhat larger than lysozyme but smaller than lactoferrin (data not shown).

To investigate which step in the activation of the complement system was inhibited, various experiments were performed. No effects were noted up to the assembly of the C14 complex on the antibody coated erythrocytes (EAC14). Table II shows evidence that tear lactoferrin does not affect the EAC14 directly but inhibits the subsequent steps in the complement pathway. The results in Table III show that lactoferrin inhibits the assembly of the classical C3 convertase (EAC142) but has no effect on the subsequent steps in the complement cascade.

Discussion

The results described here indicate that tears do not contain detectable classical complement activity, but instead contain potent inhibitors of this pathway. One of the principal inhibitors of the complement sys-

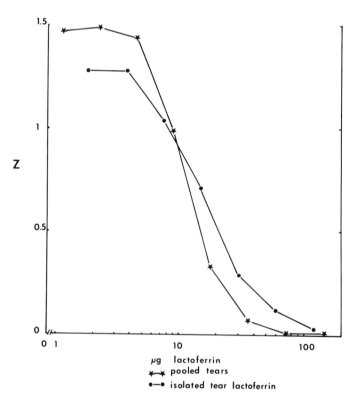

FIG. 1. The effect of different dilutions of tears or isolated tear lactoferrin on hemolytic complement activity in normal human serum. Lactoferrin or tears (100 μl) were added to 100 μl normal human serum (0.5%). After adding 100 μl antibody coated erythrocytes (10^8/ml), complement activity was measured after 60 minutes incubation at 37° C.

TABLE II
Effect of Lactoferrin on EAC14[a]

Experimental procedure				Complement Activity
EAC14 + lactoferrin	wash →	C4 deficient serum	+ buffer	2.41
EAC14 + lactoferrin	wash →	C4 deficient serum	+ lactoferrin	0.04
EAC14 + buffer	wash →	C4 deficient serum	+ buffer	2.73
EAC14 + buffer	wash →	C4 deficient serum	+ lactoferrin	0.04

[a] EAC14 were pretreated with tear lactoferrin (105 µg/ml) or buffer, and washed. C4 sites were developed by adding C4 deficient guinea pig serum (1/300) containing buffer alone or lactoferrin and incubating for 60 minutes at 37° C.

TABLE III
Effect of Lactoferrin on EAC142 Formation[a]

Experimental procedure		Complement units
EAC14 + C2 + buffer	→ C3 − C9	1.48
EAC14 + C2 + lactoferrin	→ C3 − C9	0.10
EAC14 + C2 + buffer	→ C3 − C9 + buffer	1.33
EAC14 + C2 + buffer	→ C3 − C9 + lactoferrin	1.26

[a] EAC14 were incubated with functionally purified guinea pig C2 in the presence of buffer alone or lactoferrin. EAC142 sites were developed by adding C3 to C9 (guinea pig serum 1/15 in EDTA buffer) alone or with lactoferrin.

tem in tears could be identified as lactoferrin. Due to the early inhibition of the complement system by lactoferrin,[7] the production of biologically active split products of complement activation such as C3a and C5a is prevented. This suggests that lactoferrin could have an important anti-inflammatory function during inflammatory conditions, whereby serum leaks into the tear film. The observation that high levels of lactoferrin have been found in inflammatory sites such as bronchial secretions in patients with chronic pulmonary disease[8] and in exudates of arthritic joints[9] strengthens the suggestion that lactoferrin may play a role in inflammation.

Tear lactoferrin is abundantly present in normal human tears. With a mean concentration of approximately 2 mg/ml it represents one of the main proteins in human tears.[10] The levels of lactoferrin in the tears of patients with various eye diseases are being investigated and may give more insight into the physiological role of this protein in the tear film.

In summary, the presence of hemolytic complement activity in normal human tears was investigated. Tears do not contain detectable complement activity due to the presence of strong inhibitors. One of the main inhibitors could be identified as lactoferrin. Lactoferrin was shown to block the formation of the classical C3 convertase and could thus, besides its known antibacterial properties, play an important role during external ocular inflammation.

References

1. Müller-Eberhard HJ: In *Progress in Immunology IV*. Fongerau, M and Dausset J (Eds.). 1980, pp. 1001–1024.
2. Chandler JW et al.: *Ophthalmology* **13**:151, 1974.
3. Bluestone R et al.: *Br J Ophthalmol* **59**:279, 1975.
4. Yamamoto GK and Allansmith MR: Complement in tears from normal humans. *Am J Ophthalmol* **88**:758, 1979.
5. Kijlstra A and Veerhuis R: *Am J Ophthalmol* **92**:24, 1982.
6. Veerhuis R and Kijlstra A: *Exp Eye Res* **34**:257, 1981.
7. Kijlstra A and Jeurissen SHM: *Immunology* **47**:263, 1982.
8. Zebrak J et al.: *Scand J Respir Dis* **60**:69, 1979.
9. Malmquist J et al.: *Acta Med Scand* **202**:313, 1977.
10. Kijlstra A et al.: *Br J Ophthalmol* **67**:199, 1983.

chapter 49

Studies of Complement in Corneal Tissue

Bartly J. Mondino

The complement system is an integral part of the immune system and is recognized as an important means by which immune injury and defense against invading microorganisms are mediated. Genetic deficiencies of complement in man are associated with increased susceptibility to infections and autoimmune disease, such as systemic lupus erythematosus.[1] The complement system consists of over 20 distinct plasma proteins that interact in a highly specific manner. Of the 20 individual proteins, 7 are regulatory proteins, and the remaining 13 complement components include 5 proteolytic enzymes (C1r, C1s, C2, factor B, and factor D) and 8 nonenzyme components (C1q, C3, C4, C5, C6, C7, C8, and C9).[2]

The complement system can be divided into the classical or the alternative pathway of complement activation. Antigen–antibody (IgG or IgM) complexes activate C1 (a trimolecular complex of C1q, C1r, and C1s) of the classical pathway of complement with the sequential activation of C4, C2, C3, C5, C6, C7, C8, and C9 (Fig. 1). On the other hand, endotoxin, zymosan, inulin, and aggregated IgA activate the alternative pathway, which enters the complement sequence by activating C3 instead of C1, with further reactions in the cascade being similar to the classical pathway.

As can be seen from Figure 1, C3 occupies

a pivotal position in both the classical and alternative pathways. The classical pathway C3 convertase (C4b2a) is generated when C1, activated by binding to an antibody–antigen complex, activates C4 and C2.[3] The alternative pathway C3 convertase consists of C3b, Bb (stabilized by properdin).

Activated complement is the prime mediator of tissue inflammation. *Cytolysis* results from complement-induced membrane damage to cells and can follow activation of either the classical or alternative complement pathways. The membrane attack complex consists of five proteins: C5b, C6, C7, C8 and C9. *Immune adherence* involves antibody–C3b complexes attached to antigenic cells, thereby facilitating their phagocytosis by macrophages and polymorphonuclear leucocytes that have receptors for C3b. Complement activation also results in the formation of active fragments such as C3a and C5a that have *anaphylatoxin activity* causing histamine release from mast cells or basophils. C5a also has *chemotactic activity* for polymorphonuclear leucocytes. These functions of complement are operative not only in defense against invading microorganisms but also in autoimmune tissue injury.

To date, little attention has been devoted to the role of complement in ocular diseases and especially external diseases of the eye. There have been reports demonstrating complement activity in tears[4] and aqueous humor.[5] It was believed that the cornea is probably devoid of hemolytic activity because some of the complement components

From the Department of Ophthalmology and the Jules Stein Eye Institute, UCLA School of Medicine, Los Angeles, California.

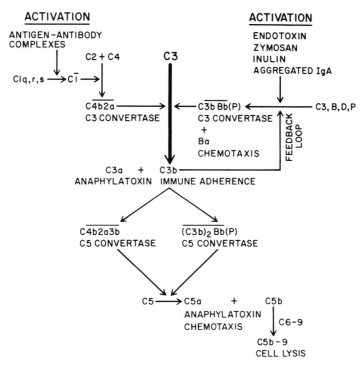

FIG. 1. Diagram of classical and alternative pathways of complement activation. A bar above a complement component indicates an active enzyme. B indicates factor B, D indicates factor D, and P indicates properdin. C3 represents the pivotal component that is acted upon by both the classical and alternative activating pathways.

are probably too large to diffuse into it. Investigators were unable to detect C3 in the cornea by radial immunodiffusion plates and concluded that C3, a relatively large molecule, probably does not enter the avascular cornea.[6]

Results and Discussion

Alternative and classical pathway components of complement in normal human donor corneas

Both classical and alternative pathway components of complement have been demonstrated in normal donor corneas using direct immunofluorescence and immunodiffusion.[7] Direct immunofluorescence was used to demonstrate C1q, C3, C4, and C5. The corneal stroma showed diffuse fluorescence for C3, C4, and C5. On the other hand, C1q (the recognition unit of the classical pathway and largest complement component) was detected only in the periphery of the cornea. C1q (molecular weight [MW] 400,000) is the largest complement component and is part of a trimolecular complex (C1q, C1r, C1s) that has an estimated MW of 647,000. The large size of this trimolecular complex may limit its diffusion into the cornea and may account for its detection by direct immunofluorescence only in the peripheral cornea.

Normal human donor corneas were also eluted in phosphate-buffered saline at 4°C for 1–4 days. Ouchterlony plates, in which the corneal eluate was reacted against antisera to complement components, disclosed the presence of C1q, C3, C4, C5, properdin, and properdin factor B. Radial immuno-diffusion was used to obtain estimates of the concentration of C3, C4, and C5 in the cornea. The mean concentrations ± S.D. in mg/ 100 g were 14.2 ± 4.9 for C3, 2.9 ± 0.9 for C4, and 1.8 ± 0.8 for C5.

Hemolytic complement activity in normal human donor corneas

The immunodiffusion technique in the previous study[7] measures breakdown products as well as intact complement and is not an assay of complement function. We sought to determine if normal human donor corneas contain functional complement with the use of 50% hemolysis (CH_{50}) of sensitized sheep red blood cells (RBCs).[8] Normal human donor corneas were minced into small fragments and eluted for 16–23 hours at 4°C. The corneal eluates were then studied for hemolytic activity of C1, C4, C2, C3, C5, C6, and C7. Sera from ten normal volunteers were also assayed for hemolytic activity. For each complement component, the mean hemolytic activity in corneas was compared with the mean hemolytic activity in sera.

These comparisons suggest that molecular weight may be a factor in determining the concentration of complement components in the cornea, because C1, the largest complement component, had the lowest hemolytic activity in corneas relative to sera, while complement components with lower molecular weights like C2 and C7 had the highest hemolytic activities in corneas relative to sera. The large size of C1 may re-strict its diffusion into the cornea, and, in the study described previously, immunoflu-orescent staining for C1q could only be demonstrated in the peripheral cornea. In summary, we showed that normal human donor corneas contain functional C1, C4, C2, C3, C5, C6, and C7 and provided normal values of hemolytic activity for further studies of complement consumption in corneal diseases (Table I).

Distribution of hemolytic complement in normal human donor corneas

Hemolytic activities in the central cornea were compared with hemolytic activities in the peripheral cornea using 50% hemolysis (CH_{50}) of sensitized sheep RBCs for each of the following complement components: C1, C4, C2, C3, C5, C6, and C7.[9] For all seven of the complement components studied, he-molytic activities in the peripheral cornea were higher than hemolytic activities in the central cornea, and the differences were statistically significant. The most striking difference was for C1, the largest complement component, which had a ratio of mean hemolytic activity in the peripheral cornea to that in the central cornea of nearly 5:1. For the other six complement components, the ratio of the mean hemolytic activity in the peripheral cornea to that in the central cornea was approximately 1.2:1.

This distribution of complement activity in the cornea suggests that the major source of complement components is the limbal vessels and that complement components diffuse from the limbus to the central cor-

TABLE I
Hemolytic Complement Activity in Normal Human Donor Corneas

	C1	C4	C2	C3	C5	C6	C7
Mean ± S.E.M. in CH_{50} units/g (N = 9)	4,113 ± 787	19,889 ± 4,763	3,996 ± 1,069	1,427 ± 271	11,040 ± 2,677	28,640 ± 9,573	45,093 ± 11,402
Percentages of hemolytic complement activity in corneas relative to sera	3.2	7.1	27.8	6.8	10.9	11.0	19.9

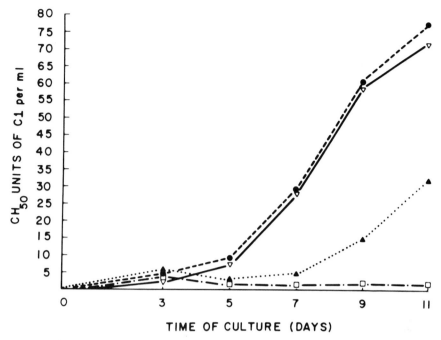

FIG. 2. Secretion of hemolytic C1 by human corneal fibroblasts into tissue culture media (ordinate) plotted as function of time of culture (abscissa). Tissue culture dishes 1 •——— —•; 2 ▽———▽; 3 ▲———▲; and 4 □—.—.□. Cycloheximide was added to tissue culture dishes 3 and 4 at day 3 but removed from dish 3 at day 5.

nea. This study was consistent with a previous study using direct immunofluorescence in which C1q was detected only in the peripheral cornea, while C3, C4, and C5 were found throughout the corneal stroma. The large size of C1 may restrict its diffusion into the cornea, and this finding may be important in peripheral inflammation and ulcers of the cornea. Because of the function of C1 as the recognition unit of the classical pathway, antigen–antibody complexes, whether formed in the cornea itself or whether derived from the tears, aqueous humor, or limbal vessels, may activate complement more effectively in the peripheral than in the central cornea.

Production of the first component of complement by corneal fibroblasts in tissue culture

Fibroblasts from lung and skin have been shown to produce C1 in tissue culture.[10] We sought to determine if corneal fibroblasts produce complement components in tissue culture and may be a potential source of complement components in the cornea. Culture media from tissue culture of human corneal fibroblasts were assayed for functional C1, C4, C2, C3, C5, C6, and C7 with the use of 50% hemolysis (CH_{50}) of sensitized sheep RBCs.[11]

The results of this study indicated that corneal fibroblasts synthesized and secreted hemolytic C1 in tissue culture. At day 0 of tissue culture, there was no detectable hemolytic C1 in the culture media, but a progressive increase was noted on days 3, 5, 7, 9, and 11 of tissue culture (Fig. 2). The increase in concentration of hemolytic C1 in the culture media corresponded to an increase in corneal fibroblast concentration that reached a peak at day 7 without a further increase thereafter. A steep increase in hemolytic C1 was noted in tissue culture media after day 7 when the corneal fibroblasts had reached confluency.

The addition of cycloheximide, an inhibitor of protein synthesis, to the culture media reversibly inhibited the increase in hemolytic C1. When the cycloheximide was removed, the progressive increase in hemolytic C1 took place. The synthesis and secretion of hemolytic C1 took place, but the synthesis and secretion of hemolytic C4, C2, C3, C5, C6, and C7 could not be detected.

The finding in this study that corneal fibroblasts have the ability to synthesize and secrete C1 in tissue culture does not prove that they do so *in vivo*. Nevertheless, corneal fibroblasts may be a potential source of C1 in the cornea. Although the large size of C1 may restrict its diffusion into the cornea, corneal fibroblasts have the potential of synthesizing this component for the central cornea, and this may account for the presence of some hemolytic activity for this large molecule in the central cornea. As the recognition unit of the classical pathway, C1 would be critical in setting off classical pathway activation of complement.

Acknowledgment

This report was supported in part by a grant from the National Institute of Health, National Eye Institute, RO1 EY02304.

References

1. Colten HR et al.: *N Engl J Med* **304**:653-656, 1981.
2. Müller-Eberhard, HJ: In *Molecular Basis of Biological Degradative Processes*. Berlin RD et al. (Eds.).Herrmann, Academic Press New York, 1978, pp. 65-104.
3. Fearon DT and Austen KF: *N Eng J Med* **303**: 259-263, 1980.
4. Yamamoto GK and Allansmith MR: *Am J Ophthalmol* **88**:758-763, 1979.
5. Chandler JW et al.: *Invest Ophthalmol* **13**:151-153, 1974.
6. Stock EL and Aronson SB: *Arch Ophthalmol* **84**:355-369, 1970.
7. Mondino BJ et al.: *Arch Ophthalmol* **98**:346-349, 1980.
8. Mondino BJ and Hoffman DB: *Arch Ophthalmol* **98**:2041-2044, 1980.
9. Mondino BJ and Brady KJ: *Arch Ophthalmol* **99**:1430-1433, 1981.
10. Reid KBM and Solomon E: *Biochem J* **167**:647-660, 1977.
11. Mondino BJ et al.: *Arch Ophthalmol* **100**:478, 1982.

chapter 50

IgE Antibodies in Ocular Immunopathology

J.H. Rockey, J.J. Donnelly, T. John, M. Khatami, R.M. Schwartzman, B.E. Stromberg, A.E. Bianco, and E.J.L. Soulsby

The role of IgE antibody in ocular immunopathology is examined in the present report in guinea pigs infected with parasites, actively immunized by systemic, intravitreal, or topical (conjunctival) routes, or passively sensitized locally or systemically, and in a population of atopic dogs.

Materials and Methods

Systemic and intravitreal injections of ascarid larvae or antigens, and passive cutaneous anaphylactic tests (7–10 day P-K reactions) were carried out in Hartley strain guinea pigs.[1] Microfilariae (mf) of *O. lienalis* were obtained from the umbilical skin of freshly killed cattle and were cryopreserved according to the method of Ham et al.[2,3] The motility of cryopreserved and fresh mf was 90% or greater. *O. lienalis* microfilarial antigen extracts were prepared from washed microfilariae homogenized in Tyrode's solution. Intracorneal injections of 10–20 μl of mf suspensions or controls were performed in inbred Strain-2 guinea pigs. Diethylcarbamazine citrate (DEC) was given orally (15

mg/kg) in 0.1–0.2 ml of sterile water as a single dose, or daily for 2 weeks.

Spleen cell suspensions were prepared in Eagle's Minimal Essential Medium and depleted of dead cells with 0.266 M sorbitol–0.042 M glucose. Recipients were given 1.3 x 10^7 cells intravenously or 2.3 x 10^6 cells intravitreally.

Hartley strain guinea pigs were immunized topically by multiple (2–4) daily instillations of ascarid antigens or fluorescein-ovalbumin (fluorescein–thiocarbamyl–ovalbumin) in the conjunctival sac. IgE antidinitrophenyl (DNP) antibody was induced in atopic dogs with dinitrophenylated ragweed pollen.[4,5] Histofluorescent microscopy of freeze-dried eyes, immunofluorescence, and histologic staining for mast cells/basophils and eosinophils were performed using previously described methods.[1]

Results

The characteristics that distinguished human, guinea pig, and canine IgE antibodies from other immunoglobulins are summarized in Table I.

Serum IgE antibodies against ascarid antigens were induced by systemic or intravitreal injection of motile *Toxocara canis* and *Ascaris suum* larvae, or ascarid ACF antigen. Intravitreal infection also induced aqueous IgE antibodies which, in some instances, occurred in animals whose undiluted serum lacked IgE antibody activity. In guinea pigs with high-titer (e.g., 1:1,000)

Department of Ophthalmology, Scheie Eye Institute, School of Medicine, and Department of Dermatology, School of Veterinary Medicine, University of Pennsylvania, Philadelphia, Pennsylvania; Department of Veterinary Pathobiology, College of Veterinary Medicine, University of Minnesota, St. Paul, Minnesota; London School of Hygiene and Tropical Medicine, Winches Farm Field Station, St. Alban's, England; Department of Clinical Veterinary Medicine, University of Cambridge, Cambridge, England.

TABLE I
Distinguishing Characteristics Shared by Human, Canine, and Guinea Pig IgE Antibodies

Characteristic	Human	Canine	Guinea Pig
Homocytotropic	PK	PK	PK
Duration local fixation	> 28d	> 14d	> 28d
Heat lability	+	+	+
Sulfhydryl sensitivity	+	+	+
Sedimentation coefficient	8S	8S	8S
G-200 Gel filtration	Intermediate	Intermediate	Intermediate
Electrophoretic mobility	γ_1	γ_1	γ_1
IgE-Specific antigens (human)	+	+	+
Ratio aqueous:serum (normal eye)		< 1:640	< 1:1,000

serum IgE antibody induced by systemic infection with ascarid larvae, acute anaphylactic activity also was demonstrated by topical (conjunctival) challenge with ascarid antigens. Acute conjunctival anaphylactic reactions also were induced in normal guinea pigs by repeated topical (conjunctival) instillation of fluorescein-ovalbumin or ascarid antigens.

A single intracorneal or intravitreal infection of inbred Strain 2 guinea pigs with 5,000 mf of *O. lienalis* did not elicit detectable IgE antibody in serum or aqueous humor. Intracorneal challenge infections with fresh or cryopreserved mf of *O. lienalis* in Strain 2 guinea pigs, previously sensitized by three subcutaneous injections of mf, induced serum and aqueous IgE antibody. Intradermal infection with mf suppressed the IgE antibody responses of guinea pigs subsequently challenged with subcutaneous and intracorneal infections. Oral administration of diethylcarbamazine (DEC) after priming with intradermal and subcutaneous injections of mf followed by intracorneal challenge with mf, resulted in increased serum and aqueous IgE antibody. Aqueous IgE antibody was present in two guinea pigs that lacked serum IgE antibody.

A pooled suspension of spleen cells from syngeneic animals sensitized by subcutaneous, intracorneal, and subconjunctival infections with *O. lienalis* mf was transferred intravenously or intravitreally into age-matched syngeneic recipients. The mean $\log_2$ serum IgE antibody P-K titer of

the donors at the time of transfer was 4.7. Serum and aqueous IgE antibodies were induced following intracorneal challenge in recipients of intravenously transferred cells, while eight recipients of normal spleen cells did not respond.

Guinea pigs were sensitized for passive anaphylactic reactions by intravenous or subconjunctival injection of high-titer IgE antisera. IgE antibody activity disappeared from the serum within 48 hours after intravenous injection, but dermal and conjunctival sensitivity persisted for more than 28 days. Intravenous antigen challenge of passive periocular anaphylactic reactions produced degranulation of mast cells, infiltration of eosinophils and neutrophils, and vascular leakage demonstrated with Evan's blue-albumin complexes, fluorescein-albumin and -IgG, in the palpebral, periocular, and episcleral tissues. Systemic anaphylaxis in addition produced degranulation of choroidal and ciliary body mast cells, a choroidal eosinophil infiltrate, and vascular leakage in the choroid but not in the iris. The reactions of animals challenged topically (conjunctivally) were similar to the passive periocular P-K reactions except that repeated topical instillation of antigen also stimulated eosinophil infiltrations in the conjunctival epithelium. Mast cell degranulation (Fig. 1) also occurred following intravitreal infection with ascarid larvae (Fig. 2), and in the bulbar conjunctiva and ciliary body of eyes infected with Onchocerca microfilariae.

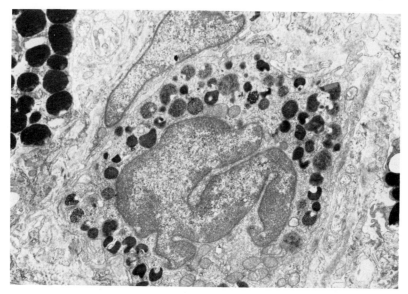

FIG. 1. Degranulating guinea pig ocular mast cell located between two mast cells with intact granules.

Discussion

IgE antibodies may play a significant role in a number of ocular diseases, particularly by means of their ability to induce the release of mast cell and basophil mediators, which may have direct actions or may pro-mote the delivery of other immunopathologic effectors to tissue sites through increased vascular permeability, the release of chemotactic agents, and kinin formation. IgE antibodies in the circulation and in and about the eye have been induced by a number of routes and by diverse antigenic challenges. The simplest method was by re-

FIG. 2. Intraocular mast cells during the course of a primary intraocular infection with ascarid larvae. Mast cells in 7–8 μm sections were stained with Unna's reagent. Ten eyes were examined for each reaction, at each time period. The numbers are the means per histologic section. Ascarid infection was produced by intravitreal injection of 3,000 second-stage *Ascaris suum* larvae on day 0.

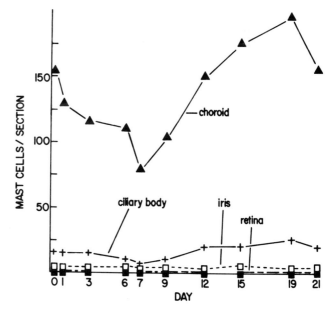

peated topical administration of antigen in the conjunctival sac,[6] a process that occurs commonly in man.

A simple correlation between the level of IgE antibody in the circulation and the role of the IgE antibody *in vivo* in an ocular immunopathologic reaction in an immunized animal may not be found. The systemic passive sensitization reactions, in which dermal and ocular hypersensitivity persisted long after the intravenously administered IgE antibody had disappeared from the serum of the recipient, emphasize that IgE-mediated reactions may be occurring in actively immunized animals even when no IgE antibody is detected in their serum.[1] Local ocular production of IgE antibody also may occur in the absence of circulating IgE antibody, as shown by the presence of aqueous IgE antibodies in some instances in eyes infected with Ascaris or Toxocara larvae, or Onchocerca microfilariae, at times when IgE antibodies were absent from the serum. On the other hand, substantial levels of IgE antibody may be found in the serum but may have a limited role in the ocular or periocular immunopathologic reaction observed because blocking antibodies (e.g., IgG antibodies with the same specificity, present in large excess in comparison to the relatively low concentrations of IgE antibodies), which do not interfere with the P-K or immunoassay for IgE antibody, may effectively reduce the antigenic stimulus for IgE-mediated mast cell degranulation *in vivo*. The use of passively sensitized animals to study IgE-mediated acute ocular and peri-ocular immunopathologic reactions following a single challenge with antigen has largely overcome these problems. However, if multiple antigenic challenges are given to the same recipient animal, to study chronic effects of repeated IgE-mediated reactions, the possibility that blocking antibodies and/or other immunopathologic mechanisms are being brought into play must again be considered. Specific pharmacologic manipulation may be required to define explicitly the role of repeated IgE-mediated reactions in chronic ocular and periocular immunopathologic processes.

Acknowledgments

This work was supported by USPHS Grant EY03984, by a Fight for Sight Postdoctoral Research Fellowship from Fight For Sight, Inc., New York City, by Grants from the UNDP/World Bank/WHO Special Program for Research and Training in Tropical Diseases and from the Wellcome Trust, by an unrestricted Grant from Research to Prevent Blindness, and by the Gretel and Eugene Ormandy Teaching and Research Fund.

References

1. Rockey JH et al.: *Arch Ophthalmol* **99**:1831-1840, 1981.
2. Donnelly JJ et al.: *Invest Ophthalmol Vis Sci* (Suppl) **22**:211, 1982.
3. Ham P et al.: *Parasitology* **83**:139-147, 1981.
4. Schwartzman RM et al.: *Clin Exp Immunol* **9**:549-569, 1971.
5. Halliwell REW et al.: *Transplant Proc* **7**:537-543, 1975.
6. Meisler DM et al.: *Invest Ophthalmol Vis Sci* (Suppl) **22**:135, 1982.

chapter 51

Langerhans' Cells

Role in Ocular Surface Immunopathology

Yoshitsugu Tagawa, Tsutomu Takeuchi, Tokuhiro Saga, Hidehiko Matsuda, and Arthur M. Silverstein[a]

Langerhans' cells have been a curiosity for over a century, since Paul Langerhans originally described them in the epidermis of the skin in 1868.[1] Recent evidence, however, suggests that epidermal Langerhans' cells share many characteristics with macrophages and may play a role in the immune phenomena of the skin.[2, 3] It has also been reported that Langerhans' cells were identified in the epithelium of the human corneal limbus by electron microscopy.[4] However, the exact distribution and function of this cell in the ocular surface still remains obscure.

In the study described herein, we attempt to clarify the distribution and surface antigens of Langerhans' cells in the guinea pig and murine ocular surfaces.

Materials and Methods

Animals. Hartley strain guinea pigs and C3H/He, BALB/c, and A/J mice were used. Prior to use, all eyes were carefully screened clinically to rule out preexisting disease.

Electron microscopy. Enucleated eyes were fixed in 2% glutaraldehyde and 1% osmium, embedded in Epon 812, sectioned, and stained with uranyl acetate and lead nitrate according to conventional methods. The sections were examined in a Hitachi 12-A electron microscope.

ATPase staining. ATPase staining was performed by the method of Juhlin and Shelley.[5] Tissues were incubated in EDTA solution at 37°C for 2 hours; then corneal and conjunctival epithelial sheets were separated from the underlying stroma. Epithelial sheets were rinsed in physiological saline for 20 minutes, fixed in cacodylate-formaldehyde at 4°C for 20 minutes, incubated in an ATP substrate at 37°C for 30 minutes, developed in ammonium sulfate for 10 minutes, then picked up on glass slides, dried, and mounted. Cryosections were prepared according to conventional methods, then fixed and stained by ATP substrate as mentioned above.

Immunofluorescence. EDTA-separated epithelial sheets were stained by immunofluorescence by the method of Nordlund and Ackles.[6] Epithelial sheets were fixed in acetone for 20 minutes, washed in PBS for 60 minutes, immersed in specific antisera at 4°C for 16 hours, washed in PBS for 60 minutes, then immersed in FITC-conjugated antimouse Ig at 37°C for 60 minutes, washed in PBS, and mounted. The specimens were observed under a Nik microscope (VFD-R).

Antisera. The following antisera were used: monoclonal antiguinea pig Ia "5S2"

Department of Ophthalmology, Hokkaido University School of Medicine, Sapporo, Japan; and [a]The Wilmer Ophthalmological Institute, Johns Hopkins University School of Medicine, Baltimore, Maryland.

(kindly provided by Dr. E. Shevach of NIH); anti-Ia^k alloserum (Cedarlane Lab., Ontario, Canada; CL 8701); antimouse I-A^k (Ia.2) (Cedarlane Lab.; CL 8709); antimouse I-E^k (Ia.7) (Cedarlane Lab.; CL 8705); antimouse I-J^k alloserum (Cedarlane Lab.; CL 8703); antimouse Thy-1.2 (Cedarlane Lab.; CL 8600); antimouse macrophage (Mac-1)[7] (Sera-Lab., Sussex, England; MAS 034); anti-TL (specificity 1.3.5) (kindly provided by Dr. Tada of Sloan-Kettering Cancer Institute); FITC-conjugated antimouse Ig (E.Y. Lab., San Mateo, CA; FA 2306).

Immunogenic keratitis. Arthus-type reactions were induced by injection of 0.01 ml of 2% ovalbumin (OA) into the central cornea 4–6 weeks after sensitization with 2% OA in complete Freund's adjuvant subcutaneously. Eyes were enucleated at 24 and 48 hours, on the third and fifth day, and 2 and 4 weeks after antigenic challenge. Delayed hypersensitivity reactions were induced by injection of 0.01 ml of 2% OA into the cornea 7 days after sensitization with 2% OA in incomplete Freund's adjuvant subcutaneously. Eyes were enucleated at 24 and 48 hours, and on days 5 and 7. Epithelial sheets were separated and stained by ATP-ase as mentioned above.

Results

Distribution of Langerhans' cells by electron microscopy

In the guinea pig, Langerhans' cells were found in both the epithelium and the substantia propria of the corneal limbus and the conjunctiva, with Birbeck granules of typical structure (Fig. 1). Langerhans' cells were usually located in the suprabasal layer of the epithelium. The ultrastructure of this cell is characterized by dendritic processes, a clear cytoplasm, the lack of desmosomal attachments to other epithelial cells, the lack of premelanosomes and melanosomes, a lobulated nucleus, and Birbeck granules. Langerhans' cells were seen only in smaller numbers in the substantia propria, usually located around blood vessels or immediately beneath the epithelium. The number of Birbeck granules in the guinea pig was less than

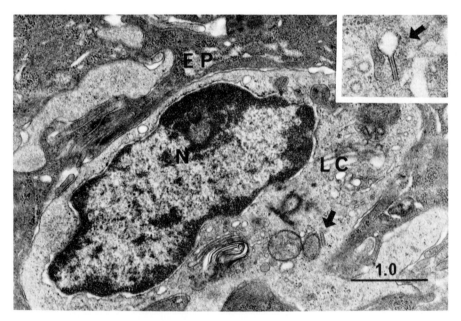

FIG. 1. Langerhans' cell (LC) in the epithelium of the normal guinea pig corneal limbus. Arrow indicates a Birbeck granule. EP: epithelium; N: nucleus (× 22,800). *(Inset)* Arrow indicates a Birbeck granule in another section (× 40,600).

that found in human Langerhans' cells. In contrast, no Langerhans' cells were identified in the central cornea.

Distribution of ATPase-positive dendritic cells

In cryosections of guinea pig ocular surface tissues, ATPase-positive dendritic cells were observed to be abundant in the suprabasal layer of the corneal limbal epithelium, and were less frequent in the conjunctival epithelium. No ATPase-positive cells were found in the central cornea. The distribution of ATPase-positive dendritic cells was more easily observed in the epithelial sheets from the ocular surface; i.e., ATPase-positive cells were numerous at the limbus, sparse in the bulbar conjunctiva, and absent in the central cornea.

Surface membrane antigens on Langerhans' cells

Guinea pig. In epithelial sheets, Ia antigen-positive dendritic cells were found to be numerous in the corneal limbus, sparse in the conjunctiva, and absent in the central cornea by immunofluorescence. The density and distribution of Ia-positive cells were almost identical to that observed with ATPase staining.

Mouse. In epithelial sheets of C3H/He mice, Ia^k antigen-positive dendritic cells were numerous in the limbus and conjunctiva, but absent in the central cornea by immunofluorescence (Fig. 2). In C3H/He mice, $I-A^k$ and $I-E^k$ antigen-positive cells were demonstrated in a distribution identical to that observed with immunofluorescence of Ia^k antigens, but $I-J^k$ antigen-positive cells were not found. In contrast, in epithelial sheets of C3H/He and BALB/c mice, no fluorescent cells were demonstrated with anti-Thy-1.2, anti-Mac-1, or antimouse Ig reagents. In A/J mice, TL antigen-positive cells were not observed in the epithelial sheets.

Immunogenic keratitis

In Arthus-type inflammation of guinea pig cornea, the migration of ATPase-positive dendritic cells occurred from the limbus into the peripheral cornea within 24 hours after an antigenic challenge. From 48 hours to day 5 after the antigenic challenge, ATPase-positive cells increased in number in the

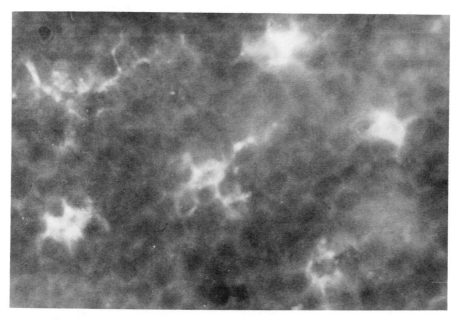

FIG. 2. Ia^k antigen-positive dendritic cells in the ocular epithelial sheet of C3H/He mouse, detected by immunofluorescence ($\times$ 1,400).

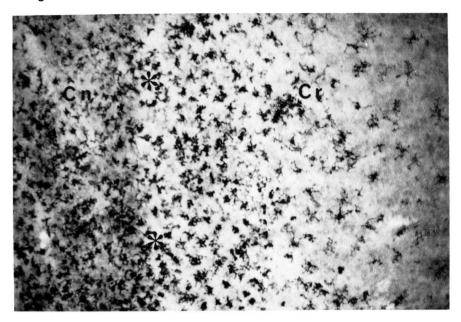

FIG. 3. ATPase staining of an ocular epithelial sheet of a guinea pig with immunogenic keratitis, 48 hours after antigenic challenge. ATPase-positive dendritic cells are noted in the limbus (∗), cornea (Cr), and conjunctiva (Cn) (× 120).

limbus and conjunctiva, and migrated into the central cornea (Fig. 3). Following the subsidence of inflammation, ATPase-positive dendritic cells gradually decreased in the central cornea, but persisted there for at least several weeks. In delayed hypersensitivity reactions, ATPase-positive dendritic cells also increased in number in the limbus and migrated into the peripheral third of the cornea, but not to the central cornea, even during the first several days after antigenic challenge. In contrast, only slightly increased numbers of ATPase-positive cells were found in the limbus and peripheral cornea of nonsensitized control animals 24–48 hours after antigenic challenge; they gradually returned to normal density and distribution within the next several days.

Discussion

Recent immunologic studies on Langerhans' cells in the epidermis indicate that they bear Fc and complement receptors,[8] and are the only cells within the epidermis that express Ia (Class II) histocompatibility antigens.[9, 10] These cells may therefore play a role in the immune phenomena of the skin, functioning as macrophages.[2, 3] Electron microscopic study has clearly shown the presence of Langerhans' cells identical to those of the epidermis in the epithelium of the normal human corneal limbus,[4, 13] indicating that Langerhans' cells are a normal cell component of the ocular surface epithelium as well as of the epidermis. However, studies on Langerhans' cells in the ocular surface have been few.[4, 11, 13, 14] We attempted, therefore, to clarify the distribution and function of this cell type in the ocular surface. Our results obtained by electron microscopy confirmed the presence of Langerhans' cells with Birbeck granules in both the epithelium and the substantia propria of the normal guinea pig corneal limbus and conjunctiva. Birbeck granules in Langerhans' cells of the guinea pig ocular surface were much less frequent than those found in the human ocular surface. This is consistent with the finding in the epidermis, though the reason for it is not clear.[12]

ATPase staining, recognized as a specific

marker for epidermal Langerhans' cells, clearly revealed the distribution of ATPase-positive cells in the guinea pig ocular surface epithelium. They were found to be numerous in the corneal limbus, sparse in the bulbar conjunctiva, but absent in the central cornea.

We confirmed the presence of Ia antigen-positive dendritic cells in epithelial sheets from the guinea pig ocular surface, employing immunofluorescent analysis. The distribution and density of these fluorescent cells were identical to that observed with ATPase staining. These results are consistent with the recent studies of Langerhans' cells of the ocular surface epithelium of the human,[13, 14] the guinea pig,[15] and the mouse.[14] However, all Ia-positive cells in the ocular epithelium may not be Langerhans' cells; other Ia-positive candidates include B-cells, certain T-cells, and macrophages.[16] Therefore, various cell surface antigens such as Thy-1 marker of T-cells, surface immunoglobulins of B-cells, Mac-1 marker of macrophages, and TL marker of thymocytes were examined in murine epithelial sheets. No fluorescent cells could be detected by any of these reagents, indicating that the Langerhans' cells are distinct from macrophages and lymphocytes, and are the only cells within the ocular surface epithelium which express Ia antigens. Further, a study of Ia antigens on Langerhans' cells in the murine ocular surface revealed that the Ia antigens on those cells were encoded for by at least the I-A and I-E subregions of the H-2 complex. These findings are consistent with results obtained with the epidermis.[17, 18]

The observation of increased numbers of ATPase-positive Langerhans' cells in the corneal limbal epithelium and their migration into the corneal epithelium during the course of immunogenic keratitis in the guinea pig suggests that Langerhans' cells may play a role in immune or other inflammatory conditions of the cornea and conjunctiva. It seems likely, therefore, that Langerhans' cells may be essential to the immune system of the ocular surface, and may contribute to various external eye diseases.

References

1. Langerhans P: *Virchows Arch Path Anat Physiol* **44**:325–337, 1868.
2. Stingl G et al.: *Immunol Rev* **53**:149–174, 1980.
3. Silberberg-Sinakin I et al.: *Immunol Rev* **53**:203–232, 1980.
4. Sugiura S and Matsuda H: *Jpn J Opthalmol* **13**:197–202, 1969.
5. Juhlin L and Shelly WB: *Acta Derm Venereal* **57**:289–296, 1977.
6. Nordlund JJ and Ackles A: *Tissue Antigens* **17**:217–225, 1981.
7. Springer T et al.: *Eur J Immunol* **9**:301–306, 1979.
8. Stingl G et al.: *Nature* **268**:245–246, 1977.
9. Rowden G et al.: *Nature* **268**:247–248, 1977.
10. Stingl G et al.: *J Immunol* **120**:570–578, 1978.
11. Brown J et al.: *Invest Ophthalmol* **7**:668–671, 1968.
12. Aberer W et al.: *J Invest Dermatol* **76**:202–210, 1981.
13. Rodrigues MM et al.: *Invest Ophthalmol* **21**:759–765, 1981.
14. Gillette TE et al.: *Ophthalmology* **89**:700–711, 1982.
15. Klareskog L et al.: *Invest Ophthalmol* **18**:310–313, 1979.
16. Hammerling GJ: *Transplant Rev* **30**:64–82, 1976.
17. Rowden G et al.: *Immunogenetics* **7**:465–478, 1978.
18. Tamaki K et al.: *J Immunol* **123**:784–787, 1979.

chapter 52

Secretory Immune Cellular Traffic between the Gut and the Eye

Hugh R. Taylor, Nathaniel F. Pierce,[a] Zhang Pu, Julius Schachter,[b] Arthur M. Silverstein, and Robert A. Prendergast

Following local antigenic stimulation, the precursors of IgA-forming cells migrate and then home to secretory tissues such as the lamina propria of the intestinal tract.[1] A few IgA plasma cells also migrate to other sites of the mucosa-associated lymphoid tissue.[2] It has been suggested that the conjunctiva is also a member of this system.[2, 3] The finding in the conjunctiva of competent lymphocytes committed to antigens first encountered in the gut would confirm this.

Cholera toxin efficiently stimulates local secretory IgA responses in the intestinal mucosa.[1] Enteric mucosal IgA response involves the systemic migration of IgA plasmablasts with organ-specific homing and local and recirculating antigen-specific memory cells. We report a study of the secretory immune system traffic between the gut and the conjunctiva using cholera toxin as the antigenic probe.

Methods

Cynomolgus monkeys were examined using a standard protocol.[4] To maintain active trachoma,[4] eyes were inoculated each week with 20μl of a $10^{3.2}$ ELD_{50} suspension of *Chlamydia trachomatis* (Bour strain). For

The Wilmer Institute, The Johns Hopkins Medical Institutions, Baltimore, Maryland; [a]Department of Medicine, Baltimore City Hospitals, Baltimore, Maryland; [b]Department of Laboratory Medicine, University of California at San Francisco, San Francisco, California.

enteric immunization, 10 ml of crude cholera toxin (CCT) (NIH lot 001) was given via a gastric tube after gastric contents were neutralized with sodium bicarbonate. The concentration of CCT varied as indicated below. For conjunctival immunization, two doses of purified cholera toxin (PCT) (NIH lot 0972, 20 μg in 20μl) were dropped in the eye one hour apart.

Immunizing and biopsy schedule

Enteric immunization. Two monkeys with active trachoma were given 50 mg of CCT enterically on days 0 and 4; 100 mg on days 5 and 6; 200 mg on day 7; 300 mg on day 11; 400 mg on day 12; 500 mg on day 13; and 1000 mg on days 25 and 52. None of these doses caused diarrhea or otherwise adversely affected the monkeys. One animal (monkey 040) was sacrificed on day 58. The second animal (monkey 041) received a further 1000 mg of CCT on day 73. The right eye was challenged with PCT on day 87. Superior tarsal conjunctivae were biopsied after 1 week (day 94), 3 weeks (day 108), and 8 weeks (day 142).

Another monkey (032) with active trachoma was given 1000 mg of CCT on days 0 and 5. The right eye was challenged with PCT on day 82, and biopsies were taken from both eyes 1 week (day 89) and 6 weeks (day 124) later.

A fourth, normal monkey (010) was given 1000 mg of CCT on days 0 and 14. On day 28, the right eye was challenged with PCT,

TABLE I
Immunization and Biopsy Schedule, and the Number of Specific Antitoxin-Containing Plasma Cells (ACC) Found

Monkey number	Enteric priming (day)	Ocular challenge, right eye (day)	ACC per mm³			
			Day	Right eye	Left eye	Jejunum
[a]040	0,4,5,6,7,11, 12,13,25,52	none	58	0	0	37,200
[a]041	0,4,5,6,7,11, 12,13,25,52,73	87	94	23,100	0	n.d.[c]
			108	2,900	0	n.d.
			142	3,300	n.d.	n.d.
[a]032	0,5	82	89	2,100	0	n.d.
			124	400	n.d.	n.d.
010	0,14	28	35	6,300	0	n.d.
[a]074	none	0	7	300	0	n.d.
		14	21	3,400	0	n.d.
			42	1,100	n.d.	n.d.
W2	none	0	7	0	0	n.d.
		14[b]	21	5,500	1,500	n.d.

[a] Animals with chronic conjunctival inflammation from weekly chlamydial inoculation.
[b] Ocular challenge to both right and left eyes.
[c] n.d., not done.

and biopsies were taken from both eyes on day 35.

Ocular immunization. Two animals were given cholera toxin to the eye without enteric immunization. One (074) with active trachoma was given PCT in the right eye on days 0 and 14. Biopsies were taken from both eyes again on days 7, 21, and 42. The other, a normal monkey (W2), was given PCT in the right eye on day 0 and in both eyes on day 14. Both eyes were biopsied on day 7 and 21.

Assay for antitoxin-containing cells (ACC)

Five micron-thick frozen sections were fixed with methanol and stained with an indirect fluorescent antibody technique by incubation with purified cholera toxoid followed by an immunopurified fluorescein-conjugated rabbit antitoxin.[1] The number of ACC and the number of high-power fields in each section were counted. ACC responses were expressed as cells per mm³; but because some biopsies were small and ACC were not evenly distributed, these

values only reflect the relative magnitude of each response.

Results

Enteric immunization alone

There was a vigorous ACC response in the jejunum of monkey 040 on day 58 (Table I), but there were no ACC in the conjunctiva despite the presence of active trachoma with large germinal centers.

Enteric immunization followed by ocular challenge

Three monkeys were given different enteric immunizing schedules with CCT followed by ocular challenge. Two had active trachoma. One monkey (041) received a total of 5.7 G of CCT over 73 days and was challenged with PCT one week after the last enteric dose. There was a vigorous ACC response in the challenged eye but none in the nonchallenged eye (Table I). Over the next 8 weeks, the number of ACC in the challenged conjunctiva declined rapidly. None appeared in the other eye.

The second animal (032) was given 2.0 G of CCT and then PCT in the right eye 77 days later. One week later there were a moderate number of ACC in the challenged eye only, and fewer ACC were present 4 weeks later.

The third animal (010) did not have inflammation of the conjunctiva. It was given 2.0 G of CCT and challenged with PCT 14 days later. Only the challenged eye showed an ACC response 7 days after challenge.

Ocular immunization alone

One monkey with active trachoma (074) was given PCT to the right eye. There was modest ACC response in the immunized eye 1 week later but none in the nonimmunized eye. One week later, the right eye was given more PCT, and biopsies 7 and 42 days later again revealed moderate numbers of ACC in the immunized eye but none in the other.

Another normal monkey (W2) was given PCT in the right eye. Conjunctival biopsies of both eyes taken 1 week later showed no ACC. Both eyes were then challenged with PCT; there were moderate numbers of ACC in the right eye 1 week later, and some ACC were also present in the left, although they were fewer in number.

Discussion

This study shows that specific antibody-producing plasma cells can be stimulated in the conjunctiva either by enteric priming followed by conjunctival challenge or by repeated conjunctival immunization alone. As conjunctival immune responses are affected by immunization at distant mucosal sites, it would seem that the conjunctiva is an integral part of the mucosal immune system.

Conjunctival challenge was required to envoke a conjunctival ACC response; repeated enteric immunization caused a vigorous intestinal ACC response but no conjunctival response. It is interesting that preexisting, chronic ocular inflammation had no apparent effect on the induction or expression of the ocular ACC response. It is likely that ocular challenge caused a con-

junctival ACC response by arresting migrating memory cells within the conjunctiva and stimulating their expansion and differentiation to antitoxin-producing plasmablasts and plasma cells at that site.

These results are in general agreement with prior studies in rats on the migration of toxin-sensitized lymphocytes between different mucosae; however, the marked localization of the conjunctival ACC response to the challenged eye differs from the intestinal immune response to cholera toxin.[1] Repeated duodenal immunization caused cells to enter the circulation and home selectively to the small intestine, including parts of the intestine not exposed to antigen. This suggested that migrating mucosal ACC precursors homed in an organ-specific but antigen-independent manner. Failure of an ACC response to appear in the unchallenged eye suggests that circulating plasmablasts did not contribute significantly to the ocular ACC response. Whether or not conjunctival priming causes plasmablast migration, sensitized memory cells must migrate from the primed to the nonprimed eye where they can support a secondary-type response to local antigen application. This finding, supported by observations in monkey W2, accords with earlier studies showing that sensitized lymphocytes convey priming from the site of mucosal immunization to distant, nonimmunized mucosae.[1]

The role played by secretory immunity in protecting against chlamydial infection is unknown. Specific antichlamydial secretory IgA antibodies are found in tears.[4, 5] Further, although trachoma is primarily an eye infection, chlamydia can be isolated from the respiratory and gastrointestinal tracts in children with ocular infection.[6]

Almost all trachoma vaccines have been prepared for parenteral administration, and although some have provided a small measure of protection, others have produced a deleterious sensitization.[7] An appropriate trachoma vaccine that stimulates secretory immunity without producing systemic sensitization may be useful in preventing chlamydial infection. This study might aid such efforts since it demonstrates that enteric immunization can prime the conjunctiva even though enteric immunization alone does not provoke a spontaneous plasma cell response in the conjunctiva. Further studies are needed to determine whether enteric priming might be more efficient than ocular priming and, therefore, have practical value for efforts at ocular protection against trachoma.

Acknowledgments

Supported by NIH-NEI grant EY 03324 to Dr. Taylor, by grant EY 02650 to Dr. Silverstein, by grant EY 01205 to Dr. Schachter, and by grant EY 03521 to Dr. Prendergast.

References

1. Pierce NF and Cray WC: *J Immunol* **128**:1311–1315, 1982.

2. Franklin RM et al.: *Invest Ophthalmol Vis Sci* **18**:1093–1906, 1979.

3. Axelrod AJ and Chandler JW: In *Immunology and Immunopathology of the Eye,* Silverstein AM and O'Connor GR (Eds.). Masson Publishing USA, New York, 1979, pp. 292–301.

4. Taylor HR et al.: *Invest Ophthalmol Vis Sci* **23**:507–519, 1982.

5. Jawetz E et al.: In *Trachoma and Related Disorders,* Nichols RL (Ed.), Excerpta Medica, Amsterdam, 1971, pp. 232–242.

6. Malaty R et al.: *J Infect Dis* **143**:853, 1981.

7. Grayston JT and Wang S: *J Infect Dis* **132**:87–105, 1975.

chapter 53

Autoimmunity and Retinal Dystrophy

Susan M. Chant, John Heckenlively, and Roberta H. Meyers-Elliott

In an effort to define the mechanisms involved in the pathogenesis of retinitis pigmentosa (RP), studies have been carried out in Royal College of Surgeons (RCS) rats[1] (which suffer from retinal dystrophy) and in man in order to assess the role played by the immune system in the degenerative process. Direct and indirect immunofluorescence were used to screen for *in vivo* bound and circulating autoantibodies, and *in vitro* transformation assays were used to detect sensitized lymphocytes. Results in both the animal model and in man indicate that an autoimmune mechanism is occurring, most probably as a secondary phenomenon following retinal degeneration.

Materials and Methods

Animal studies

Direct immunofluorescence. Frozen sections of retina from dystrophic and congenic control rats (age range: 11 days to 1 year) were incubated with FITC-labeled antirat IgG or antirat IgM to detect *in vivo* bound antibodies.

Indirect immunofluorescence. Frozen sections of retina from normal rats were incubated with sera from dystrophic or congenic control rats followed by incubation with FITC-labeled antirat IgG or antirat IgM to detect circulating autoantibodies.

Transformation assay. Splenic lymphocytes from dystrophic rats at various ages

Jules Stein Eye Institute, UCLA School of Medicine, Los Angeles, California.

and age-matched congenic control animals were cultured *in vitro* in the presence of bovine rod outer segments (ROS) (10, 25, 50, 100, and 250 μg/ml; 20 μl/well) or bovine rhodopsin antigens; (3, 5, 10, and 25 μg/ml; 20 μl/well); culture medium: RPMI 1640, 5% normal rat serum; cell concentration: 2×10^5 cells/well. After 4 days in culture (37°, 5% CO_2) the cells were pulse-labeled with 1 Ci/well ^{3}H-thymidine, and harvested 4 hours later using a MASH unit. ^{3}H-thymidine uptake was measured by liquid scintillation spectroscopy and the stimulation index (SI) calculated.

$$S.I. = \frac{\text{Mean CPM of cells cultured in presence of antigen}}{\text{Mean CPM of cells cultured in absence of antigen}}$$

Human studies

Subjects for the study consisted of (a) patients with defined forms of RP, (b) patients with non-RP eye disorders, and (c) normal subjects with no known eye disorders.

Immunofluorescence. Paraffin sections of normal human eyes were deparaffinized, fixed in 95% ETOH, washed, incubated in a moist chamber 30 minutes with test serum, washed, incubated 30 minutes in a moist chamber with FITC-antihuman IgG or FITC-antihuman IgM, washed, mounted, and examined with a UV microscope.

Transformation assay. Peripheral blood lymphocytes were cultured *in vitro* in the presence of human retinal antigens (50, 100, and 250 μg/ml; 20 μl/well) for 6 days or mi-

TABLE I
Percentages of Rats Showing Sensitization To Retinal Antigens in Transformation Assays

	Antigen		ROS		Rhodopsin
S.I.[a]	>2	RCS	27% (n = 55)	RCS	15% (n = 55)
		CC	6% (n = 35)	CC	3% (n = 35)
S.I.	>1.5	RCS	89% (n = 55)	RCS	64% (n = 55)
		CC	23% (n = 55)	CC	14% (n = 35)

[a] S.I. = Stimulation index.
RCS = Royal College of Surgeons rats (dystrophic).
CC = Congenic control rats (normal).

togens (Con A 100 µg/ml; 20 µl/well and PHA 100 µg/ml, 20 µl/well) for 3 days and pulse-labeled with ^{3}H-thymidine 4 hours prior to harvesting. Thymidine incorporation was assessed by liquid scintillation counting and results were expressed as the stimulation index.

Results

Animal studies

Although no *in vivo* bound antibody was seen by direct immunofluorescence, indirect staining demonstrated the presence of autoantibodies of the IgM class with specificity for the retina in the sera of dystrophic, but not congenic control rats. Further evidence for sensitization to retinal antigens was indicated by the *in vitro* transformation assay. Splenic lymphocytes from dystrophic but not congenic control rats demonstrated sensitization to ROS and rhodopsin antigens (Table I).

Human studies

Initial studies of *in vitro* transformation carried out on peripheral blood lymphocytes

TABLE II
Immune Responses to Retinal Antigens (Summary)

RP[a] patients		Non-RP[b] patients	Normal controls
Autoantibodies	30/79	9/19	2/9
Sensitization	26/64	6/18	3/9

[a] RP = Retinitis pigmentosa.
[b] Non-RP = Eye disease other than retinitis pigmentosa.

from RP patients and patients with other eye disorders indicated *in vivo* sensitization to retinal antigens in many patients. Sera from patients were examined by indirect immunofluorescence, and in many cases antibodies of the IgG or IgM class with specificity for retinal antigens were detected (Table II). Two normal controls also had antibodies to retinal antigens, and three controls showed evidence of sensitization in the transformation assays.

Discussion

The presence of IgM in the sera of dystrophic rats, with activity directed against photoreceptor cells, indicates that the rats have become sensitized to their own retinal antigens. The inability to demonstrate *in vivo* bound antibody by the direct method may be due to the lower sensitivity of this method. The absence of staining on rod outer segments suggests that the antibodies may be directed at precursors of disc materials or other cytoplasmic proteins.

Further evidence that dystrophic rats have become sensitized to retinal antigens comes from the *in vitro* transformation assays. Lymphocytes from dystrophic, but not those from congenic control rats, showed an enhanced response to retinal antigens *in vitro*.

Recent studies in RP have suggested a role for the immune system in human retinal degeneration. Our results from screening of sera for autoantibodies and lymphocytes for sensitized cells indicate that an autoimmune mechanism is present in almost half of the patients studied so far, and both the humoral and cell-mediated arms of the immune system are involved.

Our preliminary studies thus provide evidence of an autoimmune mechanism at play in retinal degenerative disorders and confirm earlier evidence by other investigators.[2–7]

Acknowledgments

This research was supported by USPHS grants EY 1309 and Ey 00331 from the National Eye Institute, a center grant from the National Retinitis Pigmentosa Foundation, a Research to Prevent Blindness, Inc. Research Manpower Award (R.M.E.), and a grant-in-aid from the National Society to Prevent Blindness (S.M.C.).

References

1. Chant SM and Meyers-Elliott RH: *Clin Immunol Immunopathol* **22**:419, 1982.
2. Rahi AHS: *Br J Ophthalmol* **57**:904, 1973.
3. Char DH et al.: *Invest Ophthalmol Vis Sci* **13**:198, 1974.
4. Rocha H and Antunes L: *Metab Ophthalmol* **1**:153, 1977.
5. Spalton DJ et al.: *Br J Ophthalmol* **62**:183, 1978.
6. Brinkman CJJ et al.: *Invest Ophthalmol Vis Sci* **19**:743, 1980.
7. Heredia CD et al.: *Br J Ophthalmol* **65**:850, 1981.

chapter 54

Unilateral Inoculation of Herpes Virus into the Anterior Chamber Produces a Bilateral Uveitis with a Pathogenesis Potentially Related to Anterior Chamber-Associated Immune Deviation (ACAID)

Judith A.Whittum, James P. McCulley, Jerry Y. Niederkorn, and J. Wayne Streilein

Herpes simplex-induced ocular disease may occur as keratitis, uveitis, retinochoroiditis, or some combination of the three. Despite extensive studies of herpetic ocular disease in rabbits,[1-3, 5] the relationship between virus- and immune-mediated mechanisms in the pathogenesis of herpes-induced ocular disease remains unresolved. In this report we describe our pathologic and immunologic characterization of the herpetic ocular infection induced in inbred mice by intracameral inoculation of herpes simplex virus Type 1 (HSV-1).

Materials and Methods

Mice. BALB/c mice were obtained from Cumberland View Farms (Clinton, TN).

Virus. Herpes simplex virus Type 1 (KOS strain) and the rabbit skin cells in which it was propagated were kindly provided by Dr. Lewis Pizer (University of Colorado Medical School, Denver, CO). Virus concentrations were determined by standard tech-

niques. The stock of virus used in these experiments had a titer of approximately 2 × 10^8 pfu/ml.

Clinicopathological evaluation of ocular disease. Mice were examined with a slit lamp biomicroscope 1, 3, 5, 7, 10, 14, and 21 days after unilateral inoculation of virus (2–4 × 10^4 pfu live HSV-1 in 2 μl) into one anterior chamber per mouse. Alternating control eyes received either medium or no injection at all. On each examination day, 2–3 mice were selected randomly for histopathological study. Eyes from these mice were fixed individually and processed for hematoxylin and eosin staining.

Assay for active suppression of delayed hypersensitivity (DTH) responses. In order to test for the presence of an active suppressor cell population in mice that received HSV-1 by the intracameral (IC) or intravenous (IV) routes, adoptive transfer experiments were performed in which spleen cells from mice that had been inoculated with HSV-1 IC or IV were tested for their ability to suppress DTH responses to a subsequent SC inoculation of HSV-1. Seven days after IC or IV inoculation with HSV-1 (4 × 10^4 pfu), spleen cells were harvested and injected intravenously into naive BALB/c recipients

Departments of Cell Biology and Ophthalmology, University of Texas Health Science Center at Dallas, Dallas, Texas.

(one splenic equivalent/recipient). One to two hours later, recipient mice were immunized subcutaneously with 4×10^4 pfu of live virus. Control groups received either the SC dose of virus and no cells (positive control), or nothing (negative control). Six days later, all mice were challenged in their footpads with UV-HSV or mock-infected supernatant, and footpad swelling was measured 24 hours later. The percent suppression was determined by the following formula:

$$\left(1 - \frac{Exp. - Neg.}{Pos. - Neg.}\right) \times 100$$

Results

Clinicopathological findings after unilateral intracameral inoculation of HSV-1

An intense keratouveitis developed in eyes that received HSV-1. Within 1–3 days, maximal corneal edema and neovascularization occurred. The cornea, iris, and anterior chamber contained an acute inflammatory infiltrate which persisted through day 10 (Figs. 1A, B; 2A, B). The posterior segment (vitreous, retina, and choroid) remained normal throughout the 21-day examination period with the exception of a minor inflammatory infiltrate in the ganglion

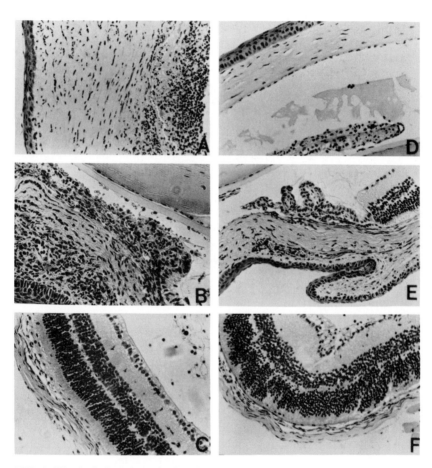

FIG. 1. Histological sections of paired injected *(A–C)* and contralateral uninjected *(D–F)* BALB/c eyes 7 days after intracameral inoculation of 4×10^4 pfu of live HSV-1. Cornea *(A, D)*, ciliary body *(B, E)* and retina *(C, F)*. Note an intense keratouveitis in the injected eye. A milder uveitis is present in the contralateral eye. A small focus of inflammation/proliferation is visible in the ganglion cell layer of the control retina (F).

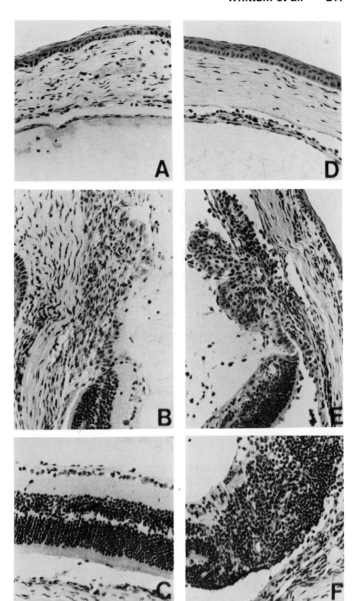

FIG. 2. Typical views of paired BALB/c eyes 10 days after unilateral inoculation of HSV-1 into the anterior chamber. Paired injected *(A–C)* and uninjected *(D–F)* eyes are shown: cornea *(A, D)*, ciliary body *(B, E)*, and retina *(C, F)*. In addition to the pathologic changes evident at day 7 is focal necrosis of the contralateral retina *(F)*.

cell layer of the retina, which was present by day 3 and persisted for 21 days (Fig. 1C). The acute inflammatory infiltrate in the anterior chamber and iris was gradually replaced by mononuclear cells, and by day 14 fibrovascular tissue filled the anterior chamber, leaving no normal iris or ciliary body tissue detectable (not shown).

Pathologic changes were observed in contralateral eyes regardless of whether they had received an injection of medium or no injection at all. Over the initial 5–6 days postinfection (p.i.), uveitis developed in these control eyes (albeit to a milder extent than that observed in eyes that received the virus inoculation) (Cf. Fig. 1B and 1E). Although the cornea and anterior chamber of control eyes remained normal throughout the examination period, dramatic pathologic changes developed in the posterior segments

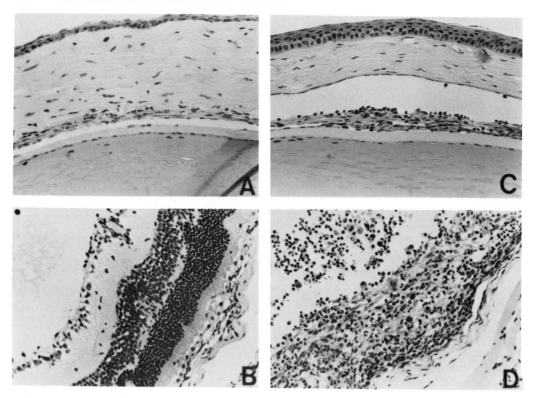

FIG. 3. Typical views of injected *(A, B)* and uninjected eyes *(C, D)* 14 days after intracameral inoculation of virus. Cornea *(A, C)* and retina *(B, D)* are shown. Note necrosis of the entire contralateral retina, and the mixed inflammatory infiltrate extending from the vitreous through the choroid *(D)*.

of these eyes. By day 7, 1–2 pinpoint foci of infiltrating cells were observed histologically in the ganglion cell layer of the retina (Fig. 1F). As shown in Figure 2F, localized necrosis through all layers of the retina was obvious by day 10 p.i. At this time, necrotic areas (usually 1/eye) were bordered by normal retina. Infiltration of the choroid was limited to areas underlying necrotic retina. By day 14, necrosis encompassed the entire contralateral retina and choroiditis was no longer focal (Fig. 3D). The inflammatory infiltrate was composed of mononuclear cells, including numerous plasma cells. By day 21, fibrovascular tissue had replaced the entire retina and choroid (not shown).

Histopathological findings following bilateral inoculation of HSV-1 into the anterior chamber

The differential retinal pathology observed after inoculation of HSV-1 into one anterior chamber raised the possibility that a local protective mechanism (e.g., interferon, antibody) was operative in inoculated eyes. To address this possibility, HSV-1 was inoculated into both anterior chambers of a group of BALB/c mice. As Figure 4 demonstrates, *both* retinas were preserved following bilateral inoculation of HSV-1 18 days previously. Similar results were obtained in another experiment in which mice were sacrificed on day 42 p.i. These results suggest that a local protective mechanism occurs in inoculated eyes to prevent retinal destruction. The nature of this protection is currently under study.

Active suppression produces depressed T-cell responses to intracameral HSV

In studies reported elsewhere,[4] we found that IC (and IV) inoculation of HSV-1 resulted in depressed T-cell-mediated responses despite intact (or enhanced) B-cell-

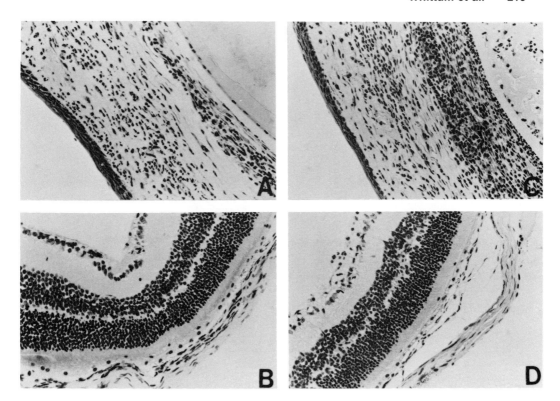

FIG. 4. Typical views of corneas *(A, C)* and retinas *(B, D)* 18 days after bilateral inoculation of HSV-1 into each anterior chamber. Despite an intense keratouveitis in each anterior segment, both retinas are preserved.

mediated responses in comparison to responses to SC inoculation of the same dose of virus. The deviant immune responses observed following inoculation of HSV-1 by the IC route are similar to those observed previously after intracameral inoculation of P815 tumor cells into BALB/c mice.[6-8] In an effort to determine the mechanism behind the impaired T-cell responses to HSV-1, we tested in adoptive transfer experiments the ability of spleen cells from mice that had received HSV-1 intracamerally to suppress the DTH responses of naive mice to a subsequent SC priming dose of virus. As Table I demonstrates, cells from donors that had received virus intracamerally were capable of suppressing DTH responsiveness by 63.7%. Of note were the results of another set of experiments in which cells from donors that had received HSV-1 by the IV route were transferred: Equivalent suppression of DTH responsiveness to HSV-1 inoculated SC was evoked by cells

induced by IC or IV inoculation of virus (74.3% vs 89.9% suppression, respectively) (Fig. 5).

These results indicate that a suppressor cell population is responsible for depressed T cell responses induced by IC or IV inoculation of HSV-1. Further characterization of the suppressor population will be described elsewhere.

Discussion

We have described a murine model of herpes simplex-induced ocular disease following intracameral inoculation of virus. This model is characterized as follows: (1) a reproducible distinctive keratouveitis develops when virus is injected into the anterior chamber of the mouse eye; (2) uninjected control eyes develop a milder uveitis after a delay of 5–7 days and a severe retinochoroiditis by day 14; (3) virus can be isolated from homogenates of either eye at

TABLE I

In vivo Evidence of Active Suppression of DTH Responses Following Intracameral Inoculation of HSV-1[a]

Group	Spleen cells transferred (IC) (day 0)	Immunizing dose (SC) (day 0)	Footpad response[b] ($\times 10^{-4}$ in) (day 7)	Percent of positive control[c]
1	0	0	18.8 ± 39.4[b]	
2	0	4×10^4 pfu	393.5 ± 122.1	
3	4.4×10^8	0	59.0 ± 18.6	
4	4.4×10^8	4×10^4 pfu	154.8 ± 10.6	36.3

[a] Spleen cell donors were inoculated IC on day (-7) with 4×10^4 pfu live HSV-1(KOS). On day 0, spleen cells were harvested and transferred to recipients (1 spleen equivalent/recipient). One to two hours later, recipients were immunized SC with 4×10^4 pfu live HSV-1; on day 6, one hind footpad was challenged with 10^6 pfu UV-inactivated HSV-1, and 24 hours later, footpad swelling was determined.

[b] Footpad swelling response = Mean (24–0 hr experimental) − Mean (24–0 hr control) footpad $\times 10^{-4}$ inches.

[c] Mice that received no cells but were subcutaneously immunized served as positive controls; mice that received neither served as negative controls.

$$\text{Percent of positive control} = \frac{\text{Experimental} - (-)\,\text{Control}}{(+)\,\text{Control} - (-)\,\text{Control}}$$

least until day 10 p.i. (data not shown); (4) anti-HSV T-cell-mediated immune responses are depressed; (5) a suppressor cell population is responsible for depressed T-cell responses; (6) anti-HSV neutralizing antibody levels are enhanced compared with levels elicited by subcutaneous priming; and (7) the deviant immune responses elicited by intracameral inoculation of HSV-1 are elicited also by intravenous inoculation of the same dose of virus.

Oh and his colleagues have described the pathogenesis of herpetic infection in rabbits after bilateral inoculation into their anterior chambers.[1] However, use of an immunologically well-defined species such as the inbred mouse will allow more precise differentiation between the viral and immunologic aspects of herpes-induced ocular disease, and may aid in finding immunological means to alter the course of active ocular infections. The presence of high levels of circulating anti-HSV antibody in conjunction with suppressed T-cell responses

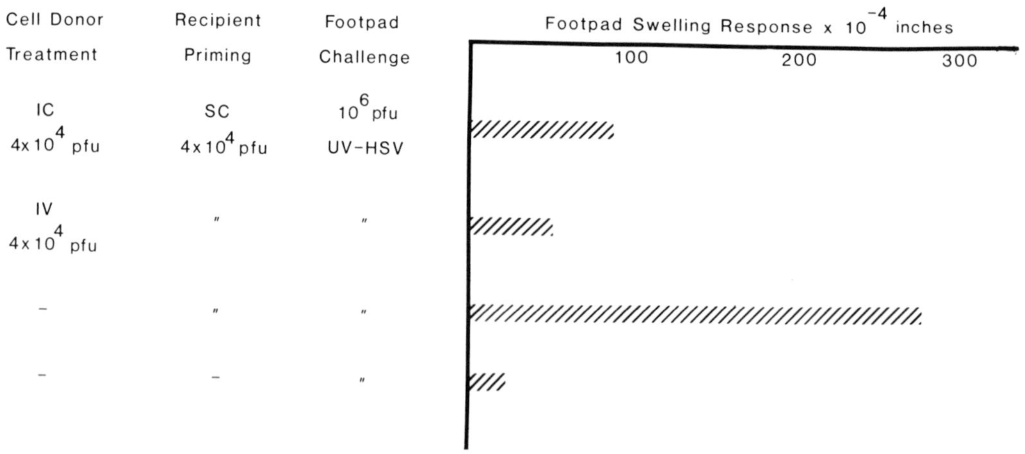

FIG. 5. Suppression of the delayed hypersensitivity response to subcutaneous (SC) priming with live HSV-1 (4×10^4 pfu) by adoptive transfer of spleen cells from donor BALB/c mice inoculated 7 days previously with HSV-1 either intracamerally (IC) or intravenously (IV).

to virus may explain why herpetic ocular infections display a pattern of progressive infection alternating with latent infections rather than resolution. When antigen reaches the anterior chamber and/or anterior uveal tract, effector T-cell mechanisms are suppressed. At this point, antibody alone, or in concert with inflammatory cells and their mediators, appears to control viral infections inadequately. The dichotomy of B- and T-lymphocyte responses to intracameral herpes simplex virus also demonstrates that the presence of circulating antibody does not denote an appropriate antiviral immune response with respect to T-cell function.

Acknowledgments

Supported in part by a Fight For Sight Postdoctoral Research Fellowship, Fight For Sight, Inc., New York; a grant-in-aid from The National Society to Prevent Blindness; and NIH grants EY-03119 and CA-09082.

References

1. Oh, JO: In *Immunology and Immunopathology of the Eye*, Silverstein AM and O'Connor GR (Eds.). Masson, New York, 1979, pp. 248–255.
2. Martenet AC: *Arch Ophthalmol* 76:858–865, 1966.
3. Meyers-Elliott RL and Chitjian PA: In *Immunology and Immunopathology of the Eye*, Silverstein AM and O'Connor GR (Eds.). Masson, New York, 1979, pp. 241–247.
4. Whittum JA et al.: *Curr Eye Res* 2:691–697, 1983.
5. Nagy, RM et al.: *Invest Ophthalmol Vis Sci* 19:271–277, 1980.
6. Niederkorn JY et al.: *Invest Ophthalmol Vis Sci* 20:355–363, 1981.
7. Streilein JW et al.: *J Exp Med* 52:1121–1125, 1981.
8. Whittum, JA et al.: *Transplantation* 34:190–195, 1982.

chapter 55

Analysis of the Induction of Anterior Chamber-Associated Immune Deviation (ACAID)

Jerry Y. Niederkorn and J. Wayne Streilein

Recently, we have explored immune privilege in the anterior chamber (AC) of the mouse eye.[1] Presentation of allogeneic P815 mastocytoma (DBA/2) cells into the AC of BALB/c mice produces specific suppression of anti-DBA/2 cell-mediated immunity with three distinct manifestations: (1) progressive growth of intraocular allogeneic tumors; (2) specific suppression of systemic allograft immunity against DBA/2 alloantigens; and (3) transient growth of subcutaneously injected P815 tumor cells. The generic term anterior chamber-associated immune deviation (ACAID) was used to describe these multidimensional effects.[1]

The spleen plays a pivotal role in the induction of ACAID, since premature removal of this organ prevents the development of immunosuppression and, in fact, permits full development of specific, cell-mediated immunity following intracameral (IC) inoculation of allogeneic cells.[2] These findings suggest that a unique immunologic interplay between the eye and spleen is necessary for the induction of ACAID.

In the present study we examined the antigen-specific signal of ACAID that originates in the eye and is transmitted to the spleen.

Departments of Ophthalmology, Cell Biology, and Internal Medicine, The University of Texas Health Science Center at Dallas, Dallas, Texas.

Materials and Methods

Mice. Adult female BALB/c (H-2^d), DBA/2 (H-2^d), LP/J (H-2^b) and C57BL/6 (H-2^b) mice were purchased from Jackson Labs (Bar Harbor, ME). BALB/c and DBA/2 mice share similar H-2 haplotypes but differ at multiple minor histocompatibility loci. C57BL/6 and LP/J mouse strains share the H-2^b haplotype but are disparate at multiple minor histocompatibility loci.

Tumor cells. B16F10 melanoma (C57BL/6 origin) and P815 mastocytoma (DBA/2 origin) were cultured as described previously.[3]

Procedures. Techniques for intracameral (IC) tumor cell inoculation, skin grafting, enucleation, and detection of extraocular tumor metastases have been described elsewhere.[1–3]

Results

Effect of injection route on the alloimmune response to DBA/2 alloantigens

Intracameral inoculation of 2×10^5 P815 mastocytoma cells fails to induce anti-DBA/2 transplantation immunity in BALB/c mice.[1] Since this dose of tumor cells is relatively small, we wished to determine if this failure to respond was due to the small dose of cells or because alloantigen presentation via the AC was a unique process. Panels of BALB/c mice received 2×10^5 P815 cells inoculated intravenously (IV), intraperitoneally (IP), IC, or subcutaneously (SC). Fourteen days later all mice were challenged with full-thickness DBA/2 skin grafts. The results show that mice receiving IC inoculations of P815 cells failed to reject DBA/2 skin grafts but were able to reject "third party" (C57BL/6) skin grafts in a normal, first-set fashion (Table I). By contrast, recipients of P815 cells administered SC, IP, or IV were able to briskly reject DBA/2 skin grafts. Thus, the route of alloantigen presentation determines the nature of the host's systemic alloimmune response; antigen presentation by extraocular routes fails to evoke ACAID. Induction of ACAID requires that alloantigen presentation occur via the AC.

Induction of ACAID requires the temporary presence of an anatomically intact eye

The unique role of the eye in the induction of ACAID might be explained by one of the following. The AC might act as an antigen depot that provides a continual, low amplitude tolerogenic stimulus delivered by vascular routes. Conversely, tissues within the AC might "process" alloantigens in situ and render them tolerogenic prior to systemic release and dissemination. A third possibility is that the eye itself functions as an autonomous immunoregulatory organ and elaborates suppressor elements that modulate the activities of peripheral lymphoid organs such as the spleen and lymph nodes.

To explore these possibilities, P815 cells were inoculated IC into the left eyes of BALB/c mice and the eyes were enucleated 2, 7, and 10 days later. All mice received orthotopic DBA/2 skin grafts 14 days after the initial IC inoculations. The results show that enucleation 2 days post-IC inoculation successfully prevented the induction of ACAID: DBA/2 skin grafts were rejected in first-set tempos with a median survival time (MST) = 11.8 ± 0.8 days (N = 9). However, the induction of ACAID was established by day 10 post-IC inoculation, since enucleation of the tumor-containing eye at this time did not prevent the development of impaired anti-DBA/2 immunity: these

TABLE I

Influence of Intracameral Inoculation of P815 Cells on the Alloimmune Response BALB/c Mice to DBA/2 Minor Alloantigens

Primary exposure to DBA/2 alloantigens[a]	Strain of skin graft donor[b]	Median survival time of subsequent skin grafts[c]
None (first set)	DBA/2	10.8 (10.2–11.8)
P815 cells injected subcutaneously	DBA/2	7.0 (6.5–7.5)
P815 cells injected intraperitoneally	DBA/2	8.3 (7.9–8.8)
P815 cells injected intravenously	DBA/2	11.5 (10.2–12.9)
P815 cells injected intracamerally	DBA/2	>30
None (first set)	C57BL/6	11.0 (10.5–11.5)
P815 cells injected intracamerally	C57BL/6	11.5 (9.2–12.8)

[a] P815 cells (2×10^5/mouse).
[b] Skin grafts applied 14 days after primary exposure to DBA/2 alloantigens.
[c] Days + 95% confidence limits.

mice failed to reject DBA/2 skin grafts (MST > 30 days; n = 9). Enucleation at 7 days had an intermediate effect: a considerable number of mice retained their skin grafts beyond 30 days (MST = 15.0 ± 3.4 days; n = 8). Thus, induction of ACAID requires the presence of an anatomically intact eye; however, the presence of the intact eye is unnecessary beyond 10 days of IC inoculation.

Splenic metastasis of intraocular P815 mastocytomas

We have previously demonstrated that induction of ACAID requires the presence of an intact spleen.[2] Since P815 cells are known to metastasize rapidly to the spleen following SC inoculation,[4] we wished to determine whether a similar pattern of splenic metastasis followed IC inoculation and might therefore play an integral role in the induction of ACAID.

IC-injected hosts were examined for splenic metastases using methods described previously.[3] Briefly, panels of BALB/c mice received IC inoculations of P815 cells on day 0 and the spleens were harvested aseptically on days 2, 3, 4, 5, 6, 10, and 14. Monocellular spleen cell suspensions were cultured *in vitro* and inoculated SC into normal DBA/2 recipients who were then observed for tumor growth. Within 3 days of IC inoculation, and at every time point tested thereafter, P815 cells were detected in the spleens of IC-inoculated BALB/c mice.

Thus, the facile dissemination of IC-injected tumor cells to the spleen suggests that viable, blood-borne tumor cells might be the inductive signal of ACAID. In order to analyze this possibility, we employed an ancillary model of ACAID by using a tumor (i.e., B16F10 melanoma) that metastasizes exclusively to the lung and not to the spleen.[5–7]

B16F10 melanoma model of ACAID

B16F10 melanoma (C57BL/6 origin) cells were inoculated either IC (10^5 cells/eye) or SC (2×10^5 cells) into panels of LP/J mice. Fourteen days later, all mice were challenged with C57BL/6 skin grafts. The results show that LP/J mice receiving IC inoculations of B16F10 cells fail to reject C57BL/6 skin grafts (Table II). By contrast, recip-

ients of B10F10 melanoma cells inoculated SC were sensitized and rejected their skin grafts at a tempo indicative of specific sensitization (i.e., second set rejections). Thus, IC inoculation of B16F10 cells into the AC of LP/J mice produces systemic immunosuppression parallel to the ACAID phenomenon observed in BALB/c mice inoculated IC with P815 mastocytoma cells.

Absence of splenic metastases in LP/J mice harboring intraocular B16F10 melanomas

C57BL/6 and LP/J mice harboring IC B16F10 melanomas were examined for the presence of splenic metastases. Panels of mice received IC inoculations of B16F10 melanoma cells (10^5 cells/eye) on day 0 and the spleens were examined as described above. Unlike P815 mastocytoma, B16F10 melanoma cells were not found in spleens 2, 3, 4, 5, 6, 10, or 21 days after IC inoculation. Moreover, histologic examination failed to reveal evidence of metastatic foci in the spleens of any hosts, including mice receiving melanoma cells injected IC 21 days earlier. Even though B16F10 melanoma cells preferentially metastasize to the lung, no metastatic foci were found in the lungs of any mice, including those harboring enormous intraocular tumors.

Thus, induction of ACAID in LP/J mice does not require metastasis of intraocular B16F10 melanoma cells. The antigenic signal of ACAID, which is transmitted from the eye and is transduced within the spleen, need not be in the form of viable tumor metastases.

Discussion

The present results offer several insights into the induction of ACAID. First, the induction of ACAID requires that alloantigen presentation occur through the AC; other routes of antigen administration fail to evoke systemic unresponsiveness. The second requirement for the induction of ACAID is that the injected eye must remain intact for at least 10 days; premature removal of the eye prevents development of ACAID.

We have previously shown that the induction of ACAID requires a functional, in-

TABLE II
Induction of ACAID in LP/J Mice by Intracameral Inoculation of B16F10 Melanoma Cells

Primary exposure to C57BL/6 alloantigens	N[b]	Median survival time of subsequent C57BL/6 skin grafts[c]
None (first set)	10	8.7 (8.2–9.2)
C57BL/6 skin graft (second set control)	10	7.0 (6.8–7.2)
B16F10 cells injected intracamerally[a]	24	>30
B16F10 cells injected subcutaneously[a]	10	7.0 (6.5–7.5)

[a] B1610 cells (2×10^5/mouse).
[b] Number of mice.
[c] Days + 95% confidence limits.

tact spleen.[2] Since IC-injected P815 cells disseminated rapidly to the spleen, it seemed likely that these viable, alloantigen-bearing cells served as the inductive signal of ACAID. However, this was not found to be the case when an ancillary model of ACAID was employed. IC-injected B16F10 melanoma cells induced ACAID in LP/J mice, yet there was neither gross nor microscopic evidence of extraocular tumor metastases. Moreover, B16F10 cells could not be cultured from the spleens of intraocular tumor-bearing hosts. Thus, induction of ACAID does not require the systemic dissemination of viable tumor cells.

We conclude, although there is no direct evidence, that local antigen processing occurs within the eye and an antigen-specific signal is transmitted from the eye, via the vasculature, to the spleen where the immunosuppressive elements responsible for the development of ACAID are generated.

Since an intact eye and spleen are only temporarily required, we further conclude that the suppressor elements of ACAID eventually become systemically dispersed and are subsequently independent of splenic or ocular influence.

Acknowledgment

This work was supported by USPHS grants EY 03319 and CA 30276.

References

1. Niederkorn JY et al.: *Invest Ophthalmol Vis Sci* **20**:355–363, 1981.
2. Streilein JW and Niederkorn JY: *J Exp Med* **153**:1058–1067, 1981.
3. Niederkorn JY and Streilein JW: *J Immunol* **128**:2470–2474, 1982.
4. Haran-Ghera N: *J Immunol* **126**:1241–1244, 1981.
5. Fidler IJ: *Nature New Biol* **242**:148–149, 1973.
6. Fidler IJ et al.: *Cancer Res* **36**:3160–3165, 1976.
7. Talmadge JE et al.: *J Natl Cancer Inst* **65**:929–935, 1980.

chapter 56

Suppressor T-Lymphocytes Mediate Immunologic Tolerance after Anterior Chamber Antigen Presentation

Richard P. Wetzig, C. Stephen Foster, and Mark I. Greene

Our work concerns active cellular regulation of immune responses in the anterior chamber of the eye, and how such regulation might play a role in the so called "immune privilege" of the anterior chamber. We have adapted a system wherein the hapten azobenzenearsonate (ABA) is used in A-strain mice to induce a cascade of regulatory T-cell interactions [1,2] to study immune responses in the anterior chamber of the eye. Our results suggest that immunologic unresponsiveness following anterior chamber priming is at least in part the result of active cell-mediated suppression of the immune response.[3,4]

Materials and Methods

Induction of DTH

Inbred A/J mice (Jackson Laboratories, Bar Harbor, Maine) were used in these experiments. Details concerning the preparation of ABA derivatized spleen cells (ABA-SC) are discussed elsewhere.[1] Mice were sensitized by the subcutaneous injection of 3×10^7 ABA-SC. At day 5, DTH was assessed by injection of 25 μ of activated ABA into the left hind footpad. Twenty-four hours later, the footpad swelling was measured with an engineer's micrometer. The differ-

ence between the injected left footpad and the uninjected right footpad was used as an index of the degree of sensitization. Negative control mice were not sensitized by subcutaneous injection of ABA-SC, but were challenged by injection to the left footpad of 25 μ of activated ABA and read at 24 hours.

Induction of immune unresponsiveness (tolerance)

Mice were primed, as before, with 3×10^7 ABA-SC injected subcutaneously. Unresponsiveness to ABA was induced in subcutaneously primed mice by injecting 5×10^7 ABA-SC intravenously. Other subcutaneously primed mice received 6×10^6 ABA-SC injected subconjunctivally. While still others received 6×10^6 ABA-SC injected into the anterior chambers (bilateral injections of 3×10^6 ABA-SC into each eye). Positive control mice were primed subcutaneously only, while negative control mice were not primed at all. All experimental animals were footpad challenged at 5 days and read 24 hours later, as before. In all experiments testing immune unresponsiveness, positive and negative control groups were included in parallel with experimental groups. The formula used to calculate unresponsiveness was:

% Unresponsiveness

$$= \frac{\text{Experimental-negative control}}{\text{Positive control-negative control}} \times 100\%$$

Departments of Ophthalmology and Pathology, Harvard Medical School, Eye Research Institute of Retina Foundation, Boston, Massachusetts.

Materials and Methods

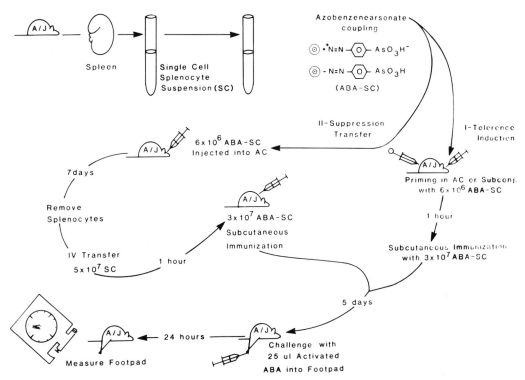

FIG. 1. Procedures for immunization and testing of A/J mice.

Each experimental group consisted of four mice. A given experiment was considered valid only if the footpad increment in the positive control group exceeded 20×10^{-2}mm in swelling and the footpad increment in the negative control group was less than 10×10^{-2}mm in swelling, which was the case in the majority of experiments run. Statistical significance was calculated using the Student's t test.

Active transfer of immune unresponsiveness (suppression protocol)

In these experiments mice were primed for the adoptive transfer of immune suppression by spleen cells. Mice received either 5×10^7ABA-SC intravenously or 6×10^6ABA-SC in the anterior chamber. Seven days later their spleen cells were harvested and 5×10^7 of these cells were injected intravenously into mice primed subcutaneously for DTH to ABA that same day. Some of the spleen cells were treated with anti-Thy 1.2 plus complement, anti-CRI (cross reactive idiotype) plus complement, or complement alone prior to transfer, as described previously.[5] Challenge with ABA and footpad measurement were done, as described earlier, 5 days following transfer and subcutaneous priming. Some mice were primed for DTH to the unrelated hapten TNP by injection of 3×10^7TNP-SC subcutaneously and 5 days later challenged by injection of 1×10^6TNP-SC into the left footpad, as previously described.[6] The influence of spleen cell transfer from mice primed in the AC with ABA-SC on DTH to TNP was studied to determine the hapten specificity of the suppressor cell activity.

TABLE I
Anterior Chamber-Induced Suppressor Cells Are Hapten Specific T-cells Lacking CRI Surface Receptors

	Splenocyte treatment prior to transfer[a]	Recipient priming and challenge	% Suppression of DTH response[b]
A	—	ABA	55
B	—	TNP	21
C	Anti-Thy 1.2 + C'	ABA	27
D	Anti-CRI + C'	ABA	85
E	C' alone	ABA	64

[a] Splenocyte donors were primed with 6×10^6 ABA-SC in the anterior chamber 7 days prior to transfer.

[b] Results represent the average % suppression in seven separate experiments that matched different experimental groups simultaneously, i.e., ABA priming vs. TNP priming of recipients or antibody treatment of transferred splenocytes vs. treatment with C' alone and no treatment.

The kinetics of AC and IV induced suppressor cell activity

In these experiments mice were primed subcutaneously for DTH to ABA as before. That same day they received IV transfer of 5×10^7 spleen cells from mice primed either with 6×10^6 ABA-SC in the AC or 5×10^7 ABA-SC intravenously 7 days previously. Other mice received the same IV- or AC-induced spleen cells 4 days following subcutaneous priming for DTH to ABA (i.e., 1 day prior to footpad challenge).

Results

Animals were rendered tolerant to ABA if they were inoculated either in the anterior chamber or intravenously with ABA-SC when subcutaneous priming was carried out. Subconjunctival priming did not significantly inhibit DTH, however.

Adoptive transfer of spleen cells from AC primed mice rendered the recipient mice unresponsive to priming for DTH to ABA (Table I). Priming for DTH to TNP, however, was not suppressed by transfer of spleen cells from animals primed in the anterior chamber with ABA-SC (Table I, Line B), establishing specificity of the AC induced suppressor cell for the ABA hapten.

Prior treatment of AC-induced suppressor cells with anti-Thy 1.2 plus complement abrogated that subpopulation of spleen cells leading to suppression (Table I, Line C).

This established the identity of AC-induced suppressor cells as T-cells (thymus derived lymphocytes).

Treatment of AC-induced suppressor cells with anti-CRI plus complement did not abrogate suppression (Table I, Line D). Therefore, the AC induced suppressor cells do not bear CRI surface receptors.

Treatment of AC induced suppressor cells with complement alone did not impair their ability to suppress DTH to ABA (Table I, Line E).

AC-induced suppressor cells as well as IV-induced suppressor cells inhibited elicitation of DTH when injected on the same day as subcutaneous priming (Fig. 2). Thus, both populations of cells are capable of blocking the afferent phase of the DTH response. In contrast, the AC-induced suppressor cells blocked the DTH response when injected 4 days after subcutaneous priming, but the IV-induced suppressor cells did not. Apparently AC-induced suppressor cells, unlike IV-induced suppressor cells, are capable of blocking the efferent phase of the DTH response.

Discussion

It has been known for some time that the nature of an immune response is in part determined by the anatomical site at which antigen is initially encountered and processed.[7,8] To account for such differences in regulation of cellular immune responses,

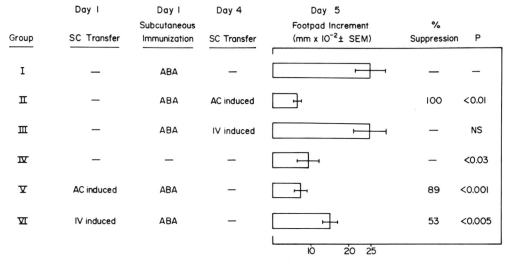

FIG. 2. Footpad increments as a measure of changes in delayed hypersensitivity.

it has been postulated that the immune system is divided between central and peripheral compartments.[9] The former would contribute to immune homeostasis through a system of suppressor/effector interactions. The latter would activate immune responses against assault from foreign antigenic insult with overt hypersensitivity and inflammation.

With subcutaneous priming, antigen is first processed through a peripheral lymphatic drainage system rich in antigen-presenting cells including macrophages and dendritic cells. This leads ultimately to activation of delayed hypersensitivity and other forms of immune response. With intravenous priming, antigen is first processed in the central reticulo-endothelial system that includes the spleen. This activates a network of interacting suppressor T-lymphocytes which dampens the response to that antigen.[5] Our observation that AC priming likewise activates hapten-specific, T-cell mediated suppression would suggest that the basis for the so-called immunologic privilege of the anterior chamber rests on its functional membership within the central division of the immune system.

It is interesting that suppressor T-cells induced by AC priming differ from suppressor T-cells induced by IV priming in both their phenotypic (cell surface antigen) and functional (mode of action) characteristics. Ts induced by IV administration of antigen bear CRI surface receptors.[10] We have shown herein that AC-induced Ts do not. Ts induced by IV administration of antigen block only the afferent limb of the immune response, during which effector T-cell clones are presumably expanded.[11] We have shown that AC-induced Ts likewise block the afferent limb. Unlike IV-induced Ts, however, AC-induced Ts also block the efferent limb of the immune response during which ABA specific effector cells express their function. Ts induced by IV injection of antigen produce a soluble factor that induces second order Ts. These cells resemble AC-induced Ts in lacking CRI surface receptors and in blocking both afferent and efferent limbs of the immune response.[11] Such suppressor T-cells differing functionally have also been demonstrated in the murine immune response to the hapten DNP.[12,13] Our studies suggest that there is perhaps a fundamental difference between the immune response to AC priming and the response to IV priming.

Acknowledgment

Supported in part by PHS Grants EY #03063 and A1 #16396 from the National Institutes of Health.

References

1. Back BA et al.: *J Immunol* **121**:1460, 1978.
2. Green MI and Sy MS: *Fed Proc* **40**(5):1458, 1981.
3. Foster CS and Wetzig RP: *Surv Immunol Res* **1**:93, 1982.
4. Wetzig RP et al.: *J Immunol* **128**:1753, 1982.
5. Sy MS et al.: *J Exp Med* **153**:1415, 1981.
6. Greene MI et al.: *J Immunol* **120**:1604, 1978.
7. Sulzbeuger MB: *Arch Dermatol Syphilol* **20**:669, 1929.
8. Chase NW: *Proc Soc Exp Biol Med* **61**:257, 1946.
9. Greene MI and Bach BA: *Cell Immunol* **45**:446, 1979.
10. Dietz MH et al.: *J Immunol* **125**:2374, 1980.
11. Dietz MH et al.: *J Exp Med* **153**:450, 1981.
12. Miller SD et al.: *J Immunol* **121**:265, 1978.
13. Miller SD et al.: *J Immunol* **121**:274, 1978.

chapter 57

Mycoplasma arthritidis-Induced Experimental Uveitis

Charles E. Thirkill and Dale S. Gregerson

Mycoplasma arthritidis is associated with the induction of pathogenic processes in rodents: primarily arthritis, and to a lesser extent conjunctivitis and uveitis. Isolation of mycoplasma from patients with rheumatoid arthritis, systemic lupus erythematosus, Reiter's syndrome,[1,2,3] and from contaminated tissue cultures,[4] prompt investigators to maintain an interest in the relevance of these ubiquitous microorganisms to disease.

Since the methods used to identify new mycoplasma isolates involve serological assays, antigenic relationships between species, such as that demonstrated here between *Mycoplasma hominis* and *M. arthritidis,* can lead to confusion.[5,6] The production of specific antisera is complicated by culture medium contamination of the organisms used to incite the response. Monoclonal antibodies against specific microbial antigens present the means of bypassing most of the immunological artifacts encountered in routine identification procedures. Accordingly, we have prepared a hybridoma that secretes monoclonal antibodies reactive with the microbial enzyme arginine deiminase. Indirect immunofluorescence with this monoclonal antibody indicates the presence of the mycoplasma enzyme within the stroma of ciliary bodies obtained from the eyes of *M. arthritidis*-infected Sprague-Dawley rats.

The uveitis-inducing characteristics of *M.*

Department of Ophthalmology and Visual Science, Yale University School of Medicine, New Haven, Connecticut.

arthritidis, together with the ability of this microorganism to exist in a form closely resembling common diphtheroids, give this animal model interesting novel aspects, which might explain some of the unanswered questions associated with idiopathic conjunctivitis and iridocyclitis.

Materials and Methods

1. Mycoplasma arthritidis was cultured as described.[7] Trypticase tellurite agar (Baltimore Biological Laboratories) contained 10% horse serum and 0.01% potassium tellurite. The specific antisera growth inhibition method [8] was used to verify the identity of *M. arthritidis*.

2. Mycoplasma-induced anterior uveitis: Four female Sprague-Dawley rats were given single tail vein inoculations of approximately 1×10^{11} cfu (colony forming units) in a volume of 0.1 ml. Two other rats used as negative controls received 0.1 ml of the mycoplasma growth medium through the same route. Five days later all six rats were killed, exsanguinated, and enucleated. Eyes were fixed in 2% glutaraldehyde in PBS at pH 7.2 for histopathology techniques.[9]

3. Arginine deiminase was extracted from *M. arthritidis* and purified by column chromatograhy.[10]

4. Swiss Webster mice were used to produce a hybridoma reactive with arginine deiminase according to the method previously described.[11] The IgG2a antibodies synthesized by this hybridoma are identified as Ma-1.

TABLE I
Standardized Protein Homogenates

Antibodies	M. arthritidis	M. hominis	M. pulmonis
Mouse anti-M. arthritidis	1:12800	<1:10	<1:10
Mouse anti-arginine deiminase	1:6400	<1:10	1:50
Ma-1	1:8000	1:25	<1:10
Rabbit anti-M. hominis	1:400	1:6400	1:50
Rabbit anti-M. pulmonis	1:10	<1:10	1:400

5. The ELISA was performed as described by Voller,[12] upon homogenates of *M. arthritidis, M. pulmonis,* and *M. hominis* standardized for equal protein content, using commercially obtained antisera as controls.

6. *M. arthritidis* was labelled with ^{125}I using the Bolton-Hunter method.

7. The SDS-Polyacrylamide gel electrophoresis (SDS-PAGE) method described by Laemmli[13] was used, applying a 4-25% polyacrylamide gradient.

8. Immunoprecipitation: Immunoglobulin, bound ^{125}I labelled antigens were adsorbed with protein-A-Sepharose (Pharmacia). Unbound material was removed by three washes in PBS-NP·40 prior to solubilization of the adherent complexes in SDS sample buffer and electrophoresis. Completed gels were stained, dried, and autoradiographed on Kodak X-AR film using Cronex intensifying screens.

9. Indirect Immunofluorescence: The ability of Ma-1 (1:800 dilution) to bind *M. arthritidis* antigen was affirmed on monolayers of kidney cells infected with this mycoplasma. Deparaffinized, glutaraldehyde-fixed sections of *M. arthritidis*-infected rat eyes were examined for mycoplasma antigens by indirect immunofluorescence using either whole mouse antisera or Ma-1.[9] The second phase "sandwich" antiserum was rabbit anti-mouse IgG2a (Litton) at 1:100, and the fluorescein labelled conjugate; goat anti-rabbit IgG (Miles) at 1:100. All reagents were reacted for 30 minutes at room temperature and washed in PBS between each phase.

10. Routine culture purity checks on *M. arthritidis* using fresh blood agar plates consistently produced small, alpha hemolytic colonies following 5 days of aerobic incubation at 37°C. Trypticase tellurite agar

plates inoculated with *M. arthritidis* produced black colonies, approximately 1 mm in diameter, under the same conditions. Colonies from fresh blood agar were transferred to Todd-Hewitt broth and incubated aerobically for one week at 37°C before being harvested and assayed for arginine deiminase activity and ELISA reactivity with Ma-1.

Results

ELISA

Anti-arginine deiminase monoclonal antibodies were compared by ELISA for reactivity with homogenates of *M. arthritidis, M. pulmonis,* and *M. hominis* standardized for equal protein content, (Table I). Mouse antisera to *M. arthritidis* homogenate and *M. arthritidis*-derived arginine deiminase reacted vigorously with their homologous antigens. No appreciable binding was evident in the reaction of these sera with *M. pulmonis* homogenate. Ma-1 exhibited strong preferential binding to the protein homogenate of *M. arthritidis* over that derived from *M. hominis*. Indications of cross-reactivity were apparent with rabbit anti-*M. hominis* on the protein homogenate from *M. arthritidis*. Rabbit antiserum directed against *M. pulmonis* was specific with low background reactivity on unrelated antigens.

A separate standardized concentrate of the possible contaminant, initially cultured from *M. arthritidis* on fresh blood agar and propagated further in Todd-Hewitt broth, gave ELISA and arginine deiminase reactions comparable to those attributed to *M. arthritidis*. Control studies utilizing an unrelated hybridoma and normal mouse serum were appropriately negative (not shown).

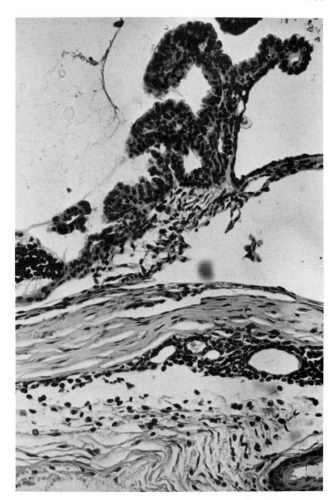

FIG. 1. Histological section of affected eye from a Sprague–Dawley rat showing polymorphonuclear neutrophil influx at base of ciliary body.

Immunofluorescence studies of rat ocular tissues

Since Ma-1 is a murine antibody, specificity problems could arise when assayed on murine tissues by indirect immunofluorescence. To circumvent this problem, an affinity-purified rabbit antibody directed against mouse IgG2a heavy chain class (Litton) was used as the second phase "sandwich" serum to examine the ocular tissues of rats infected with *M. arthritidis*. Infected Sprague-Dawley rats develop a low incidence of acute conjunctivitis followed by anterior uveitis, which strongly resemble the disease process observed in mice.[7] Sections from the eyes of three of the four infected rats exhibited early indications of an anterior inflammatory reaction consisting primarily of a polymorphonuclear neutrophil influx (Fig. 1). Biopsies from these same eyes also bound Ma-1 and fluoresced brightly within the stroma of the ciliary process, particularly around the vasculature (Fig. 2). Whole mouse anti-*M. arthritidis* serum gave much less indication of positive binding when tested using the specific antiIgG2 "sandwich" serum. Substitution with rabbit anti-mouse IgG considerably improved the intensity of the observed fluorescence within the ciliary stroma; however, additional sites of what was considered nonspecific reactivity appeared throughout the retroorbital tissues. Normal mouse serum and the immunoglobulin concentrate from a nonrelated hybridoma gave no indication of fluores-

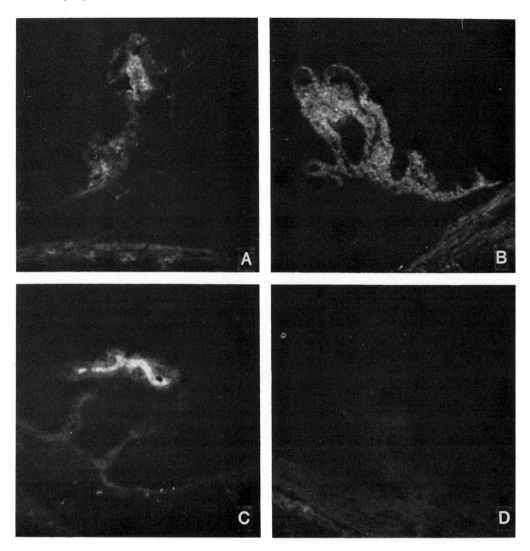

FIG. 2. Indirect immunofluorescence on sectioned eyes of *M. arthritidis*-infected rats using Ma-1 monoclonal antibodies. Localized fluorescence within the stroma of the ciliary process was observed in three (A, B, and C) of the test animals, while the fourth nonsymptomatic test rat *(D)* and the two negative controls (not shown) were nonreactive.

cence. Equivalent sections from uninfected animals tested by this same procedure were all negative.

Properties of the antigen bound by Ma-1

The antigen recognized by Ma-1 can be released from freshly harvested cells by freeze-thawing and sonication in buffered isotonic saline. Approximately one-third of this activity remained in the supernatant following centrifugation at $100,000 \times g$ for 1 hour, suggesting that the protein is not membrane bound. A single band of approximately 49,000 molecular weight was found in protein A adsorbed immunoprecipitations reacting Ma-1 with ^{125}I-labelled total detergent solubilized *M. arthritidis* (Fig. 3). A fraction having the same migratory characteristics, along with others, was also found in immunoprecipitations using mouse anti-arginine deiminase serum.

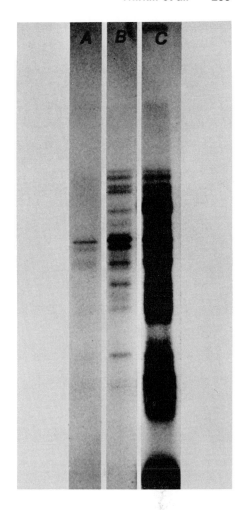

FIG. 3. Polyacrylamide gel electrophoresis: Specific immunoprecipitation of the antigen reactive with Ma-1 monoclonal antibodies, retrieved from ^{125}I labelled *M. arthritidis* homogenate by protein-A immunoglobulin adsorption. Subsequent electrophoresis and autoradiography allowed identification of an antigen that migrates with the characteristics attributed to arginine deiminase; 49,000 daltons, line *A*. Lines *B* and *C* represent the results of the same immunoprecipitation-autoradiography technique using mouse anti-arginine deiminase and mouse anti-*M. arthritidis* respectively on the radiolabelled mycoplasma homogenate.

Discussion

A monoclonal antibody, Ma-1, has been produced that is able to detect *M. arthritidis* within the tissues of infected Sprague-Dawley rats by indirect immunofluorescence. The localization of *M. arthritidis* antigens has been demonstrated to be closely associated with the stromal vasculature of the ciliary body in rats experiencing acute anterior uveitis induced by this mycoplasma. The limitations of light microscopy prevent any clear description of the exact location of the reactive antigens that could be bound to either capillary endothelium or the collagenous connective tissue of the stroma. Although at 5 days post-infection the predominant polymorphonuclear neutrophil infiltration was observed surrounding the blood vessels at the base of the ciliary body, no indication of the binding of specific antibody at this site was detectable with either Ma-1 monoclonal antibodies or whole mouse anti-*M. arthritidis* serum.

Since *M. arthritidis* induces signs in mice and rats that are similar to those seen in rheumatoid arthritis and Reiter's syndrome, the detection of this pathogenic microorganism in other host tissues and its movement through or localization within various organs and tissues could prove interesting to those who use these animal models in the study of the multiple lesions associated with the arthritides.

Acknowledgments

We thank Mary Bannon for typing the manuscript and Gary Fletcher for histology and preparation of sections for immunofluorescence. This work was supported by USPHS grants EY-03660, EY-07000, and EY-00785 from the National Institutes of Health and by unrestricted funds from the Connecticut Lions Research Foundation and Research to Prevent Blindness.

References

1. Bartholomew LE: *Arthritis Rheum* **8**:376–388, 1965.
2. Mardh PA et al.: *Ann Rheum Dis* **32**:319–325, 1973.
3. Oates JK et al.: *Br J Vener Dis* **35**:184–186, 1959.
4. Fogh J (Ed.): In *Contamination in Tissue Culture*. Academic Press, New York, 1973, p. 149.
5. Cole BC et al.: *Proc Soc Exp Biol Med* **124**:103–107, 1967.
6. Edward DG and Freundt EA: *J Gen Microbiol* **41**:263–265, 1965.
7. Thirkill CE and Gregerson DS: *Infect Immun* **35**:775–781, 1982.
8. Clyde WA: *J Immunol* **92**:958–965, 1964.
9. Bosman FT and Kruseman ACN: *J Histochem Cytochem* **27**:1140–1147, 1979.
10. Weickman JL et al.: *J Biol Chem* **253**:6010–6013, 1978.
11. Fazekas de St Groth S and Scheidegger D: *J Immunol Methods* **35**:1–21, 1980.
12. Voller A et al.: In *Manual of Clinical Immunology*, Rose and Gieden (Eds.). 1976, pp. 506–512.
13. Laemmli UK: *Nature* **227**:680–685, 1970.

chapter 58

Specific Nature of Retinal A Antigen and Its Localization

Shigeru Takano, Shingo Yajima, Toshiaki Matsushima, Kentaro Takamura, and Masahiko Usui

It has been suggested that several kinds of soluble antigens are present in the retina. Among them, it is mainly the S antigen that has been subjected to extensive studies, and knowledge of its immunological aspects has gradually been accumulated. On the other hand, studies of the other soluble antigens have been scarce, and there is only one report on the existence of A antigen by Faure.[1] The present study was undertaken to gain new knowledge concerning A antigen, especially its isolation, characterization, and localization.

Materials and Methods

Isolation of retinal soluble antigens

(1) Preparation of porcine retinal extract (PRE): Retinas obtained from pigs were homogenized and centrifuged (10,000 rpm, 15 minutes). The supernatant (PRE) was collected. Bovine retinal extract (BRE), human retinal extract (HRE), and guinea pig retinal extract (GPRE) were prepared likewise. (2) Gel filtration: PRE was filtered through Sephadex G-200 (Pharmacia Fine Chemicals Co.). The buffer solution used was 0.03M phosphate buffer solution at pH 7.6.

Preparation of antisera

(1) Rabbit antinormal-porcine-serum serum (anti-NPS) was made by MILES-YEDA

Co. (2) Rabbit anti-PRE serum (anti-PRE) was made according to the method of Takano et al.[2] (3) Rabbit antiporcine S antigen serum (antipig S) was obtained by repeated injections into rabbits of porcine S antigen (1 mg protein) emulsified in an equal volume of complete Freund's adjuvant (Difco Co.) (C.F.A.). (4) Rabbit antiporcine-retinal-A antigen-gel filtration-fraction II serum (antiFra. II) was obtained by repeated injections into rabbits of freshly prepared porcine retinal A antigen purified by gel filtration. Fraction II (Fra. II) (500μg protein), emulsified in an equal volume of C.F.A., was used as the immunizing antigen. (5) Rabbit antiporcine-retinal-A-antigen-PAG disc electrophoresis serum (antipig A) was obtained by repeated injections into rabbits of freshly prepared porcine retinal A antigen PAG disc electrophoresis fraction (PAG Antigen A) (100μg protein) emulsified in an equal volume of C.F.A.

Presence of retinal soluble antigens

Protein fractions I–V, obtained by gel filtration, and PRE were reacted with anti-PRE and anti-NPS by immunoelectrophoresis.

Isolation of A antigen

As the presence of A antigen was confirmed in Fra. II among the various fractions obtained by gel filtration, Fra. II was subjected to PAG disc electrophoresis. After electrophoresis, the gel was cut into seg-

---

Department of Ophthalmology, Tokyo Medical College, Tokyo, Japan.

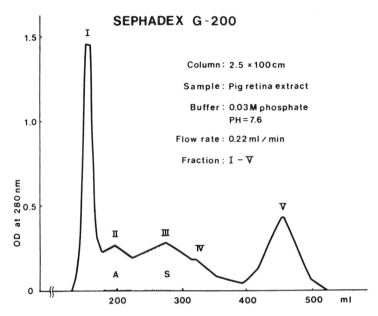

FIG. 1. Isolation by gel filtration (PRE).

ments 5 mm wide (Segments 1–14), and the protein in the segments was extracted with a small amount of 0.85% saline solution. Each extract was subjected to immunodiffusion against anti-NPS, anti-PRE, antipig S, or antipig A serum. The segment found to contain A antigen was reacted with anti-PRE and antipig A by immunoelectrophoresis.

Qualitative analysis of A antigen

(1) PAG disc electrophoresis: The A antigen in various stages of its purification was studied by PAG disc electrophoresis. (2) Isoelectric focusing electrophoresis: Fra. II and PAG antigen A were studied by isoelectric focusing electrophoresis using Ampholine PAGE plates (LKB Co). (3) Organ specificity: (a) Anti-Fra. II was reacted with PRE, BRE, HRE, and GPRE by immunoelectrophoresis. (b) Extracts of porcine brain, liver, and kidney as well as porcine serum were reacted with anti-PRE and anti-NPS by immunoelectrophoresis.

Inoculation of A antigen into animals

Guinea pigs were inoculated with PAG antigen A, and 0.85% saline solution was injected into controls. Antigen (50μg/0.1 ml)

was combined with an equal volume of C.F.A., and the emulsion was inoculated subcutaneously into each of the hind foot pads on one occasion only. Beginning 2 weeks after the inoculation of antigen, guinea pig eyes were examined under a slit lamp. The eyeballs were subsequently enucleated and examined histologically by light microscopy.

Localization of A antigen examined by a direct enzyme-labeled antibody method

Antipig A was labeled with horseradish peroxidase, and the normal rabbit eye was studied by a direct enzyme-labeled antibody method to clarify the localization of A antigen.

Results

Isolation of retinal soluble antigens

Figure 1 shows the plotting of the optical density (O.D.) at 280 nm for each fraction obtained by gel filtration. Five protein fractions, Fra. I–V, were obtained.

Presence of retinal soluble antigen

Of the proteins in Fra. I–V, three exhibited strong antigenicity, one of which oc-

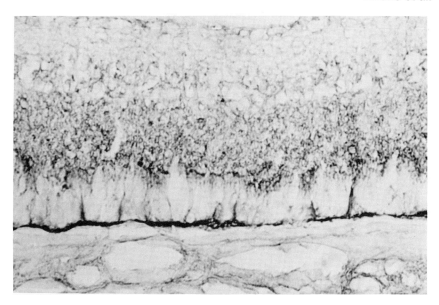

FIG. 2. Light microscopy findings of direct enzyme-labeled antibody method for the localization of A antigen. A positive response is detected in the apical portion of the retinal pigment epithelium.

curred in Fra. II, and two in Fra. III. The fraction II protein was what was called A antigen, while one of the Fra. III proteins was albumin. It formed a precipitate with anti-NPS. The remainder of fraction III was S antigen.

Isolation of A antigen

In the immunodiffusion study of the segmented gels following the PAG disc electrophoresis of Fra. II, A antigen was detected in the 5th–8th segments from the cathode end. The presence of A antigen was confirmed by immunoelectrophoresis.

Qualitative analysis of A antigen

(1) PAG disc electrophoresis: As the isolation process reached successively higher stages, the number of proteins decreased, until A antigen completely independent of S antigen was detected. (2) Isoelectric focusing electrophoresis: The isoelectric point for porcine A antigen was approximately pH 5.0.

Inoculation of A antigen into animals

Clinically, no lesion of the eye was detected. Histologic examination (40 days after inoculation) revealed that cellular infiltration of the choroid was very slightly more marked than in controls. No abnormality was detected in the iris or ciliary body.

Localization of A antigen examined by a direct enzyme-labeled antibody method

Light microscopic studies showed a positive response in the apical portion of the retinal pigment epithelium, (Fig. 2) while the blocking test was negative (Fig. 3). By electron microscopy, a positively reacting substance was seen in the areas of microvilli that had phagocytized the fragments of the outer segment (Fig. 4).

Discussion

The existence of A antigen was first suggested by Faure,[1] who stated that an antigen differing from S antigen in molecular weight could be separated by gel filtration during the process of S antigen isolation. He proposed the term "A antigen" for this antigen. With the above report in mind, the present authors subjected PRE to gel filtration with the resulting identification of A antigen, having a molecular weight around 150,000.

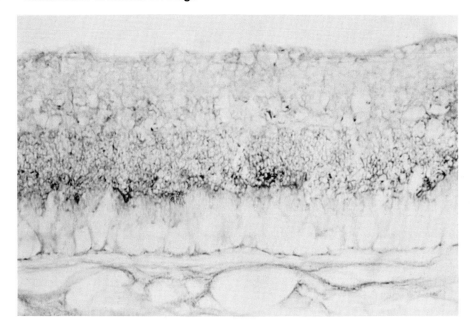

FIG. 3. Blocking test results.

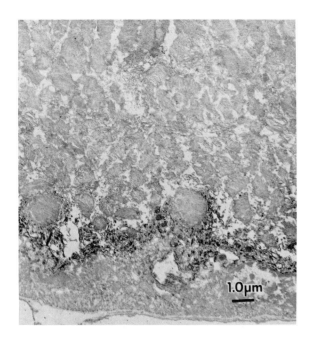

FIG. 4. Electron microscopy findings. A positively reacting substance is seen in the areas of microvilli which phagocytized the fragments of the outer segment.

At this stage, the fraction containing A antigen was contaminated with many other proteins, so that it seemed necessary to obtain a fraction free from at least S antigen for the qualitative analysis of A antigen. Therefore, PAG disc electrophoresis was performed, and this provided an A antigen fraction containing no S antigen. This fraction was used in the subsequent qualitative analysis of A antigen.

The results of the present experiment show that the characterization of newly prepared swine A antigen roughly agrees with that suggested by Faure.[1] In inoculation studies in animals, the pathogenicity of A antigen was shown to be remarkably less than that of S antigen.

The localization of A antigen appears to be in the apical portion of the retinal pigment epithelium. It is very interesting that the localization of A antigen differs from that of S antigen which has been noted to be in the outer segments of photoreceptor cells.[3]

In summary, the retinal extract prepared from swine eyeballs was subjected to gel filtration and the presence of A antigen was confirmed. The A antigen was completely isolated from the S antigen segment in order to determine experimentally its characteristics. The results were as follows:

1. A highly purified A antigen uncontaminated by S antigen was obtained by PAG disc electrophoresis.
2. The swine A antigen thus obtained had a molecular weight in the vicinity of 150,000, had an isoelectric point of 5.0, and closely resembled albumin in its electrophoresis migration pattern.
3. The swine A antigen exhibited organ specificity and was detected only in the eye.
4. Unlike S antigen, which induces severe uveitis, swine A antigen exhibited extremely low pathogenicity.
5. The localization of A antigen seemed to be in the apical portion of the retinal pigment epithelium and especially in the areas of microvilli which phagocytize the fragments of the outer segment.

References

1. Faure JP: In *Current Topics in Eye Research Vol. 2*, José A. Zadunaisky and Hugh Davson (Eds.). Academic Press, New York, 1980, pp. 215–302.
2. Takano S et al.: *Folia Ophthalmol Jpn* **32**:491–496, 1981.
3. Yajima S et al.: *Acta Ophthalmol Jpn* **86**:1558–1566, 1982.

chapter 59

Differences in Susceptibility to Experimental Autoimmune Uveitis Among Rats of Various Strains

Igal Gery,[a] W. Gerald Robison, Jr.,[a] Hitoshi Shichi,[a]
Magda El-Saied,[b] Manabu Mochizuki,[a]
Robert B. Nussenblatt,[b] and R. Michael Williams[c]

The genetic control of an experimental autoimmune disease was described a decade ago in rats in which experimental allergic encephalomyelitis (EAE) was induced. [1,2] Since then, the relationship between the genetic makeup and the susceptibility to autoimmune diseases has been well established in both man [3–5] and experimental animals.[3,6,7]

Experimental autoimmune uveitis (EAU) is an ocular inflammatory disease that can be induced in a variety of experimental animals by immunization with small amounts of the retinal S antigen (S-Ag).[8–10] The disease resembles certain human uveitic conditions and has been considered an animal model for these conditions.[9,10] The present study was aimed at determining the susceptibility of rats of different strains to EAU.

Materials and Methods

Animals. Lewis, BN, MAXX and LBNF$_1$ rats were purchased from M.A. Bioproducts, Walkersville, MD. CAR, RCS-PETH (dystrophic) and RCS-rdy$^+$ (heterozygous, unaffected) rats were supplied by the NIH breeding facility. AVN rats were the offspring of breeding pairs, kindly provided by Dr. David Gasser, University of Pennsylvania. Rats with typed RT1 locus were bred from BN and Lewis-derived ancestors and defined by one of us (RMW) at the Northwestern University facilities. Typing was performed by serology and mixed lymphocyte reactivity. Male or female rats between 7 weeks and 1 year of age were used.

Immunization. Bovine S-Ag was prepared as described by Wacker et al.[8] and emulsified (1:1) in complete Freund's adjuvant (GIBCO, Grand Island, NY), enriched with *Mycobacterium tuberculosis* H37Ra to a concentration of 2.5 mg/ml. Rats were injected into the hind footpads, with a total volume of 0.2 ml/rat, containing 30–50 μg of S-Ag. Treatment with *Bordetella pertussis* consisted of 10^{10} bacteria, injected intravenously,simultaneously with the S-Ag emulsion. The *B. pertussis* was of batch C-507F, generous gifts from Dr. Dale McFarlin, NIH, and Dr. Scott Linthicum, University of Southern California.

EAU evaluation. Clinical signs were monitored daily, with or without a slit lamp. Rats were killed usually 21 days post-immunization. Their eyes were fixed in 2.5% glutaraldehyde and embedded in glycol methacrylate; sections were cut at 2–3 μm and stained with toluidine blue or hematoxylin and eosin. The severity of changes was evaluated as described by Wacker et al.[8]

[a]Laboratory of Vision Research and [b]Clinical Branch, National Eye Institute, National Institutes of Health, Bethesda, Maryland; and [c]Department of Medicine, Northwestern University Medical School, Chicago, Illinois.

Lymphocyte cultures. Draining popliteal lymph nodes were removed, and lymphocyte cultures were set up as described in detail elsewhere.[11] The results are presented as the stimulation indices, which were calculated by dividing the mean cpm values of stimulated cultures by the mean values of unstimulated control cultures.

Results

Table I summarizes the data concerning the development of EAU in rats of different inbred strains. The data indicate that the tested strains may be divided into three groups according to their susceptibility to EAU: highly susceptible strains with incidence of >80% (Lewis, CAR, and the hybrid LBNF$_1$); low-susceptible strains, with <30% incidence (BN, MAXX, AVN); and nonresponders (rats of the RCS strain). It is noteworthy that the severity of EAU in positive rats of the low-responder strains resembled that observed in animals of the high-responder strains (see Fig. 1).

The relationship between susceptibility to EAU and the major histocompatibility antigen RT1 was further tested by employing backcross rats in which the RT1 locus was typed (Table II). Similar levels of EAU incidence were observed in rats typed as RT1$^{1/n}$ or RT1$^{n/n}$.

The incidence of EAU was found to increase when *B. pertussis* bacteria were injected simultaneously with the S-Ag emulsion (Table III). Treatment with *B. pertussis* had, however, different effects on the EAU inducibility in rats of various strains. The bacteria had no effect on RCS rats, increased slightly the disease incidence among BN and MAXX rats, but elevated the EAU incidence to 100% in AVN rats.

The possible relationship between susceptibility to EAU and the level of lymphocyte sensitization toward the immunizing S-Ag was examined by measuring the responses of lymphocytes from rats of the various strains toward the immunizing antigens. The data of three experiments are summarized in Table IV and show no correlation between the EAU development and the lymphocyte responses. Lymphocytes

TABLE I
Development of EAU in Rats of Different Strains[a]

Strain	EAU (positive/total)	
	Clinical	Histological
Lewis	42/48	44/48
CAR	8/8	8/8
LBNF	4/6	5/6
BN	3/13	3/13
MAXX	2/11	2/11
AVN	3/14	4/14
RCS-PETH	0/2	0/2
RCS-rdy$^+$	0/7	0/7

[a] Rats were immunized with 30–50 μg of S-Ag per rat, emulsified in CFA.

from rats with no detectable EAU reacted to S-Ag similarly to those from severely affected rats. Tested lymphocytes from the various donors also reacted similarly to PPD, a component of the adjuvant, or to ConA, a polyclonal mitogen.

Discussion

The data recorded here show that rats of various strains differ in their susceptibility to induction of EAU. Most strains showed either high or low levels of susceptibility, yet a complete refractoriness to EAU induction was observed only in the RCS rats. Of particular interest was the comparison made between EAU and EAE: A good correlation was observed with the Lewis, CAR, and LBNF rats, which showed high susceptibility to EAU; they are also susceptible to EAE [1,2,12,13] (and our unpublished data). On the other hand, severe EAU was found to develop in some of the BN, MAXX, and AVN rats, in contrast with the complete resistance of these rats to EAE [1,2,12,13] (and our unpublished data). A dissociation between EAU and EAE was also observed in the rats with typed RT1 locus. Half of the animals homozygous for RT1^n developed EAU, in contrast to the established refractoriness of such rats to EAE[12] (and unpublished data). The resistance to EAE of rats of the BN or other strains homozygous for the RT1^n locus has been attributed to the genetically controlled inability of these rats

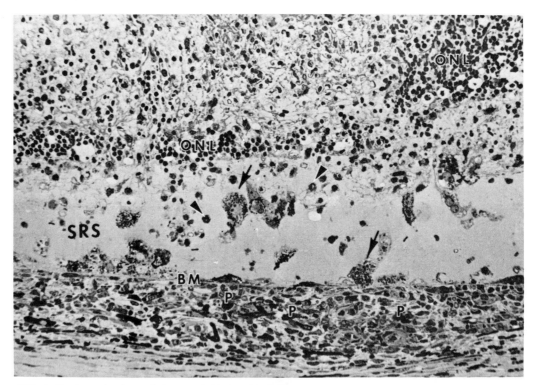

FIG. 1. Portions of retina and pigmented (P) choroid of a BN rat with severe EAU. The rat was immunized 14 days previously with S-Ag in CFA. Most of the photoreceptor cell nuclei are missing from the outer nuclear layer (ONL). Essentially no inner or outer segments are left and the retina is detached. The subretinal space (SRS) is filled with infiltrate containing leukocytes (arrowheads) and erythrocytes, as well as pigmented cells (arrows), which may be derived from invading histiocytes or from pigment epithelial cells detached from Bruch's membrane (BM).

to mount an immune response to the encephalitogenic peptide.[14] Our data thus show that the immune response to the uveitogenic antigen is controlled by genes other than those controlling the response to the encephalitogenic antigen.

Although EAU developed in some rats of the strains resistant to EAE, it is noteworthy that the majority of these animals did not develop EAU. This finding is interpreted to indicate that rats of these strains are deficient in one or more mechanisms that are essential for a full development of the pathogenic process. One such mechanism may be related to the mast cell-mediated increased permeability of the blood–retinal barrier. Recent studies have indicated the important role of this mechanism in the pathogenesis of both EAE[7] and EAU[15,16];

this notion is also supported by our results in rats treated with *B. pertussis*. Treatment with these bacteria increased the frequency of EAU in low-responder rats, in particular of the AVN strain (Table III).

Our results with the BN rats differ from those of de Kozak et al.[16] who did not detect EAU histologically in any of the rats of this strain following immunization with S-Ag in CFA. However, since only five BN rats

TABLE II
EAU in Rats with Typed RT1 Locus

RT1 type	EAU (positive/total)	
	Clinical	Histological
n/n	1/4	2/4
l/n	4/7	4/7
l/l	1/1	1/1

TABLE III

Effect of Treatment with *B. pertussis* on Development of EAU in Rats of Different Strains[a]

Strain	*B. pertusis*	Rats with EAU/total
AVN	No	4/14
	Yes	11/11
BN	No	1/4
	Yes	2/4
MAXX	No	0/4
	Yes	1/4
RCS-rdy[+]	No	0/5
	Yes	0/4

[a] Treated and untreated rats were matched for age and sex and were identically immunized with S-Ag in CFA. *B. pertussis* was given I.V., 10^{10}/rat.

TABLE IV

Lack of Correlation Between Development of EAU and Lymphocyte Responses in Culture

Exp No.	Rat strain	EAU (severity)	Stimuli added to cultures (µg/ml)		
			S-AG (5)	PPD (10)	ConA (3)
I	Lewis	+	8.3[a]	16.1	37.3
	BN	0	34.6	26.9	71.9
	LBNF	+ +	17.4	20.3	30.3
	CAR	+ + +	11.3	13.0	161.4
II	Lewis	+ + +	6.0	29.3	39.2
	BN	+ + +	16.8	31.2	48.1
	MAXX	0	7.4	18.8	17.2
III	Lewis	+ + +	13.2	74.2	130.3
	RCS	0	13.4	54.6	28.5
	AVN	0	5.0	26.9	13.2

[a] Stimulation index.

were tested in the mentioned study, it is conceivable that by increasing the number of tested animals these authors could have obtained BN rats with demonstrable susceptibility to EAU. BN rats were also examined for EAU induction by Marak et al.[17] who reported uveitis of "minimal" intensity in 40% of the tested rats.

Lymphocytes from rats of all tested strains reacted similarly in culture when incubated with the immunizing S-Ag. The finding of positive lymphocyte responses to S-Ag in rats with no EAU may be explained by the assumption that the bovine "S-Ag" preparation used in this study comprises multiple antigenic sites, not all of which are uveitogenic. This assumption is in line with the finding of McFarlin et al,[14] that BN rats immunized with the myelin basic protein (MBP) may show lymphocyte responses to the whole MBP molecule, but not to the encephalitogenic peptide. An alternative explanation to our finding of positive lymphocyte responses to S-Ag in rats with no EAU is that cellular immunity in these rats may develop without producing the pathogenic changes. More studies are needed to analyze further these hypotheses.

References

1. Williams RM and Moore MJ: *J Exp Med* **138**:775–783, 1973.
2. Gasser DL et al.: *Science* **181**:872–873, 1973.
3. Rose NR, et al. (Eds.): *Genetic Control of Autoimmune Disease*. Elsevier North Holland, New York, 1978.
4. Strober W: In *Immunology of the Eye. Workshop I: Immunogenetics and Transplantation Immunity*. Steinberg GM et al. (Eds.). (*Suppl Immunology Abstracts*) pp. 43–72, 1980.
5. Nussenblatt RB et al.: *Am J Ophthalmol* **94**:147–158, 1982.
6. Lillehoj HS et al.: *J Immunol* **127**: 654–659, 1981.
7. Linthicum DS and Frelinger JA: *J Exp Med* **155**:31–40, 1982.
8. Wacker WB et al.: *J Immunol* **119**:1949–1958, 1977.
9. Faure JP: *Curr Top Eye Res* **2**:215–302, 1980.
10. Nussenblatt, RB et al.: *Arch Ophthalmol* **99**:1090–1092, 1981.
11. Salinas-Carmona MC et al.: *Eur J Immunol* **12**:480–484, 1982.
12. Moore MJ et al.: *J Immunol* **124**:1815–1820, 1980.
13. Gasser DL, et al.: *J Immunol* **115**:431–433, 1975.
14. McFarlin DE et al.: *J Exp Med* **141**:72–81, 1975.
15. de Kozak Y et al.: *Eur J Immunol* **11**:612–617, 1981.
16. de Kozak Y et al.: *Curr Eye Res* **1**:327–337, 1981.
17. Marak GE and Rao NA: *Ophthalmic Res* **14**:29–39, 1982.

chapter 60

Natural Killer Cell Activity in Patients with Uveitis

Jung Koo Youn,[a] Jean Belehradek Jr.,[a]
Thu Cuc Lexuan Quang,[b] Jean-Claude Timsit,[a] Gilbert Hue,[a]
René Campinchi,[c] and Etienne Bloch-Michel[a,c]

Natural killer (NK) cells, a subpopulation of lymphocytes, are capable of lysing a variety of target cells *in vitro* without prior sensitization. They have recently been the subjects of many investigations, since NK cells seem to play an important role in various immune surveillance mechanisms.[1]

We have undertaken a study to elucidate whether NK cells intervene in the pathogenic process of various forms of uveitis, in particular, those forms accompanied by retinal vasculitis. The latter are generally suspected to be autoimmune diseases in which immune complexes might play a major role.[2] In this paper we report our observations that uveitis patients with retinal vasculitis displayed significantly increased levels of NK cell activity as compared with uveitis patients without such vascular lesions.

Materials and Methods

Patients and controls

Eighty one patients with a diagnosis of endogenous uveitis (35 males and 46 females) were studied. Their ages were 8–73 years. These patients were classified into two major groups according to their retinal vasculitis status: 29 patients with vasculitis

(Uv$^+$) and 52 patients without vasculitis (Uv$^-$). They were further divided into several subgroups according to the pathologic types of their diseases and the anatomical sites of the lesions. Fifty healthy adult blood donors (27 males and 23 females) 21–60 years of age served as controls.

In vitro NK assay

NK cell activity was determined by a 4-hour ^{51}Cr-release assay using NK-sensitive K562 cells, derived from a human myeloid leukemia,[3] as targets. The target cells were prelabelled with ^{51}Cr as described previously.[4] Effector cells were prepared from heparinized peripheral blood by a Ficoll-Hypaque (Pharmacia Fine Chemicals, Div. of Pharmacia Inc., Piscataway, N.J.) density gradient. The blood cells were washed three times with RPMI1640 medium supplemented with 10% heat-inactivated fetal calf serum, 2 mM L-glutamine, 25 mM Hepes-buffer and antibiotics (hereafter referred to as complete medium), resuspended in complete medium, and incubated overnight in plastic Petri dishes at 37°C in a 5% CO_2 humidified incubator. Nonadherent mononuclear cells were then collected and used as effector cells. Labelled K562 cells (10^4) in 100 μl were incubated in each well of a round-bottomed microtiter plate (Limbro, Titertek) with 100 μl of effector cells at a ratio of 1:100, for 4 hrs at 37°C in a 5% CO_2 incubator. Three to five replicate wells were

[a]Institut Gustave-Roussy, Villejuif, France. [b]Centre de Transfusion Sanguine, Hôpital, Broussais, Paris. [c]Service d'Ophtalmologie, Hôpital Lariboisière, Paris.

used for each assay. The NK activity was expressed as

$$\% \text{ of specific } ^{51}\text{Cr release} = \frac{T - S}{M - S} \times 100$$

where T is the test release in 100 μl of supernatant, M is the maximal release from target cells treated with 0.25% Triton X-100, and S is the spontaneous release from target cells alone. Percentages of spontaneous release were <5.0%.

Statistical analyses of significance were made by the Student t-test, in parallel to the distribution analysis χ^2 test.

Results and Discussion

Cells from a total of 81 patients with uveitis and from 50 healthy adult blood donors were submitted to *in vitro* NK assays during a period of 14 months. Most patients were tested once and in some cases repeatedly during the course of treatment.

The NK activities of the patients varied considerably from patient to patient. They were 12.1–89.1% of specific ^{51}Cr release and the mean percentage of ^{51}Cr release of these patients (Uv$^-$ + Uv$^+$) was not significantly different from that of the control subjects, i.e., 56.3% ± 22.4 for the patients and 53.0% ± 21.0 for the controls. The NK activities of individual control subjects also varied greatly, ranging from 2.0 to 85.9%. In our experiments, the mean NK activity of the controls seemed to be slightly higher than those reported by other authors.[5,6] This might be due to our overnight incubation of effector cells before the assays. As a matter of fact, increased lytic activity of human NK cells after relatively prolonged incubation at 37°C has been reported previously.[7] Repeated assays at a 3–7 months' interval in the same patients (a total of 18 patients) showed quite stable NK activities except for three patients who were receiving high doses of corticosteroid therapy (12 mg of prednisolone per day).

When these patients were divided for data analysis into two major groups, i.e., Uv$^-$ and Uv$^+$ (according to the retinal vasculitis status) and their NK activities were compared, we found that 29 patients suffering from retinal vasculitis (Uv$^+$), with either typical (IRV) or atypical inflammatory lesions including Behçet's disease (ARV), showed significantly increased levels of NK activity (64.4% ± 20.8) as compared with those of 51 patients without vasculitis (Uv$^-$, 51.9% ± 21.8) and 50 controls (53.0% ± 21.1). The difference between Uv$^+$ and Uv$^-$ was highly significant ($p < 0.02$). All Uv$^-$ patients with panuveitis (Pan U), intermediate uveitis (IU), or posterior uveitis (PU) showed NK activities comparable to those of the controls, whereas in eight Uv$^-$ patients with anterior uveitis (AU), NK activity was lower (41.6% ± 22.0) than in other types of the disease. However, the differences are hardly significant, probably due to the limited number of patients tested.

In order to determine whether corticosteroid therapy influences the NK activity, the patients in both Uv$^-$ and Uv$^+$ groups were further subdivided as follows: (1) patients not receiving any corticosteroid therapy or receiving less than 12 mg of prednisolone daily for at least one month before the day of assay, and (2) patients receiving more than 12 mg of prednisolone daily over a similar period.

As shown in Figure 1, seven Uv$^+$ patients receiving high doses of prednisolone showed NK activity significantly ($p < 0.02$) lower (48.7; p ± 27.7) than 22 Uv$^+$ patients not receiving corticosteroid therapy or receiving low doses (69.4% ± 15.8). In the Uv$^-$ group, decreased NK activity was also observed in patients receiving high doses. However, the difference between groups receiving high or low doses was at the limit of significance. In line with our results, depressed NK activity due to corticosteroid therapy has been reported in man[8] and in mice.[9]

To determine whether the clinical status of the patient could influence NK activity, the patients in both Uv$^-$ and Uv$^+$ groups were subdivided into two categories: (1) patients in a clinically active phase showing acute inflammatory lesions and (2) patients in a subsiding phase without such signs. Our criteria for the acute inflammatory signs were the degree of flare and the number of

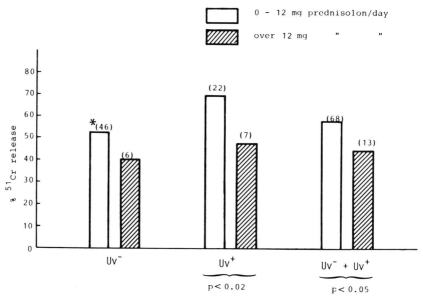

FIG. 1. Influence of corticosteroid therapy on NK cell activity of patients with uveitis. (*Numbers of patients tested.)

cells in both aqueous and vitreous humors and greatly increased leakage of fluorescein from retinal vessels on fluoroangiograms.The NK activities were compared in each group. No difference in the levels of NK activity was found between the patients in the active phase and those in the subsiding phase in both Uv^- and Uv^+ groups.

The mechanisms underlying the increased NK activity in patients with retinal vasculitis are not known. It is widely accepted that interferon or interferon-inducing agents enhance NK activity.[10] On the other hand, one can suspect a local involvement of immune complexes in retinal vasculitis. Autosensitization to retinal S antigen has been reported as one of the mechanisms of experimental retinal vasculitis. Whatever the initial triggering mechanisms might be, it can be postulated that certain immunopathologic processes lead to the enhancement not only of specific immune responses but also of the production of interferon which in turn might activate NK cells in the hosts. Further investigations will certainly be required to validate this hypothesis.

In summary, NK cell activities were studied in 81 patients with uveitis and in 50 healthy adult blood donors in 4-hour ^{51}Cr

release assay using NK-sensitive K562 cells as targets. The NK cell activity of all the patients tested showed a wide range of reactivity, and the average NK activity of the patients was not significantly different from that of the healthy control subjects. However, patients suffering from retinal vasculitis showed significantly higher levels of NK cell activity than those without such findings. High doses of corticosteroid therapy (more than 12 mg of prednisolone per day) resulted in a substantial decrease of NK cell activity in these patients as compared with those receiving low doses (less than 12 mg) or no steroids at all. Little difference in NK cell activity was found between patients in a clinically active phase and those in a quiescent phase, regardless of their retinal vasculitis status.

Acknowledgment

Supported in part by a grant from INSERM France (ATP 68.78-100, Contract No. 002).

References

1. Herberman RB and Holden HT: *Adv Cancer Res* **27**:305–377, 1978.
2. Rahi AHS et al.: *Immunology and Immunopathology of the Eye,* Silverstein AM and O'Connor

GR (Eds.). Masson Publishing USA, Inc., New York, 1979, pp. 23–28.

3. Lozzio CB and Lozzio BB: *Blood* **45**:321–326, 1975.
4. Youn JK et al.: *Cancer Res* 1982. *(in press).*
5. Antonelli P et al.: *Cell Immunol Immunopathol* **19**:161–169, 1981.
6. Nagel JE et al.: *Cancer Res* **41**:2284–2288, 1981.
7. Koren HS et al.: *Proc Natl Acad Sci USA* **76**:5127–5131, 1978.
8. Lipinski M et al.: *Transplantation* **29**:214–218, 1980.
9. Hochman PS and Cudkowicz G: *J Immunol* **119**:2013–2015, 1977.
10. Djeu JY et al.: *J Immunol* **122**:175–181, 1979.
11. Zarling JM et al.: *J Immunol* **124**:1852–1857, 1980.

chapter 61

Lymphocyte Transformation Test with Retinal S Antigen in Human Retinal and Uveal Inflammatory Diseases

Alfred Tanoé, Jean-Pierre Faure, Phuc Lê Hoang, and Etienne Bloch-Michel

Cell-mediated immunity (CMI) to retinal antigens frequently occurs in patients suffering from posterior ocular inflammation.[1] CMI testing of patients using purified retinal S antigen has confirmed that this organ-specific autoantigen is involved in retina sensitization in man.[2-4] But whether anti-S sensitization plays a pathogenetic role or not in ocular disease is at present unknown. Is it a primary factor or an accessory phenomenon? Does it influence the severity of the disease, its duration, the triggering of recurrences, the bilateralization of inflammation? We tried in the course of this work to analyze the *in vitro* reactivity of lymphocytes to S antigen with regard to the anatomical type of uveitis, the level of inflammation, and the course and treatment of the disease.

Patients

Sixty two patients with well-characterized uveitis or retinitis were tested. Fifty of them had chorioretinitis or retinitis; three had retinitis pigmentosa, and nine had intermediate and/or anterior uveitis. Of the 50 patients with posterior inflammation, 28 had only posterior involvement, and 22 had posterior uveitis associated with intermediate and/or anterior uveitis. Thirty six blood donors were used as controls.

Methods

S antigen was isolated from bovine retinas according to the method of Dorey et al.[5] The lymphocyte transformation (LT) test was performed on blood lymphocyte cultures in RPMI medium supplemented with 10% heat-inactivated human AB positive serum. Cultures were set up in triplicate in microtiter plates. Each well contained 4×10^5 cells in 0.2 ml of culture medium. Five concentrations of S antigen (3, 6, 10, 20, 40 µg/well) and 3 concentrations of phytohemagglutinin (PHA) (4, 8, 20 µg/well) were tested for each patient. Cultures were processed for 5 days, followed by 24 hour incubation with tritiated thymidine. Cells were harvested with a Scatron harvester. The results are expressed as stimulation index:

$$SI_{(SAg)} \text{ or } SI_{(PHA)} = \frac{\text{Mean CPM with S antigen or PHA}}{\text{Mean CPM with medium alone}}$$

In each experiment, the test was performed simultaneously on two to four patients and one control.

Laboratoire d'Immunopathologie de l'Oeil, CNRS ER 227, INSERM U 86, Hôtel-Dieu, Paris, France.

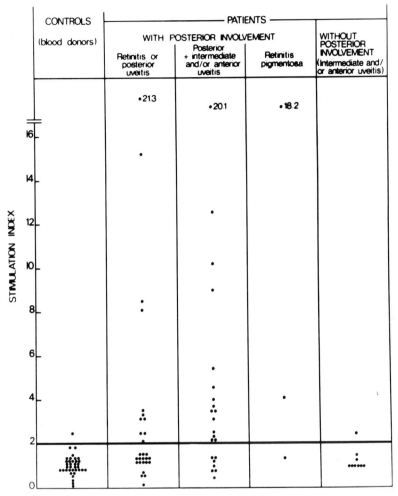

FIG. 1. Response to S antigen in controls and in patients.

Results

Response to PHA

Lymphocytes from all patients and controls were stimulated by PHA. The mean SI_{PHA} was slightly higher in patients

$$(X \pm \frac{s}{\sqrt{n}} = 84.3 \pm 12.7)$$

than in controls (65.4 ± 10.4), but the difference was not significant. The mean SI_{PHA} was significantly higher in patients with a recurrence of uveitis (105 ± 17.7, n = 42)

than in first attacks (41.5 ± 5.8, n = 20), $p < 0.05$. No difference in the response to PHA was noted in other clinical categories.

Response to S antigen (Fig. 1)

$SI_{S\ Ag}$ varied from 0.14 to 2.51 in controls. Mean SI + 2 SD of controls was 2.04 and was taken as the inferior limit of positivity of the test. Only one control of 36 (3%) was positive (SI 2.51). $SI_{S\ Ag}$ values ranged from 0.18 to 21.34 in patients. The test was positive in 29 of 62 patients (47%), the χ^2 of patients versus controls was 18.98 with $p <$

TABLE I
S Antigen Sensitization and Clinical Conditions

	No. of positive/total
Retinal vasculitis	12/26 (46.2%)
Multifocal chorioretinitis	11/18 (61%)
Focal chorioretinitis (toxoplasmic)	1/5
Behçet's disease	2/4
"Sympathetic" phenomenon	2/4
Pigment placoid epitheliopathy	1/4
Retinitis pigmentosa	2/3

0.0005. The test was positive in 28 of 53 (53%) patients with chorioretinal or retinal involvement and in only 1/9 (11%) patients without posterior involvement ($p < 0.005$). In patients with posterior uveitis or retinitis, the test was more often positive when the disease extended to the anterior uvea (15/22 = 68%) than when it was limited to the choroid and retina (11/28 = 39%) ($p < 0.05$). Sensitization was found in patients with chorioretinitis and/or retinal vasculitis, in cases of Behçet's disease, in pigment placoid epitheliopathy, in retinitis pigmentosa, and in secondarily bilateralized ocular inflammations (Table I). Regarding bilaterality, no significant difference was found between unilateral and bilateral ocular diseases: seven positive of 14 (50%) in unilateral; 22 positive of 48 (46%) in bilateral uveitis. But strong sensitization to S antigen was noted in two of four patients suffering primarily from unilateral recurrent posterior uveitis (one post-traumatic, the other toxoplasmic) with secondary contralateral intermediate uveitis, in conditions suggesting a "sympathetic" phenomenon.

Ten patients, in whom the clinical conditions are summarized below, had especially high stimulation indices.

Case reports: patients with highest SI $_{S\ Ag}$

DES. M., 40 years old. Bilateral exudative retinal detachment, disc edema, meningeal reaction, first attack (3 months). SI $_{S\ Ag}$ = 8.16.

DUC. F., 15 years old. Recurrent protracted (8 years) focal chorioretinitis in the right eye, with antitoxoplasmic antibodies in

the aqueous. Secondary intermediate uveitis with hyalitis in the left eye without antitoxo antibodies in aqueous, occurring during a recurrence of chorioretinitis. The LT test was performed 1 year after the involvement of the second eye, at a time when the toxoplasmic focus was healed and the inflammation of the left eye was still active. SI $_{S\ Ag}$ = 12.60.

GAC. F., 55 years old. Bilateral retinitis pigmentosa for 29 years and bilateral chronic hyalitis of moderate intensity. SI $_{S\ Ag}$ = 18.22.

HOU. F., 64 years old. Active intermediate uveitis with macular edema for 1 year in the left eye. Active intermediate uveitis for 2 months in the right eye. SI $_{S\ Ag}$ = 10.26.

LEB. M., 47 years old. Bilateral chronic (1 year) multifocal chorioretinitis and choroidal detachment occurring after bilateral cataract extraction. SI $_{S\ Ag}$ = 20.07.

MAT. F., 35 years old. Unilateral peripheral chorioretinitis, first attack (6 months), with diffuse active retinal vasculitis and hyalitis. SI $_{S\ Ag}$ = 15.23.

PER. M., 48 years old. Leprosy. Bilateral peripheral multifocal chorioretinitis, first attack (13 months), hyalitis, active inflammation. SI $_{S\ Ag}$ = 5.40.

SIR. F., 22 years old. Chronic recurrent posterior and intermediate uveitis after traumatic retinal detachment 12 years ago in the left eye; recurrent hyalitis for 6 years in the right eye: sympathetic uveitis. SI $_{S\ Ag}$ = 8.93.

TCH. M., 48 years old. Bilateral multifocal chorioretinitis, first attack (six months); exudative retinal detachment, macular edema, severe inflammation. SI $_{S\ Ag}$ = 8.68.

VIG. M., 41 years old. Bilateral, chronic (20 years), extensive retinal vasculitis; vitreous strands. SI $_{S\ Ag}$ = 21.34.

Relation between S-antigen sensitization and course of the disease

There was no significant difference in the results of the S antigen LT test between first attacks (9 positive of 20 = 45%) and recur-

TABLE II
S Antigen Sensitization and Chronicity of Ocular Inflammation

	Duration of the ongoing episode		
	< 3 months	> 3 months	
All patients	8/27[a] (30%)	21/35 (60%)	$p < 0.05$
First attacks	2/7 (29%)	7/13 (54%)	not significant
Recurrences	6/20 (30%)	14/22 (64%)	$p < 0.05$

[a] No. positive/total.

rences (20 positive of 42 = 48%) of uveitis. In first attacks and in recurrences the test was more often positive when the ongoing episode was of long duration rather than of recent onset (Table II). The test was more often positive when the disease was quiescent than when the inflammation was active. Systemic corticosteroid therapy strongly reduced the frequency of positive tests in active disease. On the other hand, positive tests were more frequent among treated patients when the disease was quiescent (Table III).

Discussion

This study confirms the frequent association between cell-mediated immune responsiveness to a soluble retinal antigen and uveoretinal inflammatory disorders in man. Lymphocytes from normal people were not stimulated, whereas more than 50% of patients with posterior ocular inflammation showed a positive lymphocyte stimulation with S antigen. The response of lymphocytes to S antigen did not correlate with the response to the mitogen PHA, neither in controls ($r = 0.06$) nor in patients ($r = 0.23$), a fact confirming the notion that the posi-

tivity of the LT test with S antigen cannot be related to a nonspecific polyclonal T-lymphocyte activation. In our series, the frequency of patients with posterior ocular involvement who exhibited immunologic memory to the S antigen (53%) was similar to the frequency observed by Nussenblatt[2] with the same test (43% in 40 patients).

In most of our patients no etiology of the ocular disorder has been found in spite of thorough clinical and biological examinations. We are inclined to propose "primary retinal autoimmunity" as the etiology of some of the diseases manifested by patients who showed evidence of strong S antigen sensitization. Especially suggestive of this etiology is the contralateral ("sympathetic") inflammation occurring during the course of a primarily unilateral posterior uveitis, whatever its original cause (toxoplasmosis, trauma). Certain observations suggest that the appearance of a positive LT test in recurrent unilateral uveitis might presage a bilateralization of the inflammation and might give the test some prognostic value. Lymphocyte activation with S antigen is observed less frequently in patients treated systemically with corticosteroids. We found that the suppressive effect of treatment on immune recognition was prominent in the group of patients with active inflammation. In contrast, the treatment was correlated with a high frequency of S antigen sensitization in patients with a low degree of inflammation. Most of those patients, who were undergoing systemic therapy in spite of quiescent disease (5 of 7), were in fact proven to be "treatment-dependent", i.e., they had already suffered a relapse after their treatment had been stopped. One could assume that the therapy incompletely con-

TABLE III
S Antigen Sensitization and Activity of the Ocular Inflammation

		Systemic therapy		
	All patients	No	Yes	
All patients		20/34 (59%)	9/28 (32%)	$p < 0.05$
Active	16/42[a] (38%)	13/21 (62%)	3/21 (14%)	$p < 0.01$
Quiescent	13/20 (65%)	7/13 (54%)	6/7 (86%)	$p = 0.20$
	$p < 0.05$			

[a] No. positive/total.

trols autoimmune reactivity, and one could hypothesize that residual sensitization could be a factor in the relapse of those patients upon withdrawal of treatment.

The observed correlations of S antigen sensitization with both the duration of the attack and the extent of inflammation outside the posterior segment support the hypothesis that autoimmune mechanisms do influence the severity of posterior uveitis.

References

1. Faure JP: *Curr Topics Eye Res* **2**:215, 1980.
2. Nussenblatt RB et al.: In *Immunology of the Eye II, Autoimmune Phenomena and Ocular Disorders,* Helmsen RJ et al. (Eds.). Information Retrieval Inc., Washington DC p. 49, 1981.
3. Faure JP and de Kozak Y: *Ibid.* p. 33.
4. Nussenblatt RB et al.: *Am J Ophthalmol* **94**:147, 1982.
5. Dorey C et al.: *Ophthalmic Res* **14**:249, 1982.

chapter 62

Suppressor Networks in TNP-Anterior Chamber-Associated Immune Deviation (TNP-ACAID)

J. Clifford Waldrep and Henry J. Kaplan

Mechanisms of immunoregulation within the eye have been studied following immunization via the anterior chamber (AC).[1-3] Suppressed systemic delayed-type hypersensitivity (DTH) to the hapten TNP, develops after AC immunization with TNP-modified splenocytes, a phenomenon termed "TNP anterior chamber-associated immune deviation" (TNP-ACAID).[4] Antigen specific, efferent suppressor T-cells (Ts) mediate TNP-ACAID. The induction of Ts requires a time-dependent interaction within the splenocameral axis.

Immunoregulation is an extremely complex process, involving numerous types of molecular and cellular interactions. Regulatory networks within the T-cell population have been demonstrated to encompass different subsets of helper and suppressor T-cells.[5] We have postulated that TNP-ACAID may be regulated through suppressor T-cell subsets. Here we demonstrate two distinct suppressor pathways and delineate their regulatory interactions.

Materials and Methods

TNP-ACAID was induced in Balb/c mice immunized intracamerally with hapten-modified autologous splenocytes as previously described.[4] Briefly, erythrocyte-free splenocyte suspensions in phosphate buffered saline were trinitrophenlated using 10 mM trinitrobenzenesulfonate. Mice were immunized on day -1 with 10^6 TNP-splenocytes via the AC using a Hamilton syringe and 33 gauge needle, or subcutaneously (SC). The SC mice were sensitized 24 hours later on day 0 with 100 microliters of 7% trinitrochlorobenzene in acetone-olive oil base. Adoptive transfer was carried out on day $+4$ using pooled splenocytes from AC immunized mice (suppressor cell donors) and SC immunized mice (DTH donors). Normal recipients received 2×10^7 splenocytes (experimental or control) by intravenous injection. After 24 hours (day $+5$), the recipients' basal ear thickness was measured, followed by challenge with 1% TNCB. Twenty-four hours later (day $+6$) the ear swelling response was determined.

Some mice were treated with cyclophosphamide (CY, 200 mg/kg) 48 hours prior to the initial AC or SC injection. In some experiments, fractionated splenocytes were adoptively transferred. DTH T-cells (DTH$_T$) were purified from SC immunized splenocytes passed through nylon wool. DTH accessory cells were depleted of T-cells by treatment with anti-T-cell serum plus complement. B-cells were removed by anti-Ig panning, and macrophages were removed by carbonyl iron treatment. Contrasuppressor (i.e., suppressor of suppressor cells) AC cells were fractioned in an analogous manner.

Department of Ophthalmology, Emory University, Atlanta, Georgia 30322.

TABLE I
Two Suppressor Pathways in TNP-ACAID

	Cells transferred				Ear swelling response (-10^{-4}cm)
	DTH	DTH_{CY}	AC	AC_{CY}	
Group A	+	−	−	−	53 ± 8
B	+	−	+	−	23 ± 5
C	+	−	−	+	23 ± 5
D	−	+	−	−	43 ± 8
E	−	+	+	−	59 ± 4
F	−	+	−	+	17 ± 3

Results

Two suppressor pathways in TNP-ACAID

Intracameral immunization with TNP-modified splenocytes induces systemic tolerance to the hapten TNP (TNP-ACAID). We have demonstrated that TNP-ACAID is mediated by splenic, suppressor T-cells that inhibit the efferent arm of the DTH response to TNP.[4] Regulatory mechanisms in TNP-ACAID were delineated following cyclophosphamide (CY) treatment. Table I demonstrates the existence of two suppressor pathways in TNP-ACAID. Suppression of the ear swelling response to TNP is demonstrated when AC splenocytes were cotransferred with DTH splenocytes (Group B vs. A). We have designated this the primary suppressor pathway of TNP-ACAID. CY treatment of the DTH (DTH_{CY}) population does not diminish the ability to transfer DTH (Group D). However, these cells are no longer suppressed when AC splenocytes are cotransferred (Group E vs. D). This demonstrates that a CY sensitive auxiliary cell is required for the primary suppressor pathway of TNP-ACAID. In contrast, cotransfer of DTH_{CY} and CY treated AC splenocytes (AC_{CY}) reestablished suppression (Group F vs. D). These results demonstrate a secondary (CY resistant) suppressor pathway in TNP-ACAID.

The secondary suppressor pathway requires an accessory macrophage

Passage of DTH splenocytes through nylon wool blocked the suppressive activity of AC_{CY} suggesting that the secondary pathway requires a DTH-derived accessory cell. The results shown in Table II demonstrate that the accessory cell is a macrophage. Depletion of T-cells (Group B) or B-cells (Group C) from the DTH accessory cell population does not abrogate suppression of DTH. Conversely, macrophage depletion of the DTH accessory population (Group D) does abolish suppression mediated by AC_{CY}.

Contrasuppression (suppression of suppressor cells) in TNP-ACAID

The existence of two suppressor pathways in TNP-ACAID is demonstrated by the ability of AC_{CY} but not AC splenocytes to mediate suppression when cotransferred with DTH_{CY} (Table I, Group F vs. E). The absence of the secondary suppressor pathway suppressors in non-CY treated AC splenocytes (Table I, Group E) suggests that the latter may possess regulatory properties. Table III demonstrates that the AC splenocyte population contains cells that are capable of suppressing the secondary suppressor pathway (i.e., contrasuppression). Cotransfer of AC T-cells plus AC_{CY} (Group D) abrogates the suppressive activity of AC_{CY} (Group B vs. A). Conversely, cotransfer of AC B-cells and macrophages plus AC_{CY} did not abrogate the suppression (Group C vs. A). Thus the AC splenocyte population contains a CY-sensitive contrasuppressor T-cell subset that regulates the secondary suppressor pathway.

Discussion

The immune "privilege" of the anterior chamber is thought to be mediated through

TABLE II
The Secondary Suppressor Pathway of TNP-ACAID
Requires an Accessory Macrophage

	Cells Transferred Accessory Cells					Ear swelling response (10^{-4}cm)
	DTH_{T}	T	B	MØ	AC_{CY}	
Group A	+	+	+	+	−	42 ± 3
B	+	−	+	+	+	22 ± 4
C	+	+	−	+	+	19 ± 3
D	+	+	+	−	+	55 ± 3

TABLE III

Control of the Secondary Suppressor Pathway Is Mediated by a Suppressor of Suppressor Cells (Contrasupression)

	Cells Transferred Contrasuppressor					Ear swelling response $(10^{-4}cm)$
	DTH_{CY}	T	B	MØ	AC_{CY}	
Group A	+	−	−	−	−	39 ± 6
B	+	−	−	−	+	10 ± 4
C	+	−	+	+	+	13 ± 3
D	+	+	−	−	+	47 ± 8

active suppressor mechanisms. We have demonstrated the phenomenon of TNP-ACAID following AC immunization with TNP-modified splenocytes. The hallmark of TNP-ACAID is antigen-specific suppressed systemic cellular immunity. TNP-ACAID is mediated via *two* distinct suppressor pathways (Table I). We have designated the suppression mediated by AC splenocytes on DTH cells (Group B vs. A) as the primary pathway of TNP-ACAID. This pathway requires a CY sensitive auxiliary cell that is DTH derived (Group E vs. D). This auxiliary cell is analogous to the auxiliary T-cell described by Claman et al.[6] for intravenous induced suppression. The secondary pathway of TNP-ACAID is mediated by CY resistant AC splenocytes (Group F vs. D). A DTH accessory macrophage is required for this alternate suppressor pathway (Table II, Group D). Thus, two mechanisms for efferent suppression in TNP-ACAID have been identified by their resistance or susceptibility to CY. The suppressor cells comprising each pathway constitute elements of the suppressor network regulatory system.

The secondary suppressor pathway is nonfunctional in the presence of active primary suppressors. This is demonstrated in Table I, Group E. Inactivation of the primary suppressors by CY renders the secondary suppressors active (Table I, Group F). Thus there is a subset of cells within the AC population that functions as a suppressor of suppressor cells (i.e., a contrasuppressor). The existence of an AC contrasuppressor T-cell is demonstrated in Table III, Group D. Two possible mechanisms can be proposed for contrasuppression in TNP-ACAID. Suppressor T-cells of the primary pathway (Ts_x) may exert a feedback inhibition on the suppressor T-cells of the secondary pathway (Ts_y). Alternatively, a third T-cell subset (Ts_z) may act to regulate Ts_y. The demonstration of contrasuppression in TNP-ACAID represents an example of the complex nature of the multiple regulatory pathways that have been demonstrated in immunoregulation.

Acknowledgments

This work was supported by NIH Grant EY-03723 and by a general departmental grant from Research to Prevent Blindness, Inc.

We thank Ms. Martha Johnson for excellent technical assistance and Ms. Edith Barron and Ms. Cynthia Hale for excellent secretarial assistance.

References

1. Kaplan HJ and Streilein JW: *J Immunol* **118**:809–814, 1977.
2. Streilein JW and Niederkorn JY: *J Exp Med* **153**:1058–1067, 1981.
3. Wetzig RP et al.: *J Immunol* **128**:1753–1757, 1982.
4. Waldrep JC and Kaplan HJ: *Invest Ophthmal Vis Sci* **24**: 1086–1092, 1983.
5. Germain RN and Benacerraf B: *Scand J Immunol* **13**:1–10, 1981.
6. Claman HN et al.: *Immunol Rev* **50**:105–132, 1980.

chapter 63

Determination of Protein Content in Ocular Fluids by High-performance Gel Filtration Chromatography

K. Matti Saari, Esko Aine, and Markku T. Parviainen

High-performance liquid chromatography (HPLC) is a new technique that provides for the rapid separation of various biomolecules with a high degree of resolution. We have studied the protein content of (1) tears in the normal state and in different ocular and rheumatic diseases, (2) the aqueous humor in cataract and in uveitis patients, and (3) subretinal fluid in retinal detachment.

Materials and Methods

We studied the protein content of tears in 15 normal subjects, ages 15–66 (mean 41) years, and in a 42-year-old woman with bronchial asthma. We also performed the same determinations in a 53-year-old man with seronegative rheumatoid arthritis, and in a 42-year-old man with Behçet's disease. The tears were collected along the lower lid margin by means of a sterile capillary pipette (5 μl). The protein content of the aqueous humor was studied in 17 cataract patients, ages 46–86 (mean 69) years, and in some patients with chronic anterior uveitis. These aqueous humor samples were obtained at the time of normal cataract extraction. In some patients with toxoplasmic chorioretinitis or with secondary glaucoma after occlusion of the central retinal vein, an anterior chamber tap was made in an operating room. The protein content of subretinal fluid

Department of Ophthalmology and the Biochemical Research Laboratory, Institute of Clinical Sciences, University of Tampere, Tampere, Finland.

was studied in nine patients with retinal detachment. The duration of the retinal detachment was shorter than 5 weeks in all nine patients.

We used a Perkin-Elmer Model Series 2/2 liquid chromatograph (Norwalk, Conn.) equipped with a Perkin-Elmer Model LC-15 UV-detector (280 nm) or Model LC-75 UV-detector (215 nm), a Rheodyne syringe loading sample injector Model 7125 (Berkeley, Calif.) and a Perkin-Elmer Model 561 pen recorder. Gel filtration columns MicroPak TSK 2000 SW and MicroPak TSK 3000 SW were obtained from Varian Corp. (Walnut Creek, Calif.).

Protein separations were performed by connecting the gel filtration columns in series. Used in all separations were 50 mM phosphate buffer, pH 7.4, containing 100 mM KCl and 0.6 mM sodium azide and a constant flow rate of 1.0 ml/minute. The sample volumes used were 5–50 μl. We used commercially available high and low molecular weight protein standards to plot a calibration curve for evaluating the molecular weights of the protein peaks in the chromatograms of the ocular fluids. The routine sensitivity of HPLC for a single protein was about 100 ng.

Results

Tears

A typical high-performance gel filtration chromatogram (HPGFC) of normal tears

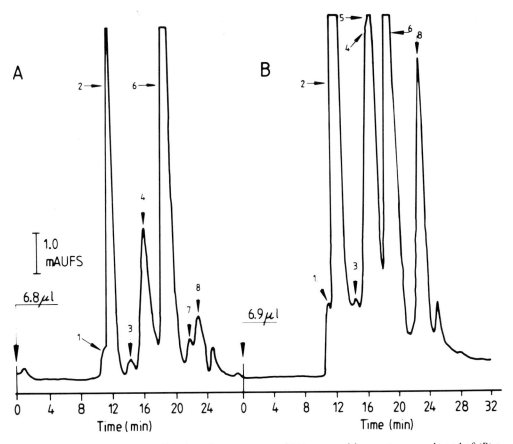

FIG. 1. High-performance gel filtration chromatograms of *(A)* a normal human tear sample and of *(B)* a patient with bronchial asthma and acne rosacea. The absorbance units of full scale (AUFS), injection point, sample volume, and peak numbering are indicated.

(Fig. 1*A*) showed eight peaks with the occurrence of IgA and lactoferrin in the second peak (range 0.43–17.1 $\mu g/\mu l$, mean 4.61 $\mu g/\mu l$), albumin in the fourth (range 0.54–9.1 $\mu g/\mu l$, mean 2.39 $\mu g/\mu l$), and lysozyme (and presumably also tear specific prealbumin) in the sixth peak (range 2.54–27.5 $\mu g/\mu l$, mean 6.99 $\mu g/\mu l$). In normal subjects the first peak containing high molecular weight proteins (IgM, lipoproteins, immune complexes) was of very low optical density and could be quantitated only in two cases (0.146 $\mu g/\mu l$ and 0.64 $\mu g/\mu l$). The third peak containing IgG was also of very low density and could be quantitated only in one normal subject (0.06 $\mu g/\mu l$).

In the 42-year-old woman with bronchial asthma and acne rosacea the HPGFC of the tears showed a concentration of materials in peaks 1–4 that was about twice that of normal tears. The most prominent findings were elevation of IgA (10.3 $\mu g/\mu l$) and other immunoglobulins. In addition, an unidentified high-shoulder peak was seen in HPGFC very close to the albumin (3.84 $\mu g/\mu l$) (Fig. 1*B*).

The 53-year-old man with seronegative rheumatoid arthritis had normal findings in the Schirmer and tear break-up-time tests, as well as in the serum electrophoresis test, whereas the HPGFC of his tears showed an elevation of peak 1 (0.33 $\mu g/\mu l$), IgA (9.83 $\mu g/\mu l$), and of IgG (0.67 $\mu g/\mu l$).

Aqueous humor

A typical HPGFC of normal aqueous humor showed 7–12 peaks (Fig. 2*A*). The first peak contained high molecular weight proteins (MW 250,000 or more; range 0.004–

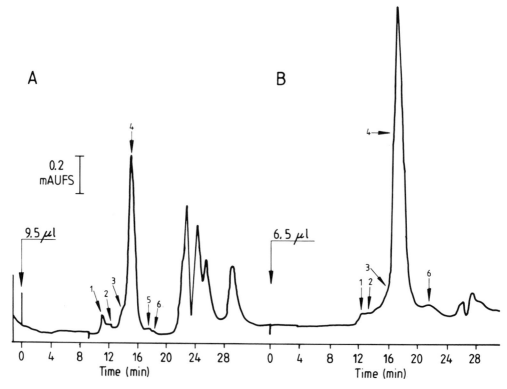

FIG. 2. High-performance gel filtration chromatograms of *(A)* a normal aqueous humor and of *(B)* aqueous humor from the left eye of a patient with bilateral chronic anterior uveitis.

0.674 μg/μl, mean 0.085 μg/μl). The second peak containing IgA was seen in three cases, but it was quantitated in only one case (0.011 μg/μl). The third peak containing IgG was found in two cases but the quantitative determination for IgG could be done only in one case (0.011 μg/μl). The fourth peak contained albumin (range 0.068–2.191 μg/μl, mean 0.6 μg/μl). The next two peaks were low; the second of them contained lysozyme, which could be quantitated in three cases (range 0.019–0.060 μg/μl, mean 0.038 μg/μl). The last six peaks matched for peptides, amino acids, and perhaps lipids.

In a 47-year-old man with bilateral chronic anterior uveitis of 6 years' duration and complicated cataract, the HPGFC showed an elevation of peak 1 (0.33 μg/μl) and lysozyme (0.22 μg/μl) in the aqueous humor of the right eye, and a high elevation of albumin (2.62 μg/μl) in the aqueous humor of the left eye (Fig. 2B).

A 78-year-old woman had an intraocular pressure of 40–52 mmHg and rubeosis iridis in her blind left eye following occlusion of the central retinal vein and hemorrhagic glaucoma. The HPGFC of the aqueous humor of the left eye showed a massive elevation of proteins: peak 1, 2.7 μg/μl, IgA 0.25 μg/μl, IgG 2.8 μg/μl, albumin 61.4 μg/μl, and lysozyme 2.3 μg/μl.

Subretinal fluid

A typical HPGFC of subretinal fluid in retinal detachment (Fig. 3) showed peaks similar to those of tears and of the aqueous humor, but IgG and albumin were present in greater concentration and peak 2 was not seen. Peak 1 ranged between 0.16–0.61 μg/μl, IgG 0.6–1.2 μg/μl, albumin 10–14.8 μg/μl, and lysozyme 0.6–1.6 μg/μl.

Discussion

The advantages of HPLC consist of rapid and accurate analyses with increased pre-

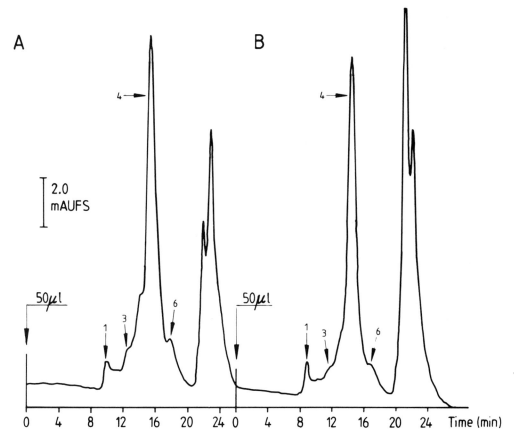

FIG. 3. High-performance gel filtration chromatograms of subretinal fluid from *(A)* a 70-year-old normal woman and *(B)* a 59-year-old woman with detachment of the retina.

cision. The sample volume may be as small as 5–25 µl. The sample components remained biologically active during the run. Thus HPLC may be convenient for the separation, partial purification and characterization of immunoglobulins and other proteins in immunological studies of the eye. However, HPGFC is not specific, since it shows only the major proteins according to their molecular weight. Lactoferrin binds to several proteins, such as albumin and IgA,[1] and IgA and lactoferrin have a similar molecular weight. Therefore, in the HPGFC of tears it was not possible to separate lactoferrin and IgA from each other.

HPLC is more precise than electrophoresis. In this study HPGFC of tears showed eight peaks, whereas electrophoresis of tears showed only four peaks.[2] IgA was the dominant immunoglobulin in the tear fluid, and high peaks of albumin and lysozyme were also seen. The concentration of lysozyme in tears is higher than in any other body fluid.[3] Diminished concentrations of lysozyme, tear-specific prealbumin, and lactoferrin have been demonstrated in patients with keratoconjunctivitis sicca.[1,4] Because cases of damage to the secretory tissues of the lacrimal gland produce a decrease of one or several peaks, and an inflammatory reaction induces an increase in the albumin and immunoglobulin fractions,[2] (as was also seen in this study) HPGFC of tears can be helpful in the diagnosis of the sicca syndrome and in revealing inflammation of the gland. HPGFC of tears may reveal a functional change, enabling the physician to follow the evolution of glandular disease and to avoid biopsy of the lacrimal gland.

HPGFC showed low concentrations of the high molecular weight proteins IgA and IgG in the normal aqueous humor. This may be due to a molecular sieving effect of the blood–aqueous barrier with a theoretical pore size of 104 Å.[5] Fielder and Rahi[6] observed a linear relationship between aqueous and serum IgG (aqueous IgG = 0.9101 + 0.00096 × serum IgG), but no significant correlation between the aqueous and serum IgA. Furthermore, they found no IgM in the aqueous samples from normal individuals. The present and earlier[7] studies showed elevated levels of immunoglobulins including IgM in the aqueous humor in patients with chronic anterior uveitis. HPGFC also showed a high elevation of albumin in the aqueous humor of a uveitis patient. This could be explained by the increased vascular permeability in uveitis. Similarly, a massive elevation of immunoglobulins, albumin, and lysozyme occurred in the aqueous humor of a patient with thrombotic glaucoma, obviously due to the increased vascular permeability seen in rubeosis. In uveitis, the vascular damage caused by antigen–antibody complexes and complement has been previously discussed.[8]

HPGFC of subretinal fluid showed high levels of albumin and IgG in our patients in whom duration of retinal detachment was shorter than 5 weeks. When the duration of the detachment increases, high levels of proteins including IgG and IgA (but not IgM) are found due to increased permeability of the choriocapillaris.[9]

References

1. Janssen PT and Van Bijsterveld OP: *Clin Chim Acta* **114**:207, 1981.
2. Liotet S et al.: *Ophthalmologica* **184**:87, 1982.
3. Lemp MA et al.: *Ann Ophthalmol* **2**:258, 1970.
4. Avisar R et al.: *Am J Ophthalmol* **87**:148, 1979.
5. Dernouchamps JP and Heremans JF: *Exp Eye Res* **21**:289, 1975.
6. Fielder AR and Rahi AHS: *Trans Ophthalmol Soc UK* **99**:120, 1979.
7. Ghose T et al.: *Br J Ophthalmol* **57**:897, 1973.
8. Allansmith MR and O'Connor GR: *Surv Ophthalmol* **14**:367, 1970.
9. Chignell AH et al.: *Br J Ophthalmol* **55**:525, 1971.

chapter 64

Cyclic AMP Enhances the Susceptibility of Human Cultured Retinoblastoma Cells to Natural Killer Cells

Yuichi Ohashi, Tetsuo Sasabe, Teruo Nishida, and Reizo Manabe

Retinoblastoma is a highly malignant tumor of the eye in infancy. It has been noted that spontaneous regression occurs in a small number of patients.[1,2] Differentiation of the tumor cell itself or destruction of the tumor cells by an immunological attack might be considered among the possible mechanisms by which this regression occurs. Nishida et al. showed that rat retinoblastoma-like tumor cells were made to differentiate *in vitro* by 8-bromo cyclic AMP (8BrcAMP) treatment.[3] Recently, we found that 8BrcAMP treatment caused growth inhibition and morphological alteration of human retinoblastoma cells *in vitro* (unpublished observation). Thus, this treatment might lead to the alteration of cell surface antigens so that the interaction of tumor cells with immunocompetent cells would be changed.

We found that NK cells in the human peripheral blood could kill several human retinoblastoma cell lines and that fresh retinoblastoma cells isolated from the patients were also killed by NK cells activated by interferon.[4] The aim of this study was to determine how the susceptibility of retinoblastoma cells to attack by NK cells might

be altered, and how this, in turn, might be related to changes in the levels of intracellular cAMP.

Materials and Methods

Cell lines

Human retinoblastoma-derived cell line, Y-79, human erythroleukemic cell line, K-562, and Epstein-Barr virus-transformed lymphoblastoid cell line (Raji) were used. These cell lines were cultured with RPMI-1640 medium plus 15% fetal calf serum. Y-79 cells were used exclusively for the cytotoxicity assay. In some experiments, Y-79 cells were pretreated with either 0.1 mM of 8BrcAMP or 10 mM of theophylline for a specified amount of time and used as target cells.

Preparation of effector cells

Effector cells were prepared from the peripheral blood of healthy donors as described previously.[4] Heparinized blood was treated with killed yeast suspension for 45 minutes and then centrifuged on Ficoll-Paque gradient to remove phagocytic cells. The cells at the top of the gradient were pooled and used for the usual cytotoxicity assay.

Further fractionation of human peripheral lymphocytes was done to purify NK cells as described previously.[4] The non-phago-

From the Transparent Ocular Tissue Laboratory, Department of Ophthalmology, Osaka University Medical School, Osaka, Japan.

cytic cell fraction was applied to the nylon-wool column and incubated for 1 hour. Then the column was washed and the cells that emerged were pooled. These cells were further centrifuged on a discontinuous Percoll gradient according to the method of Timonen et al.[5] The cells in the low density fraction (fraction 0-2) and in the high density fraction (fraction 4) were separately pooled and used for the cytotoxicity assay.

The cytotoxicity assay was performed using a ^{51}Cr-release method as described previously.[4] Specific cytotoxicity was calculated according to the following formula:

Specific cytotoxicity =
     cpm(sample) −
     cpm(spontaneous)/cpm(maximum) −
     cpm(spontaneous) × 100(%).

In the cold inhibition test, unlabeled K-562 cells or Raji cells were added to the cytotoxicity cultures at the ratio desired.

Interferon treatment

Heparinized human peripheral blood was treated with 1,000 IU/ml of human fibroblast interferon for 2 hours. Then the blood was incubated with killed yeast suspension and centrifuged on a Ficoll-Paque gradient. The cells at the top of the gradient were used as effector cells.

Results

Nonphagocytic peripheral lymphocytes from nine healthy donors killed Y-79 cells to a variable extent (Table I). When Y-79 cells were treated with 8BrcAMP to increase the level of intracellular cAMP, the same effectors lysed treated Y-79 cells more extensively than those that were untreated ($p<0.01$) (Table I). When Y-79 cells were treated with theophylline, the treated cells were also lysed more extensively than those untreated ($p<0.01$) (Table I).

Peripheral lymphocytes were fractionated and examined for their cytotoxic activity against untreated as well as 8BrcAMP-treated Y-79 cells. The cytotoxic activity became higher as the fractionation proceeded (Table II). After processing on a Percoll gradient, the cells recovered in the low den-

TABLE I

Effect of 8BrcAMP and Theophylline on the Susceptibility of Y-79 Cells to Natural Killer Cells

Donors	Y-79 Cells[a]		
	Untreated	8BrcAMP	Theophylline
A	12.7	25.2	20.5
B	25.8	45.8	36.8
C	61.4	76.2	67.7
D	25.5	36.6	35.4
E	45.2	56.9	47.5
F	49.9	54.4	52.2
G	22.8	32.8	26.3
H	13.4	39.1	28.7
I	2.7	7.4	6.1

[a] Y-79 cells were treated with 0.1 mM 8BrcAMP or 10 mM theophylline for 48 hours.

sity fraction showed much higher cytotoxicity against target cells than those in the high density fraction (Table II). The cells in this fraction were enriched with large granular lymphocytes, the size and the granules being morphological characteristics of NK cells.[6] When the effector cells were treated with interferon, the lysis of target cells was markedly enhanced (Table II). Furthermore, this cytotoxicity was inhibited by adding K-562 cells, which are standard NK target cells, but not by Raji cells, which are resistant to NK cell-mediated lysis (Table III).

The cytotoxic activity of nonphagocytic peripheral lymphocytes against Y-79 cells was increased in all four samples as the treatment time of Y-79 cells with 8BrcAMP

TABLE II

Cytotoxicity of Fractionated Peripheral Blood Lymphocytes against Y-79 Cells

Effectors	Y-79 Cells	
	Untreated	8 BrcAMP
MØ-Depleted	25.4	33.6
+IFN(1000 IU/ml)	77.5	88.8
Nylon Wool-Column Passing Cells	34.6	39.4
Percoll Gradient[a]		
Fraction 0-2	48.3	52.2
Fraction 4	12.7	15.2

[a] The effector/target ratio was 12.5.

TABLE III
Cold Inhibition Test on Y-79 Cell Lysis

Ratio (Y-79/Added Cell)	Y-79 Cells			
	Untreated		8BrcAMP	
	K-562	Raji	K-562	Raji
–	34.6	34.6	39.4	39.4
1:1	32.1	33.3	38.8	36.6
1:2	24.3	36.0	31.1	43.4
1:4	22.1	33.3	23.0	35.6
1:8	15.6	30.9	17.7	28.4

was prolonged (Table IV). At 72 hours after the initiation of the treatment, a significant increase of the cytotoxicity was observed.

Discussion

The present study clearly showed that the human retinoblastoma cell line, Y-79, became more susceptible to attack by NK cells after the treatment with cAMP analogs or with phosphodiesterase inhibitors in a time-dependent manner. We reconfirmed by several lines of evidence that this cytotoxic activity was mediated by NK cells in the peripheral blood: (1) recovery of the highest cytotoxicity in the low density fraction of Percoll gradient; (2) enhancement of the cytotoxicity by interferon treatment; (3) and marked inhibition of the cytotoxicity by K-562 cells.

Recently, we found that an increase of the intracellular cAMP level caused a morphological alteration of the cell surface of Y-79 cells. For example, scanning electron microscopy revealed that the numerous microvilli usually seen on the cell surface of Y-79 cells had almost disappeared after 8BrcAMP treatment. The results of the cold inhibition test indicate that Y-79 cells might share with K-562 cells a common antigenic determinant which is supposed to be NK cell-directed. Thus, treatment with 8BrcAMP or theophylline could alter the distribution or location of NK-directed antigens on the cell surface by increasing the intracellular cAMP level.

Stern et al. reported that embryonal carcinoma cells were highly sensitive to NK cells, unlike well-differentiated cells of the same origin.[7] Adamson et al. found that embryonal carcinoma cells lost NK cell sensitivity after the induction of differentiation by retinoic acid treatment.[8] Also, Werkmeister et al. showed that the treatment of K-562 cells with butyrate made these cells differentiate and become less susceptible to NK cells.[9] Our results do not agree with these reports, however. One explanation is that the effect of cAMP on the nature of cancer cells could differ from one cell line to another.

In conclusion, an increase of the intracellular cAMP concentration by treatment with cAMP analogs or by phosphodiesterase inhibitors leads to a unique situation *in vitro* in which retinoblastoma cells stop multiplying and, at the same time, become more sensitive to NK cell lysis. We may possibly be able to apply this phenomenon to the *in vivo* situation in ways that will permit us to combat retinoblastoma.

TABLE IV
Effect of Incubation Time with 8BrcAMP on the Susceptibility of Y-79 Cells to Natural Killer Cells

Donors	Y-79 Cells			
	Untreated	Treated with 8BrcAMP for[a]		
		24 hr.	48 hr.	72 hr.
B	25.5	31.0	34.2	44.3
D	22.0	24.0	25.2	37.6
H	10.9	14.8	18.0	24.9
I	3.0	3.5	5.9	11.2

[a] Y-79 cells were incubated with 0.1 mM of 8BrcAMP for an indicated period and used as a target cell.

References

1. Bonuik M and Girard LJ: *Trans Am Acad Ophthalmol Otolaryngol* **73**:194, 1969.
2. Khodadoust AA et al.: *Surv Ophthalmol* **21**:467, 1977.
3. Nishida T et al.: *Invest Ophthalmol Vis Sci* **22**:145, 1982.
4. Ohashi Y et al.: *Jpn J Ophthalmol* **29**:1985 (in press).
5. Timonen T, Saksela E: *J Immunol Meth* **36**:285, 1980.
6. Timonen T et al.: *J Exp Med* **153**:569, 1981.
7. Stern P et al.: *Nature* **285**:341, 1980.
8. Adamson ED et al.: *Cell* **17**:469, 1979.
9. Werkmeister J et al.: *J Cell Biol* **91**:24a, 1981.

chapter 65

Characterization of Human Corneal Antigens by Hybridoma-Secreting Antibodies

Noveen D. Das, Z. Suzanne Zam, and Pauline Jones

The potential antigenicity of corneal tissue has been amply demonstrated.[1,2] Major controversies exist in the literature concerning the presence of blood group antigens on the endothelial and stromal layers and concerning the importance of major histocompatibility typing for keratoplasty. Furthermore, it is not clear which antigenic components of the individual cell layers are responsible for triggering an immune response to the grafted tissue or how these antigens are presented and processed by the immune system. When an animal is injected with an immunogen, the animal responds by producing an enormous diversity of antibody molecules directed against different antigens, different determinants of a single antigen, and even different antibody structures directed against the same determinants. Once these are produced they are released into the circulation, and it is nearly impossible to separate all the individual components present in the serum. However, since each antibody molecule is made by individual cells, the "immortalization" of specific antibody-producing cells by somatic cell fusion, followed by cloning of the hybrid cell derivative, allows permanent production of the antibodies in culture.[3] The study to be reported here summarizes investigations on the protein antigens of human corneal epithelium and endothelium by hybridoma antibodies prepared against these antigens.

Materials and Methods

Preparation of antigen. Human research corneas were obtained from the North Florida Lions Eye Bank, from a mixed pool of individuals 65–80 years of age, in M-K medium, 36 hours post-enucleation. The epithelial and endothelial cells were scraped from these corneas into sterile phosphate buffered saline (PBS). Three separate immunizing antigens were used in this study: (1) soluble protein antigens of epithelium, (2) insoluble protein antigens of epithelium obtained after homogenization and centrifugation, and (3) endothelial whole cells.

Immunization of mice. Soluble and insoluble protein antigens from epithelium (100 μg protein) and endothelial cells (10^6 cells/ml) were individually injected, subcutaneously, into BALB/cJ mice. Two weeks later an additional amount of these antigens was injected intraperitoneally. One week later a third immunization was made intravenously with the above amounts of antigens. Three days after the last immunization the mice were sacrificed, their spleens were aseptically removed, and a small volume of their blood was saved.

Somatic cell hybridization. The immune splenocytes were fused with the mouse

Department of Ophthalmology, College of Medicine, University of Florida, Gainesville, Florida.

TABLE I
Binding Ratios for Cross-reaction Specificity of Antihuman Epithelial Water Soluble Protein Hybridoma Antibodies to Other Corneal Water Soluble Proteins

Culture supernatant tested	Human epithelial WSP (immunogen)	Human endothelial WSP	Human stromal WSP	Rabbit epithelial WSP
1E3	2.26	1.30	1.01	1.10
1E4	2.32	1.00	0.95	1.28
1E1	3.05	0.39	1.18	0.97

A binding ratio of greater than 2.0 was considered positive.

myeloma Sp2/O-Ag14 cell line[4] in the presence of polyethylene glycol-1000 (PEG-1000) according to the procedure of Koprowski.[5] Briefly, 10^7 mouse myeloma cells and 10^8 spleen cells were mixed in 1 ml of 50% PEG-1000, in DME (Dulbecco's modified Eagles Medium) at 37°C for 1 minute. The cells were then resuspended to 3×10^6 cells/ml in 50% conditioned media containing hypoxanthine/aminopterin/thymidine (DME-HAT)[6]. The cell suspension was then distributed in individual wells of a 96-well microtiter plate and allowed to incubate at 37°C in 5.0% CO_2 incubator.

Radioimmunoassay (RIA). After 7–14 days' growth, the hybrid cells were tested for antibodies specific to soluble or insoluble corneal epithelium or to endothelial whole cell by RIA.[7] The immunizing antigen was fixed to individual wells of U-bottom, flexivinyl microtiter plates. The plates were washed with PBS containing 0.1% BSA, and 50 µl of the supernatant from each hybrid colony was added and allowed to incubate at 37°C for 90 minutes. Following five washes with PBS-BSA buffer, ^{125}I labelled sheep anti-mouse IgG was added (50,000 cpm) and allowed to incubate for 90 minutes. The wells were then washed, cut, and counted in a gamma counter. Conditioned media served as the negative control. The binding ratio was obtained by dividing the average CPM of the experimental wells by that of the negative control. A value of 2.0 or more was considered positive.

Cross-reaction assays. The supernatants from colonies producing antibodies to water soluble antigens (WSP) or water insoluble antigens (WISP) of corneal epithelial cells were tested by RIA for cross-reactivity to equal amounts of WSP and WISP from various human and rabbit cells. Hybridoma antibodies to human endothelial cells were tested against equal numbers of human corneal epithelial cells, lymphocytes, and red blood cells.

Results

The immunizing antigens were examined by sodium dodecylsulfate polyacrylamide gel electrophoresis and judged to contain no evidence of immunoglobulins. The number of antibody positive hybrids obtained was as follows:

1. Soluble protein antigens of epithelium; 200 tested, 7 positive.
2. Insoluble protein antigens of epithelium; 125 tested, 5 positive.
3. Endothelium whole cell; 200 tested, 2 positive.

Table I shows that the hybridoma antibodies produced to human epithelial WSP did not cross-react with human endothelium, human stroma, or rabbit epithelial WSP. Antibodies produced to human epithelial WISP (Table II) did not cross-react with the WISP of human endothelium, RBC, or lymphocytes, nor with rabbit kidney WISP. Two of the antibodies, however, reacted with rabbit epithelial WISP (2D3 and 2A7). One of the antibodies produced to human endothelial whole cells cross-reacted with human cornea epithelial cells but neither reacted with human RBC or lymphocytes (Table III).

TABLE II
Binding Ratios for Cross-reaction Specificity of Antihuman Epithelial Water Insoluble Protein Hybridoma Antibodies to Other Corneal Water Insoluble Proteins

Culture supernatant	Human epithelial WISP (immunogen)	Human endothelial WISP	Human RBC WISP	Human lymphocyte WISP	Rabbit epithelial WISP	Rabbit kidney WISP
Mouse antiserum	12.26	3.47	N.T.	N.T.	5.44	N.T.
2D3	6.68	1.83	1.26	1.15	2.68	1.10
2A7	2.86	1.23	0.93	0.96	2.23	1.80
4H2	3.10	1.12	0.94	0.97	1.82	0.99
3D6	3.83	1.04	0.93	1.00	1.89	1.00
4F5	2.79	1.17	0.72	1.00	1.63	1.06

N.T. = Not tested.
A binding ratio of greater than 2.0 was considered positive.

Discussion

These results indicate that it is possible to produce hybridoma antibodies to human corneal cells and cellular extracts although the number of positive cultures obtained per fusion is low (less than 5%). Of course, if more hybrid cultures had been screened, we might have detected a greater number of specific antibody-producing hybridomas. However, the amount of antigen obtained from human corneas limited the number of screening assays. None of the hybridomas produced in this study secretes antibody to human blood group or histocompatibility antigens. This finding does not preclude the presence of these antigens on human corneal cells. Preliminary data suggest that there is a common antigen shared between human corneal epithelial and endothelial cells, and also a common antigen(s) between human epithelial and rabbit epithelial WISP preparations.

The immunologic characteristics of ocular tissues have been the subject of much investigation,[8–10] but the functional importance of antigens other than HLA has not yet been defined. It is possible that immune responses to ocular antigens are involved in the etiology of certain chronic ophthalmic disorders.[11,12] Previous studies, however, have been based on reactions to complex mixtures of antigens, with a few exceptions.[13,14] The technique of somatic cell fusion and the production of monoclonal antibodies have great potential in the further characterization of individual corneal antigens and their role in autoimmune reactions.

Acknowledgments

This study was supported in part by a grant from the National Society to Prevent Blindness, Inc., NIH grant EY-02681, and by an unrestricted departmental grant from Research to Prevent Blindness, Inc.

TABLE III
Binding Ratios for Cross-reaction Specificity of Antihuman Endothelium Whole Cell Hybridoma Antibodies to Other Human Cells

Culture supernatant	Human endothelial cells (immunogen)	Human epithelial cells	Human lymphocytes (mixed pool)	Human RBC		
				Type A	Type B	Type O
1B10	16.3	3.57	0.87	0.83	1.23	1.21
3G9	7.46	1.19	1.14	1.13	1.18	1.22

A binding ratio of greater than 2.0 was considered positive.

References

1. Billingham RE and Boswell T: *Proc R Soc Lond (Biol)* **141**:392–406, 1953.
2. Maumenee AE: *Am J Ophthalmol* **34**:142–152, 1951.
3. Milstein C; *Proc R Soc Lond* **B211**:393–412, 1981.
4. Svasti Y and Milstein C; *Biochem J* **128**:427–444, 1972.
5. Koprowski H et al.: *Proc Natl Acad Sci USA* **74**:2985–2988, 1977.
6. Littlefield JW: *Science* **145**:709–710, 1964.
7. Bechtol KB: In *Monoclonal Antibodies, Hybridomas: A New Dimension in Biological Analysis,* Kennet RH et al. (Eds.). Plenum Press, New York, 1980, pp. 381–384.
8. Hall JM et al.: *Invest Ophthalmol* **13**:304–307, 1974.
9. Whiteside TL et al.: *Exp Eye Res* **16**:413–420, 1973.
10. Nussenblatt R et al.: In *Immunology and Immunopathology of the Eye,* Silverstein AM and O'Connor GR (Eds.). Masson Publishing USA, New York, 1979, pp. 145–150.
11. Wong VG et al.: *Am J Ophthalmol* **72**:960–966, 1971.
12. Shore B et al.: *Am J Ophthalmol* **73**:62–67, 1972.
13. Holt WS and Kinoshita JH: *Invest Ophthalmol* **12**:114–126, 1973.
14. Alexander RJ et al.: *Exp Eye Res* **32**:205–216, 1981.

chapter 66

An Experimental Model of Immunogenic Pseudotumor of the Orbit

Sammy H. Liu, Robert A. Prendergast, and Arthur M. Silverstein

An orbital pseudotumor is a concentrated aggregation of chronic inflammatory cells, whose cytology may range from purely lymphoid hyperplasia to granuloma. Numerous clinical and histopathologic reports have provided the basis for our current understanding of idiopathic inflammatory pseudotumors,[1] but specific data regarding the etiology and pathogenesis of this disease are still lacking. To the present time, however, no experimental model of ocular inflammatory pseudotumors has been available. Recent work in other fields on the development of chronic lymphoproliferative and granulomatous inflammation of immunogenic origin offers the opportunity to develop an experimental model of this type of ocular disease. These studies utilize insoluble particles coated with a variety of soluble antigens, to induce chronic inflammatory reactions with lymphoproliferative and granulomatous complications in the lungs[2–4] or skin[5] of immunized animals. In the present study, we use a similar approach to explore the possibility of producing an experimental immunogenic pseudotumor of the orbit.

Materials and Methods

Animals. Hartley strain guinea pigs of either sex weighing approximately 300 g were used throughout the experiments.

The Wilmer Ophthalmological Institute, The Johns Hopkins University School of Medicine, Baltimore, Maryland.

Preparation of protein-coated beads. Sepharose 4B beads (Pharmacia, Piscataway, NJ) were sieved through Spectramesh (Fisher Scientific Co., Springfield, NJ) to select the range of 40–60 μ in diameter. Bovine gamma globulin (BGG) or ovalbumin (OA) (Sigma Chemical Co., St. Louis, MO) was separately coupled to the beads by the method of Axen et al.[6] Sepharose 4B beads were activated with cyanogen bromide. The pH of the suspension was maintained at 10.5 for 15 minutes by the addition of 1.0 M sodium hydroxide. The activated Sepharose 4B beads were washed extensively with 0.1 M sodium borate–boric acid buffer, pH 8.5. A 2 g/ml slurry of the activated beads in borate buffer was mixed with a 50 mg/ml solution of BGG on a magnetic stirrer for 18 hours at 4°C. The beads were washed extensively with sterile saline and adjusted to 2000 beads/ml with the aid of a white blood cell-counting chamber.

Immunization and challenge. Guinea pigs were sensitized with 200 μg of BGG emulsified in an equal volume of complete Freund's adjuvant (CFA) containing 2 mg/ml of *Mycobacterium tuberculosis* H37Ra (Difco Lab., Detroit, MI). The emulsion (0.4 ml) was distributed equally among the four footpads. Animals were challenged intraorbitally with BGG-coated beads in their right (test) eyes, and with OA-coated beads in their left (control) eyes 4 weeks after immunization. Each challenge was delivered in a volume of 0.05 ml with 100–120 beads and a total quantity of 10 μg of BGG or OA. Groups of three animals were sacrificed at various times after the injection of beads. The eyes were enucleated, fixed in 10% phosphate-buffered formalin, processed by standard histologic methods, and stained with hematoxylin and eosin.

Results

Histological findings in lacrimal and Harderian glands

Guinea pigs immunized 4 weeks previously with BGG in CFA were challenged with BGG-coated beads in their right (test) eyes, and with OA-coated beads in their left (control) eyes. The orbital tissues were examined at 2, 4, and 8 days after injection of the beads.

At 2 days, intense inflammatory infiltrates were seen around the beads in the right eyes of the immunized animals (Fig. 1a). There were large collections of mononuclear cells that formed confluent masses around groups of BGG-coated beads; these cells were mainly lymphocytes and macrophages. In some cases lymphocytes predominated, whereas in others macrophages were more numerous. In contrast, in the left eyes there was either no reaction or a very mild inflammatory reaction around the beads coated with OA (Fig. 1b).

At 4 days, the histologic picture had changed in the right eyes (Fig. 2a). There was a marked shift to granuloma formation. The BGG-coated beads were surrounded by multinucleated giant cells, epithelioid cells, lymphocytes, and macrophages. A severe focally necrotizing, but mainly granulomatous inflammation, was observed. The inflammatory cellular reaction extended into and around a duct in the lacrimal gland, as Figure 2a indicates. In the left eyes, the inflammatory changes were still minimal at this time (Fig. 2b). A few inflammatory cells, consisting mainly of polymorphonuclear leukocytes, were seen surrounding the OA-coated beads, with an early mild foreign-body reaction.

At 8 days, the inflammatory changes in the right eyes underwent a continuing resolution to granuloma formation. The lesion was accompanied by an early fibrous scar which was infiltrated with giant cells, epi-

thelioid cells, lymphocytes, and large numbers of plasma cells (Fig. 3a). In contrast, the left eyes showed a mild foreign-body and granulomatous reaction around the OA-coated beads (Fig. 3b).

Histologic findings in retrobulbar connective tissue

Histologic examination of orbital connective tissue in the right (test) eyes at 4 days after injection of beads showed the formation of granulomas around the BGG-coated beads (Figs. 4a, b). Most of the granulomas were fully developed and tended to be more organized, with a circumscribed lesion consisting mainly of giant cells and epithelioid cells, but also containing variable numbers of lymphocytes and macrophages. The inflammatory response appeared to be widespread. The most severe foci of granulomatous inflammation were noted where there were small clusters of the BGG-coated beads, and the inflammation extended into the adjacent adipose tissue. There was also severe granulomatous inflammation in the choroid. In contrast, the intensity of inflammatory change in the left (control) eyes was much less than that observed in the experimental eyes. This foreign-body type reaction, consisting mainly of polymorphonuclear leukocytes, was typical for the control eyes at 4 days after injection of the OA-coated beads.

Discussion

An experimental model of immunogenic pseudotumor of the orbit has been developed in the present study. The basic experimental design consists of insoluble beads coated with antigen injected retrobulbarly into sensitized animals. These beads induce chronic inflammatory reactions, which progress through the several characteristic histologic stages with time. The initial reactions produce characteristic lymphoproliferative inflammation, and then continue to a granulomatous transformation. Various models of pulmonary granulomatous inflammation[2-5] have created the conceptual framework needed for an understanding of the immunopathogenesis of

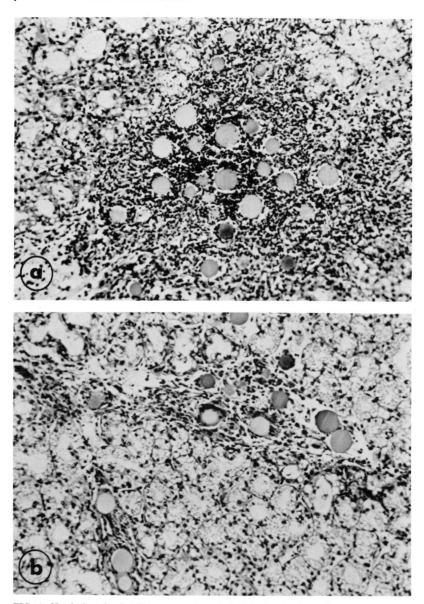

FIG. 1. Harderian gland at 2 days after injection of the beads. *(a)* Right (test) eye. Note intense infiltrate surrounding the BGG-beads. This is predominantly a mononuclear cell infiltrate of lymphocytes and macrophages. *(b)* Left (control) eye. Foreign-body type reaction to the OA-beads can be seen. This mild inflammatory reaction, consisting mainly of polymorphonuclear leukocytes, is typical for the control eye.

infectious or allergic granulomatous inflammations. This concept suggests that the delayed-type of hypersensitivity results from the interaction of locally retained antigen and sensitized T-lymphocytes, and that it triggers antigen-specific T-lymphocytes to release a variety of lymphokines. These, in their turn, recruit and activate additional lymphocytes and macrophages, contributing to granuloma formation in an accelerated manner.[2–5] Thus, the key factor in lymphoproliferative inflammation appears to be the

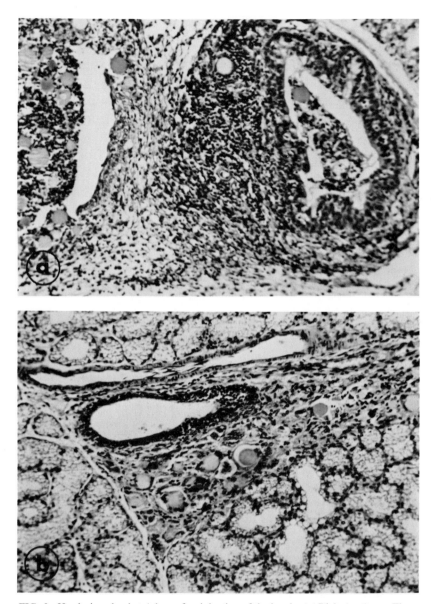

FIG. 2. Harderian gland at 4 days after injection of the beads. *(a)* Right (test) eye. There has been a marked shift to granuloma formation. The BGG-beads are surrounded by multinucleated giant cells, epithelioid cells, lymphocytes, and macrophages. Note focal necrosis in and around a duct. *(b)* Left (control) eye. The inflammatory changes are still minimal. A small number of inflammatory cells are seen around the OA-beads, with early foreign-body reaction.

specific and chronic triggering of T-cells by antigen, leading to lymphokine release and the nonspecific attraction and activation of other inflammatory cells. Such recruitment is probably responsible for the cellular polymorphism of the response. The major element in any granulomatous response seems to be the extent and degree of macrophage activation. Lymphokines have been shown to affect macrophages by inducing a rapid cell accumulation, maturation into epithelioid cells, and enhanced phagocytic and

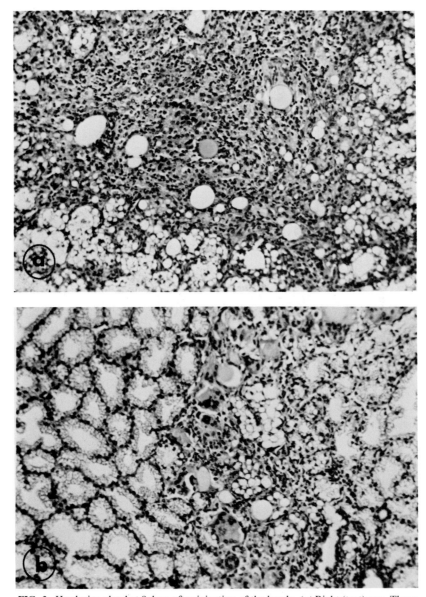

FIG. 3. Harderian gland at 8 days after injection of the beads. *(a)* Right (test) eye. There is a continuing resolution to granuloma formation, with plasma cells and early scar formation. *(b)* Left (control) eye. A mild foreign-body and granulomatous reaction around the OA-beads is observed.

digestive activity due to elevated lysosomal enzyme levels.[7–9] High levels of macrophage activation may then result in lysosomal enzyme-mediated tissue damage and reparative fibrosis.

An important question relating to this animal model is whether it has histopathologic similarities to human inflammatory pseudotumors. It is well known that the histologic picture of inflammatory pseudotumor has been found to have considerable variation in different cases. Blodi and Gass[10] studied the clinical and pathologic pictures in 140 cases. They found that no two lesions

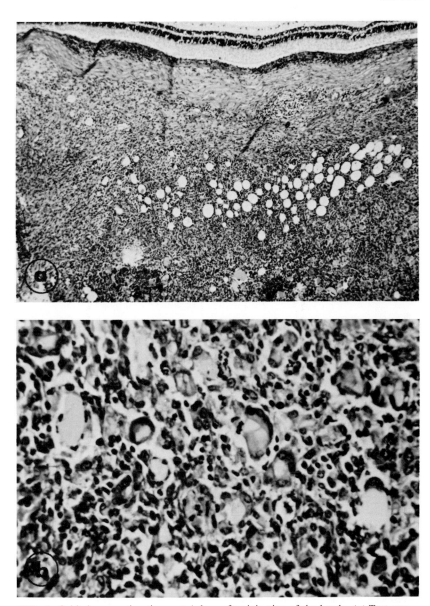

FIG. 4. Orbital connective tissue at 4 days after injection of the beads. *(a)* Test eye—low power. Granulomatous infiltrate is seen in the retrobulbar tissue. *(b)* Same eye—high power. Epithelioid and giant cells can be recognized around the BGG-beads. The inflammatory response also involves the choroid.

appear exactly alike because of the many stages and combinations possible. The histologic differences may involve variations in cellular polymorphism, true granulomatous infiltration, or involvement of a particular orbital tissue. These workers suggest that the inflammatory pseudotumor is prob-ably a single entity, the varied histologic picture depending upon the stage of the disease when the biopsy is taken. The histologic results seen in the present animal model support their hypothesis. Furthermore, the human disease is characterized by an extension of reactive lymphoid hyperplasia from the orbit to the intraocular tissues in a high percentage of cases.[11] In our model, we have demonstrated a similar extension of the orbital reaction into the choroid. Therefore, the histologic findings in the guinea pig model emphasize its potential relevance to the human disease, and point the way for investigations of the immunopathologic mechanisms involved in the formation of orbital pseudotumor.

Acknowledgments

Supported in part by U.S. Public Health Service Research Grants EY-04444, EY-00279, and EY-03521 from the National Institutes of Health, and by an Independent Order of Odd Fellows Research Professorship.

References

1. Garner A: *J Clin Pathol* **26**:639–48, 1973.
2. Boros DL et al.: *J Immunol* **114**:1437–41, 1975.
3. Kasdon EJ and Schlossman SF: *Am J Pathol* **71**:365–74, 1973.
4. Unanue ER and Benacerraf B: *Am J Pathol* **71**:349–64, 1973.
5. Kasdon EJ et al.: *J Invest Dermatol* **63**:411–14, 1974.
6. Axen R et al.: *Nature (London)* **214**:1302–04, 1967.
7. Galindo B et al.: *J Reticuloendothel Soc* **18**:295–304, 1975.
8. Dumonde DC et al.: In *Mononuclear Phagocytes in Immunity, Infection and Pathology,* van Furth R (Ed.). Blackwell, Oxford, 1975, p. 675.
9. Nathan CF et al.: *J Exp Med* **137**:275–90, 1973.
10. Blodi FC and Gass JDM: *Trans Am Acad Ophthalmol Otolaryngol* **71**:303–23, 1967.
11. Ryan SJ et al.: *Trans Am Acad Ophthalmol Otolaryngol* **76**:652–71, 1972.

chapter 67

The Role of T-Lymphocytes in the Induction of Experimental Autoimmune Uveitis in Rats

Mario-Cesar Salinas-Carmona,[a] Robert B Nussenblatt,[b] Waldon B. Wacker,[c] and Igal Gery[b]

Experimental autoimmune uveitis (EAU) is an inflammatory eye disease that may be induced in various experimental animals by immunization with the soluble retinal component, S-antigen (S-Ag).[1,2] EAU resembles certain human pathologic conditions.[2–4] Furthermore, the notion that autoimmunity to S-Ag plays an active role in some human ocular diseases has been supported by the finding of specific immune responses to S-Ag in a large proportion of patients with posterior uveitis.[4]

Little is known about the pathogenic mechanisms that bring about EAU. Recently, we showed that EAU may be prevented by treating the S-Ag immunized rats with cyclosporin A.[5] This drug is known to affect T-lymphocytes[6] selectively; thus, our finding supports the hypothesis that T-cells play a mandatory role in the pathogenesis of EAU. The study reported here further supports this hypothesis by showing that the disease does not develop in athymic rats but can be transferred to these animals by T-lymphocytes from S-Ag immunized donors.

Materials and Methods

Female Lewis rats were purchased from M.A. Bioproducts (Walkersville, MD), while athymic (rnu/rnu) rats and their heterozygous (rnu/+) controls were kindly provided by Dr. C. Hansen, NIH. These rats were of the original colony of Rowett Research Institute, Bucksburn, Aberdeen, Scotland.[7]

Bovine S-Ag was emulsified in complete Freund's adjuvant (CFA) and injected as described,[5] at 30 μg per rat. Rats tested for lymphocyte responses were sacrificed 10–12 days after immunization, their draining popliteal lymph nodes were removed, and lymphocyte suspensions were prepared as described.[5] Lymphocyte responses to the immunizing antigens (S-Ag or tuberculin [PPD]), or to concanavalin A (Con A) were determined in culture as described[5] and the results are presented both as mean cpm values ± S.E. and stimulation index values.

Rats used as lymphocyte donors for transfer experiments were sacrificed 7–10 days after immunization. Lymph node cells, prepared as described,[5] were injected intraperitoneally immediately, or following treatment with antibodies (to eliminate T-cells), or culturing with antigens.

Treatment with antibodies against rat T-lymphocytes (M.A. Bioproducts, diluted 1:4) was carried out at 37°C for 45 minutes

[a]Laboratorio de Immunologia, Apdo. Postal 4355-H Facultad de Medicina, U.A.N.L., Monterrey, Mexico; [b]Clinical Branch and Laboratory of Vision Research, National Eye Institute, National Institutes of Health, Bethesda, Maryland; [c]University of Louisville, Louisville, Kentucky.

TABLE I
Development of EAU in Athymic and Control Rats[a]

Rat strain	EAU (positive/ total in group)
Homozygous rnu/rnu (athymic)	0/4
Heterozygous rnu/+	10/12
Lewis	9/10

[a] Rats were immunized with 30µg S-Ag emulsified in CFA.

and was followed by the addition of complement for 60 more minutes. The dead cells were removed by centrifugation on an Isolymph gradient. The specificity of the antibody effect was established by the incapability of the remaining lymphocytes to react in culture against a T-cell mitogen, Con A.

Lymphocytes cultured prior to injection were incubated for 3 days with S-Ag or PPD (2µg/ml), in aliquots of 2 ml containing 4×10^6 cells. After incubation, the cells were washed and injected into recipients.

Rats tested for EAU were examined daily for clinical signs of uveitis. Actively immunized rats were sacrificed 12–14 days after immunization while recipients of cells were killed 7 days after transfer. Preparation of histological sections and evaluation of severity of EAU were carried out as described.[5]

TABLE III
Transfer of EAU to Athymic Rats

Transferred cells[a]		EAU in recipients (positive/total)
Donor rats	Treatment in vitro	
rnu/+	none	2/7
rnu/+	Incubation with S-Ag	4/5
rnu/+	Incubation with PPD	0/3
Lewis	Incubation with S-Ag	0/3
rnu/+	Antibodies to T-cells + complement	0/5

[a] 10^8 Nucleated cells/recipient.

Results

Immunization with S-Ag emulsified in CFA induced severe bilateral uveitis in most control rats of both the Lewis strain and the heterozygous rnu/+ line (Table I). On the other hand, no disease developed in the athymic nude rats (rnu/rnu).

Athymic rats differed from the Lewis and rnu/+ controls also by the incapability of their lymphocytes to react in culture to the immunizing antigens, S-Ag and PPD, or to the polyclonal mitogen Con A (Table II). The specificity of the lymphocyte responses is depicted here by the selective response to PPD of rats injected with CFA alone, as

TABLE II
Mitotic Responses of Lymphocytes from Athymic and Control Rats

Rat strain	Immunizing agent	None	Stimulant in culture		
			S-Ag (5µg/ml)	PPD (10 µg/ml)	Con A (5µg/ml)
Lewis	C F A	737 ± 42	848 ± 110 (1.1)	30,409 ± 306 (41.3)	137,280 ± 537 (186.3)
Lewis	S Ag + C F A	205 ± 11	11,069 ± 868 (54.0)	20,542 (100.2)	125,978 ± 5128 (614.5)
rnu/+	C F A	332 ± 19	423 ± 17 (1.3)	18,190 ± 396 (54.8)	69,179 ± 840 (208.4)
rnu/+	S-Ag + C F A	297 ± 13	23,396 ± 16 (78.8)	33,542 ± 1122 (112.9)	83,064 ± 2186 (279.7)
rnu/rnu	C F A	229 ± 36	222 ± 41 (1.0)	198 ± 24 (0.9)	477 ± 56 (2.1)
rnu/rnu	S-Ag + C F A	268 ± 23	351 ± 17 (1.3)	370 ± 42 (2.7)	652 ± 37 (2.4)

The values are mean cpm ± S.E. In parentheses, stimulation indices.

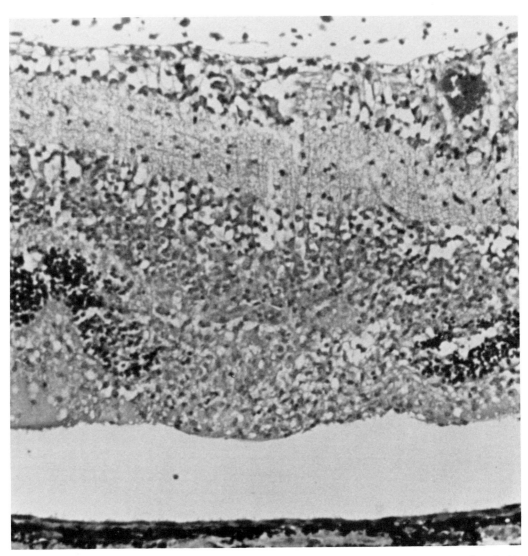

FIG. 1. Passively transferred EAU in an athymic rat. Lymph node cells from a rnu/+ donor were cultured with S-Ag for 3 days before being injected, and the recipient was sacrificed 7 days later. Note the severe inflammation in the retina with a partial loss of photoreceptor cell layer.

compared with responses to both PPD and S-Ag of rats immunized with S-Ag in CFA.

In contrast to the failure to induce EAU actively in athymic rats, these animals were found susceptible to passively induced disease. As shown in Table III, EAU developed in a proportion of athymic rats when injected with lymph node cells from S-Ag immunized rnu/+ donors. The efficacy of EAU transfer was markedly increased when the injected cells were preincubated for 3 days in the presence of S-Ag. Incubation with PPD, however, did not produce this effect. Also, no disease could be induced by transferring lymphocytes from the allogeneic Lewis rats to the rnu/rnu recipients.

The role of T-lymphocytes in the transfer of EAU was further depicted by the finding that treatment of the lymph node cells with specific antibodies against T-lymphocytes (plus complement) abolished the capacity of the injected cells to transfer the disease (Table III).

The histological changes found in athymic recipients of cultured lymph node cells resembled those observed in control rats with the actively induced EAU. A typical section is shown in Figure 1. The changes included mainly intense inflammatory infiltration and destruction of the photoreceptor region.

Discussion

The data presented here provide more supporting evidence for the notion that T-lymphocytes play a major role in the pathogenesis of EAU. Athymic rats, deficient in their T-cell capacity, were incapable of developing EAU, in contrast to their heterozygous rnu/+ controls. However, the disease could be induced in the athymic rats by transferring lymphocytes from S-Ag immunized rnu/+ donors, in particular following incubation of the cells with the S-Ag. Moreover, the active role of T-lymphocytes in the induction of EAU was established by the finding that the capacity of lymph node cells to transfer EAU was abolished by the selective lysis of the T-cell population by specific antibodies and complement.

The exact role of T-cells in the pathogenesis of EAU is not yet clear. It is conceivable, however, that these cells mediate the disease by more than one way. T-lymphocytes can manifest direct cytotoxic effects on target cells, but more important, perhaps, may be the capacity of T-cells to recruit other cells into the pathogenic inflammatory process. The recruited cells, mainly polymorphonuclears, macrophages, and other lymphocytes, are attracted and activated by a battery of released T-cell products that include chemotactic and activating factors.[8] T-cell-derived helper factors are crucial for antibody production, and thus any humoral involvement in the pathogenesis of EAU is also regulated by T-cells.

References

1. Wacker WB et al.: *J Immunol* **119**:1949–1958, 1977.
2. Faure JP: *Curr Top Eye Res* **2**:215–302, 1980.
3. Nussenblatt RB et al.: *Arch Ophthalmol* **99**:1090–1092, 1981.
4. Nussenblatt RB et al.: *Am J Ophthalmol* **89**:173–179, 1980.
5. Nussenblatt RB et al.: *J Clin Invest* **67**:1228–1231, 1981.
6. Britton S and Palacios R: *Immunol Rev* **65**:4–22, 1982.
7. Festing MFW et al.: *Nature* **274**:365–367, 1978.
8. Cohen S et al. (Eds.): *Biology of the Lymphokines.* Academic Press, New York, 1979.

chapter 68

Murine Susceptibility to Herpes Simplex Keratitis Is Influenced by the Igh-1 Gene Locus

C. Stephen Foster,[a,b] **Yvonne Tsai,**[a] **Richard Wetzig,**[a]
Mark I. Greene,[c] **and David Knipe**[c]

Herpes simplex keratitis is a major cause of corneal blindness in the world today. The available evidence suggests that both viral factors and recipient host factors, most notably immune, influence the clinical expression of herpes simplex virus infection. The purpose of this report is to describe the results of our studies in a murine model of the influence of certain genes responsible for controlling immunologic responses on the development of clinical lesions after HSV corneal inoculation.

Materials and Methods

Mice. Genetically defined, inbred congenic murine strains were employed in these studies. The background genetics and the H-2, Igh-1, and Igk phenotype of each of these congenic strains are shown in Table I. Female mice, 6–8 weeks of age, from each of the strains studied were randomly selected for corneal inoculation of herpes simplex virus or placebo control. Six mice from each strain were employed for each experimental run, and at least three replicates of each experimental run were performed.

Virus. Herpes simplex virus Type-1, KOS strain, was grown on Vero cell layers. An inoculum of 10^7 plaque forming units (PFU) was employed for these studies.

Inoculation and clinical observations. The right cornea of each mouse was scratched eight times (four vertical and four horizontal scratch marks) with a 25 gauge needle under binocular microscopic observation. Fifty microliters of a suspension of KOS herpes simplex virus Type-1 at a concentration at 10^7 PFU/ml was inoculated into the cul-de-sac, and the lids were compressed over the cornea for five seconds. Daily masked biomicroscopic observations were performed and clinical parameters scored. HSV lid lesions, conjunctival inflammation, epithelial keratitis, stromal keratitis, and anterior chamber cellular reaction were each graded on a scale of 0–4 +. Representative members of each study sample were randomly killed at various times; blood was harvested for herpes-specific antibody determinations and for mononuclear cell isolation, and eyes were harvested for histopathology and for immunofluorescence. The specimens for histopathology were fixed in Karnovsky's fixative, embedded in JB4 plastic, sectioned at 1 μm, and stained with alkaline Giemsa or with hematoxylin and eosin. Tissue for immunologic studies was snap frozen immediately after being obtained, embedded in Tissue Tek II OCT embedding compound (Lab Tex Products,

---

[a]Department of Cornea Research, Eye Research Institute of Retina Foundation, Boston; and the Departments of [b]Ophthalmology and [c]Pathology, Harvard Medical School, Boston, Massachusetts.

TABLE I
Congenic Murine Strain Genetics

Murine Congenics	Background	Chromosome 17 H-2	Chromosome 12 Igh-l-linked	Chromosome 6 Igk-l-linked
Balb/c	Balb/c	d	a	b
Cal-20	Balb/c	d	d (e)	b
C58-AL20	Balb/c	d	d (e)	a
A/J	A/J	a	e	b
ABY	A/J	b	e	b
AKR	AKR	k	d (e)	a
B10	C57BL10	b	b	b
B10.A	C57BL10	a	b	b
B10.D2	C57BL10	d	b	b
B6	C57BL6	b	b	b
B6 Igh-le	C57BL6	b	e	b

Inc., Naperville, IL), and sectioned at 4 μm. Direct immunofluorescence staining was performed with fluorescein-conjugated rabbit antisera to mouse IgG, IgA, IgM, third component of complement and antiserum directly against herpes simplex virus.

Antiherpes antibody determination. A modified, direct enzyme-linked immunosorbent assay was used to measure anti-HSV antibody titers. HSV (KOS strain) was inactivated under ultraviolet light for 20 minutes. A dilution of 10^7 PFU/ml was made in 0.1 M sodium carbonate, pH 9.6, with 0.2% NaN_3. This antigen solution was plated into 96-well flat-bottom polystyrene microtitration plates (0.35 ml well capacity) at 0.2ml/well. The plates were then incubated for 4 hours at room temperature. The reactions in the plates were then washed 3 times with a 0.9% NaCl with 0.05% Tween 20. The plates were then blocked for 1 hour with 0.2ml/well of 1% bovine serum albumin in phosphate-buffered saline with 0.05% Tween 20 and 0.02% NaN_3; the plates were washed again 3 times. Then 0.1ml of the appropriate dilution of experimental serum in phosphate-buffered saline with 0.05% Tween 20 and 0.02% NaN_3 was added, and the plates were incubated for 2 hours at room temperature. The plates were washed three times once again. Alkaline phosphatase-conjugated goat anti-mouse immunoglobulins (Fab2 fragment specific, Dynatech Diagnostics, Inc, South Windham, ME), at 1:100 in phosphate-buffered saline with 0.05% Tween 20 and 0.02% NaN_3 was added at 0.1 ml per well. The plates were incubated overnight at room temperature and again washed three times. Then 0.2 ml of 1 mg/ml p-nitrophenylphosphate (Sigma Chemical, St. Louis, MO) in 0.05 M sodium carbonate, pH 9.8, with 10^{-3} $MgCl_2$ was added to each well. After 45 minutes incubation at room temperature, the reaction was stopped by adding 0.1 ml of 1 N NaOH to each well. The substrate absorbance at 405 nm was measured with a Flow microtiter plate reader. A well was read as positive if the absorbance was more than twice that of a control well containing a comparable dilution of nonimmune serum.

Results

Clinical Observations. Major differences were observed in the clinical disease pattern between the various murine strains studied. C57 Black mice, both with the B10 background and with the B6 background, were relatively protected from development of severe keratitis after HSV corneal inoculation. On the other hand, A/J mice routinely developed severe keratopathy after similar inoculation. Balb/c mice regularly exhibited a pattern of initial resolution of keratitis with subsequent relapse and development of se-

TABLE II
Background Genetics and H-2, Igh and Igk
Phenotypes of A/J, ABY, and B10 Mice

Mice	Background	H-2	Igh	Igk
A/J	A/J	a	e	b
ABY	A/J	b	e	b
B10	B10	b	b	b

vere keratopathy. Cal20, ABY, and AKR mice developed severe keratopathy. The presence of the H-2b phenotype, a phenotype carried by the B10 strain on the A/J background, does not confer protection onto ABY mice (Table II, Fig. 1).

To explore further the influence of the H2 phenotype on development of keratopathy, we compared B10 congenics differing only at the H-2 locus. If the B10 background and the Igh and Igk loci were held constant, variability in the H-2 phenotype had no obvious effect on clinical disease expression (Table III). In contrast, Table IV and Figure 2 show an apparent influence of the Igh locus on clinical disease expression. Thus, since the genetic background, the H2 phenotype, and the Igk locus are identical for Balb/c and Cal20 mice, the development of severe keratitis is much more rapid in the Cal20, which has the Igh-1^d phenotype as opposed to the Igh-1^a phenotype of the Balb/c mice. Clinical examples of these differences are seen in Figures 3 and 4. The Igk

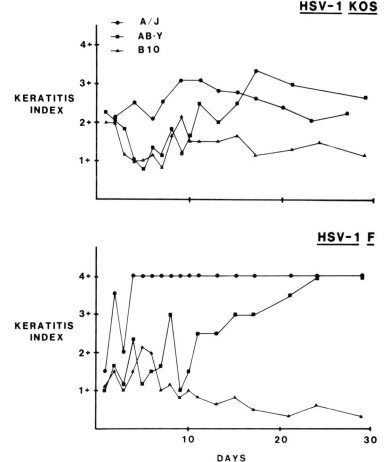

FIG. 1. Keratitis scores for A/J, ABY, and B10 mice after corneal inoculation with HSV-1, KOS and F strain. Note that in these experiments, while the keratitis scores for the A/J and ABY mice were consistently higher than those for the B10 mice, the degree of difference between the keratitis scores is not impressive. After corneal inoculation with the F strain of HSV-1, however, note that the differentiation between keratitis scores of B10 mice, compared to A/J's and ABY's is impressively enhanced. A/J and ABY mice developed severe degrees of keratopathy that persist indefinitely; the keratitis of B10 mice in these experiments resolves to nearly control levels.

TABLE III
Genetics of Three B10 Congenic Murine Strains

Congenics	Background	H-2	Igh	Igk
B10	B10	b	b	b
B10.A	B10	a	b	b
B10.D2	B10	d	b	b

TABLE IV
Genetics of Balb/c Congenic Murine Strains

Congenics	Background	H-2	Igh	Igk
Balb/c	Balb/c	d	a	b
Cal-20	Balb/c	d	d (e)	b
C58-AL20	Balb/c	d	d (e)	a

locus has no apparent influence on disease susceptibility; both Cal20 and C58AL20 mice, although differing at the Igk locus, develop similar degrees of ocular destruction after HSV inoculation into the cornea (Table IV). Antiherpes serum antibody responses are not significantly different between these murine strains, regardless of the corneal pathology that develops.

Our lymphocyte proliferation data would suggest that there is a more efficient, more rapid development of a cellular response in the murine strains that resist severe keratopathy after HSV corneal inoculation, and *in vivo* skin test results tend to corroborate this difference in the development of a DTH response to HSV corneal inoculation. Thus, we find an impaired *in vitro* lymphocyte proliferative response to HSV antigen in A/J mice compared with the response of lymphocytes from B10 mice; the *in vivo* DTH responses, though more global as a

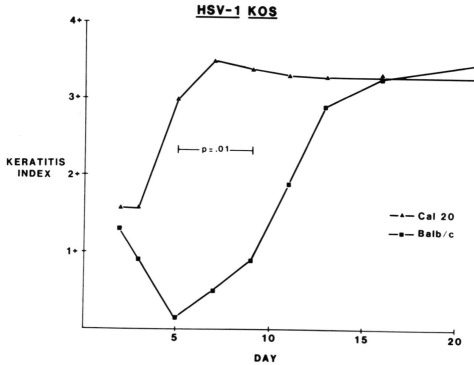

FIG. 2. Keratitis in Balb/c and in CAL20 mice after HSV-1, KOS strain corneal inoculation. These congenic murine pairs differed dramatically in their corneal response to HSV inoculation. CAL20 mice rapidly developed impressive keratitis that persists indefinitely. Balb/c mice, as seen before, had a keratopathy pattern characterized by initial resolution of keratitis, with subsequent relapse and development of significant corneal pathology. The only genetic difference between these two murine strains is in the Igh-1 phenotype.

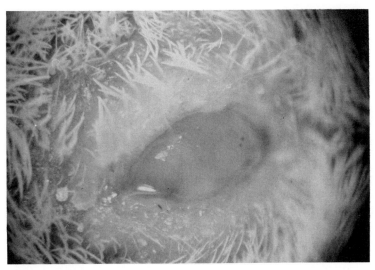

FIG. 3. Clinical example of the keratopathy of a CAL20 mouse 6 days after HSV-1 inoculation.

measure of T-cell-mediated activities, show the same trend. Preliminary studies with T-cell-mediated cytotoxicity and natural killer activity are suggestive of a similar trend in Igh-associated strain differences.

Immunofluorescence and histopathology. The immunofluorescence and histopathology patterns that develop in corneas from the various murine strains are dramatically different between "susceptible" and "resistant" strains. The initial response (up to 2 days after HSV corneal inoculation) in all strains is characterized by the appearance of the third component of complement (C3) in the superficial corneal stroma and by a neutrophil and lymphocyte peripheral corneal stromal infiltrate; A/J mice, however, subsequently fail to develop a prominent

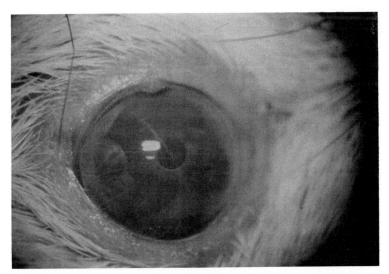

FIG. 4. The cornea of a Balb/c mouse, 6 days after HSV-1 inoculation. Note the impressive difference in corneal pathology, relative to the preceeding figure.

mononuclear response in the cornea. Instead, neutrophils continue to constitute the major cell type migrating into A/J corneas.

Immunofluorescence studies show that, compared to the patterns in B10 mice, excessive IgG, IgM, IgA, and C3 commonly accumulate in the corneal stroma of A/J mice; also herpes persistence in the epithelium past day 16 after inoculation is common. B10 mice, on the other hand, rapidly develop a prominent mononuclear cell response in the HSV-inoculated cornea, with few neutrophils seen by day 5 after inoculation. Macrophages and lymphocytes are the predominant cells. By immunofluorescence and by HSV culture we find B10 mice to harbor few HSV particles after day 7 postinoculation; this is in distinct contrast to A/J mice, which frequently continue to have HSV in the cornea 16–26 days after inoculation. Balb/c mice, which tend to exhibit clinical keratitis severity intermediate to that of A/J and B10 mice, usually develop a mixed mononuclear cell/neutrophil corneal response; these animals usually show no virus in the cornea by day 16, but frequently still have the virus in the cornea past day 7. Exceptions to the typical responses are sometimes seen, in that an occasional B10 mouse will develop clinically severe keratitis. Histopathology of these corneas always shows intense neutrophil infiltration of the corneal stroma, and immunofluorescence always demonstrates large amounts of IgG, IgM, IgA, and C3 in these corneas. HSV persistence is sometimes demonstrable.

Discussion

Our results suggest that one factor governing the clinical response to HSV encounter is the genetic control, from a locus on chromosome 12, of T-lymphocyte receptor activity for herpes simplex virus recognition. These observations of a VH (Igh-1)-associated clinical response to HSV are especially cogent (1) in view of the well-documented work showing that VH (Igh-1)-encoded structures may be used by T-cells as a portion of their antigen binding sites[1] and (2) in view of our findings relating to the powerful influence on the immune response that the site of antigen encounter in the eye has.[2]

The important work of Lopez et al. has shown that there are major differences in murine susceptibility to the lethal effects of intraperitoneally inoculated herpes simplex Type 1.[3] This genetically governed "natural" resistance seems to be mediated by a bone marrow-dependent cell (apparently a natural killer or NK-like cell) which rapidly clears the virus, thereby preventing viral dissemination.[4] The natural immunity governing such resistance or susceptibility appears to be controlled by two major, independently segregating, non-H2 genetic loci.[5] These findings are consistent with the experimental findings described by our group.

Acknowledgments

This study was supported in part by NIH Research Grants EY-03063 (Dr. Foster) and EY-04092 (Drs. Greene and Foster).

References

1. Owen FL et al.: *Eur J Immunol* **12**:94, 1982.
2. Wetzig RP et al.: *J Immunol* **128**:1753, 1982.
3. Lopez C: *Nature* **258**:152, 1975.
4. Lopez C et al.: *Infect Immun* **28**:1028, 1980.
5. Lopez C: *Immunogenics* **11**:87, 1980.

chapter 69

Immunoglobulins and Secretory Component in Rat Tears and Ocular Adnexa

Olafur G. Gudmundsson,[a,b,c] **David A. Sullivan,** [a,b] **Kurt J. Bloch,**[d] **and Mathea R. Allansmith** [a,b]

The tissues of the ocular adnexa are part of the secretory immune system. This observation is based on the identification of immunoglobulin (Ig)-containing cells and secretory component (SC) in the conjunctiva and/or lacrimal gland of several species, as well as the finding of Igs and SC in tears.[1–10] The contribution of the ocular secretory immune system of the rat to protection of the eye has not been evaluated. We have chosen the rat as a model to examine the secretory immune system of the eye and report here the ocular tissue distribution of Ig-containing cells and SC. In addition, we have measured the levels of Igs and SC in rat tears.

Material and Methods

Adult male Sprague-Dawley rats were used: 12 (pool) animals were involved in the tear Ig studies; 9 individual rats for the SC study; and 5 for the immunofluorescence studies. Tears were collected with microcapillary pipettes. Blood was obtained from the tail. Bile was collected by cannulation of the bile duct.[11] Exorbital gland, Harder's gland, conjunctiva, and a segment of the ileum were excised from each of five ether-anesthetized and exsanguinated rats. Specimens were fixed in absolute alcohol–acetic acid, embedded in paraffin, and cut into 4 micron sections.

Antisera to IgG1, IgG2a, IgG2b, IgG2c, and IgM (all antisera produced a single precipitin line with normal rat serum on Ouchterlony analysis) were obtained commercially (Miles Laboratories, Cappel Laboratories). Antisera to IgA and IgE were prepared by Kurt J. Bloch, M.D. Antiserum to rat secretory component was a gift from Charles Wira, Ph.D. (Dartmouth Medical School, Hanover, NH) and was made by Brian Underdown, Ph.D. (Institute of Immunology, University of Toronto, Toronto, Ontario, Canada). Sera from nonimmunized rabbits, goats, and sheep were used as controls in immunofluorescent experiments. The presence and quantity of Ig in the rat fluids was determined by single radial immunodiffusion (RID)[12] in 0.75% agarose containing antiserum at a final dilution of 1:20. Standard curves were constructed by plotting the diameter of precipitin rings versus the dilution of a reference serum whose Ig content was determined by the manufacturer (Miles). The results for IgA were expressed without taking into account the possible difference in the degree of polymerization of IgA in the various fluids. The error of the RID determinations was ± 5%

[a]Department of Ophthalmology, Harvard Medical School, Boston; [b]Department of Cornea Research, Eye Research Institute of Retina Foundation, Boston; [c]Fellow of the World Health Organization; [d]Clinical Immunology and Allergy Units, Medical Services, Massachusetts General Hospital, Boston, Massachusetts.

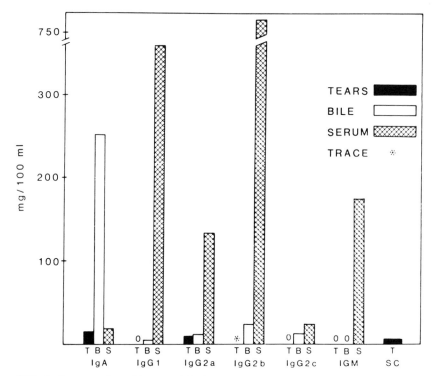

FIG. 1. Concentration of immunoglobulins in rat tears, bile, and serum and the concentration of SC in rat tears.

at one standard deviation (SD). The secretory component (SC) level of rat tears was measured by radioimmunoassay (RIA).[13] This assay detects primarily free SC.[13] The molecular size of IgA in tears was analyzed by fractionating a tear pool (220 µl) on a Biogel A 1.5 (Bio-Rad) column (0.9 × 60 cm) and assaying eluates for IgA by RIA (14).

Tissue sections for immunofluorescence studies were examined with a Leitz Orthoplan microscope. Photomicrography was accomplished with a Leitz Orthomat camera and Kodak Ektachrome film (400 ASA).

Results and Discussion

In this paper we report the Ig and SC content of rat tears and the distribution of Ig and SC in rat lacrimal glands, conjunctiva, and gut lamina propria. As shown in Figure 1, we were able to measure the concentration of IgA, IgG2a, and IgG2b in rat tears. We were unable to detect IgG1, IgG2c, or IgM. The predominant Ig in rat

tears was IgA, which had a concentration similar to that found in serum. Preliminary characterization of the molecular size of tear IgA showed that this Ig is primarily polymeric. In contrast to tear IgA levels, the concentrations of IgG2a and 2b in tears were considerably less than their respective serum concentrations. The distribution and levels of Igs in tears were different from those of bile (Fig. 1).

The interstitium of the exorbital gland was found to contain a considerable number of IgA-containing cells (Table I). The number was approximately one-fifth the number of IgA-containing cells in the lamina propria of the gut (ileum). The cytoplasm of the acinar epithelial cells typically contained only a trace of IgA. Occasionally, an epithelial cell or an entire acinus stained brightly. These structures were usually close to IgA-containing cells. The distribution of IgA in the rat exorbital gland appears to be different from that of the human lacrimal gland in which IgA has been found in the acinar collecting ducts and interstitial plasma

TABLE I

The Number of Immunoglobulin-containing Mononuclear Cells in the Rat Lacrimal Glands, Conjunctiva, and Gut Lamina Propria (average number of cells per 500x field)[a]

Location	IgA	IgG1	IgG2a	IgG2b	IgG2c	IgM	IgE
Exorbital gland	2.4	0.1	0.1	0.2	0.1	0.2	0
Harder's gland	0	0	0	0	0	0	0
Conjunctiva	0	0	0	0	0	0	0
Gut lamina propria	10.6	3.5	3.1	1.9	3.1	1.9	0

[a] Ten fields counted for each immunoglobulin for each specimen.

cells.[2,9] In the rabbit lacrimal gland, IgA is primarily located both in interstitial plasma cells and in the interstitial space.[4] In the rat, however, the interstitium was found to be essentially devoid of staining for IgA (Fig.2).

In the present study Ig of all IgG subclasses was found in the interstitium of the exorbital gland. These proteins were observed primarily as diffuse extracellular deposits, but IgG-containing cells were found infrequently (Table I). This observation is similar to the findings for the human lacrimal gland[2] and the rabbit lacrimal gland.[4] The number of Ig-containing cells in the rat exorbital gland is similar to that of the human lacrimal gland; the IgA/IgG cell ratio is about 20:1. The number of IgM-containing cells in the rat exorbital gland is slightly greater than that in human lacrimal gland; no IgM-containing cells were found in the rabbit lacrimal gland.[4] No IgE was detected in the rat exorbital gland, whereas it has been found in the human lacrimal gland.[2,10]

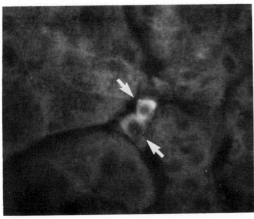

FIG. 2. Rat exorbital gland stained for IgA. Notice IgA-containing cells in the interstitium (arrows).

We were unable to detect any IgA staining in Harder's gland, but Igs of all subclasses of IgG were present. These proteins were evident primarily in extracellular areas of the interstitium. Low levels of IgG were also observed in the acinar epithelial cells and collecting ducts. A trace of IgM was detected in the interstitium.

The substantia propria of the conjunctiva contained a rare IgA-containing plasma cell. Extracellular IgA was present only in a trace amount. Igs of all IgG subclasses were found extracellularly in the conjunctiva (Fig.3) but only a trace of IgM was detected (Table II). Our findings with respect to IgA differ from those reported for rabbit[4] and human conjunctiva.[15] In the rabbit, IgA was present both in plasma cells and extracellularly. In the human, IgA was found only extracellularly. IgG was present in the substantia propria in both human and rabbit conjunctivae. IgM was detected (weak staining) in the human conjunctiva, but only as a trace in the rabbit conjunctiva.

The lamina propria of the gut was utilized as a positive control tissue in the immunofluorescent studies (Table I). In agreement with the results of others,[16,17] there was a predominance of IgA-containing cells, relatively few IgG- and IgM-containing cells, and no IgE-positive cells. Because all of the tissues that we had examined were negative for IgE, we tested our rabbit anti-rat IgE antiserum on a mesenteric lymph node from a rat infected with Nippostrongylus brasiliensis. Many IgE-containing mononuclear cells were detected in this tissue.

Secretory component (SC) was identified in the exorbital gland but not in the conjunctiva or Harder's gland. SC appeared both in the acinar cell cytoplasm and in the lumina of collecting ducts. This pattern of

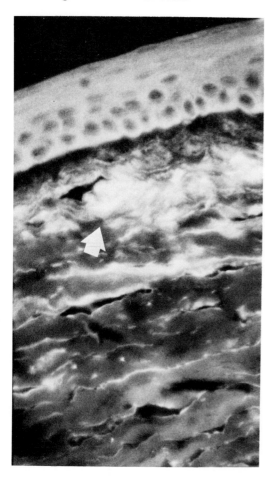

FIG. 3. Rat conjunctiva stained for IgG2a. Notice diffuse staining of substantia propria (arrow).

TABLE II

Location of Extracellular Immunoglobulins and Intraepithelial SC in the Rat Exorbital Gland, Harder's Gland, Conjunctiva, and Gut Lamina Propria

Location	IgA	IgG	IgM	IgE	SC
Exorbital gland					
acinar cells	*	–	–	–	+
interstitium	–	+	+	–	–
collecting ducts	–	–	–	–	+
Harder's gland					
acinar cells	–	*	–	–	–
interstitium	–	+	*	–	–
collecting ducts	–	*	–	–	–
Conjunctiva					
substantia propria	*	+	*	–	–
Gut lamina propria	–	+	–	–	+

* Trace.

Acknowledgments

This work was supported by grant EY-02882 from the National Institutes of Health, and by grants from the Massachusetts Chapter and the National Arthritis Foundation and the HOR Foundation.

References

1. Josephson AS and Weiner R: *J Immunol* **100**:1080–1092, 1968.
2. Franklin RM et al.: *J Immunol* **110(4)**:984–992, 1973.
3. McClellan BH et al.: *Am J Ophthalmol* **76(1)**:89–101, 1973.
4. Franklin RM et al.: *Invest Ophthalmol Vis Sci* **18(10)**:1093–1096, 1979.
5. Dohms JE et al.: *Avian Dis* **22**:151, 1978.
6. Kowlasky WJ et al.: *Immunology* **34**:663–667, 1978.
7. Killinger AH et al.: *Am J Vet Res* **39**:931, 1978.
8. Porter P: *Adv Vet Sci Comp Med* **23**:1, 1979.
9. Allansmith MR and Gillette TE: *Am J Ophthalmol* **89**:353–361, 1980.
10. Allansmith MR et al.: *Am J Ophthalmol* **82**:819–826, 1976.
11. Lambert R: In *Surgery of the Digestive System in the Rat.* Charles C Thomas, Springfield, IL, 1965, chapter 13. (Quoted by Lemaitre-Coelho I et al. *Eur J Immunol* **8**:588, 1977).
12. Mancini G et al.: *Immunochemistry* **2**:235–254, 1965.
13. Sullivan DA and Wira CR: *J Immunol* **130**:1330–1335, 1983.
14. Sullivan DA and Wira CR: *Endocrinology* **112**:260–268, 1983.
15. Allansmith MR and Hutchison D: *Immunology* **12(2)**:225–229, 1967.
16. Mayrhofer G and Fisher R: *Eur J Immunol* **9**:85–91, 1979.
17. Tourville DR et al.: *J Exp Med* **129**:411–429, 1969.

staining is almost identical to that described for both human and rabbit lacrimal glands.[2,5,9] SC was also found in rat tears (Fig. 1).

The predominant Ig in rat tears was polymeric. Lesser amounts of IgG 2a, IgG 2b, and SC were also present. Immunofluorescent analysis of various ocular tissues suggested that the exorbital gland of the rat is the major site of Ig production. This gland, but not the conjunctiva or Harder's gland, had a relatively high density of IgA-containing cells, few IgG and IgM cells, and SC. Our results suggest that, in the rat as in other species, the ocular adnexae constitute a portion of the secretory immune system.

chapter 70

Studies of Antinuclear Antibody-Positive Serum from Patients with Juvenile Rheumatoid Arthritis and Iridocyclitis Using Ocular Substrate

E. Lee Stock, Mark A. Rosanova, and Lauren M. Pachman

The association between the presence of antinuclear antibody (ANA) in the sera of patients with juvenile rheumatoid arthritis (JRA) and iritis has been well documented.[1,2] To date, various tissues such as thymus, mouse liver, mouse kidney and human leukocytes have been used as substrates for antinuclear antibody testing.[3] However, the end organ of inflammation, the human uveal tract, has not been studied as a substrate. In the present study we used human uveal tract to study reactivity of antinuclear antibody from sera of patients with JRA and iritis.

Methods and Materials

Ten patients with JRA and iritis were chosen from the Immunology Clinic of Children's Memorial Hospital, Chicago. Sera were stored at $-70°C$ until used for testing. Five patients with systemic lupus erythematosus (SLE) without a history of eye disease and with positive antinuclear antibody tests were used as positive controls. Sera were likewise frozen at $-70°C$ until used.

Donor eyes were obtained from the Illi-

nois Eye Bank, and each whole globe was frozen in liquid nitrogen. The globe was partially thawed and the anterior segment was divided with a razor blade at the equator. The remaining anterior segment was then quarter-sectioned through the center of the cornea. Neither vitreous nor aqueous was allowed to thaw. Each quarter-section was then placed in plastic vials, subsequently dipped in liquid nitrogen for at least one minute, and stored at $-70°C$ until used. One millimeter sections of cornea, iris, ciliary body, and sclera were cut with a razor blade, frozen in O.C.T. compound, and cut at $-20°C$ on a cryostat into 4-micron thick sections. One section was stained with H&E in order to allow for orientation with the fluorescent microscope.

A Kallested Quantafluor® mouse kidney substrate slide was run with all of the eye substrate slides. All slides were placed in electron microscope grade formalin, then in chilled acetone 5–10 minutes, then placed in a moist chamber at room temperature. A positive control Kallested Quantafluor ANA positive control (homogeneous pattern) was applied to one eye substrate slide and one section of mouse kidney. A negative control Quantafluor test solution and ANA positive serum from SLE patients, diluted 1:10, 1:20, and 1:80 were applied to mouse kidney and eye substrates. Patient serum samples diluted 1:10 were similarly applied to eye and mouse kidney substrates.

Division of Immunology, Children's Memorial Hospital; The Cornea and External Eye Disease Laboratory, VA Lakeside Medical Center; and the Department of Ophthalmology, Northwestern University Medical School, Chicago, Illinois.

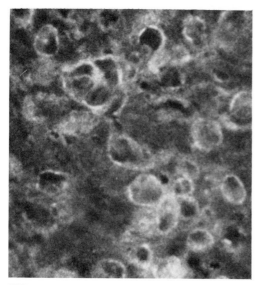

FIG. 1. JRA test serum showing positive nuclear fluorescence against mouse kidney.

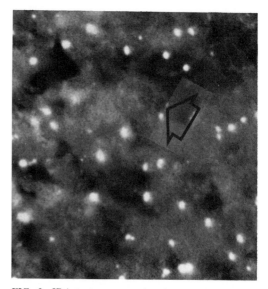

FIG. 2. JRA test serum against human ciliary body. Arrow shows absence of fluorescence of nuclei.

All slides were incubated in a moist chamber at room temperature for 30 minutes. Sections were removed from the moist chamber and rinsed with phosphate buffered saline (PBS). The slides were then washed with gentle stirring in PBS for 10 minutes; for a second time they were washed as above with fresh PBS. The slides were placed in a moist chamber. Fluorescein-conjugated IgG fraction of goat antihuman IgG, IgM, and IgA (Kappel Laboratories) diluted to 1:15 was placed on all sections. Sections were incubated for 30 minutes in a moist chamber and rinsed with PBS. Slides were then washed with gentle stirring in PBS for 10 minutes and washed for a second time with fresh PBS. Phosphate buffered glycerol mounting medium (Zeus Scientific) was applied to each section and the sections were covered with glass slips. All sections were viewed with a fluorescent microscope after the H&E stained slides were viewed with light miscroscopy in order to orient the ocular substrate.

Results

Of the 10 patients with JRA and iritis, three patients had polyarticular disease and seven patients had pauci-articular disease.

Nine of 10 patients had the onset of their JRA at less than 4 years of age. The age of onset of the eye disease was between 3 and 16 years of age, although six of the 10 patients had the onset of their eye disease before the age of 5.

Nuclear fluorescence of ciliary body nuclei as found with Kallested positive control demonstrated our ability to detect positive ANA with this method. In addition, known systemic lupus erythematosus serum when diluted 1:10, 1:20, and 1:80 showed equal fluorescence whether done on mouse kidney or ciliary body substrate (Fig.1). However, none of our patients with JRA and iritis showed nuclear fluorescence of their serum against ciliary body (Fig.2). Nine of 10 of these patients showed nuclear fluorescence against mouse kidney done under the same treatment schedule in parallel.

Discussion

In the present study, ANA-positive sera from patients with SLE were reactive against human ciliary body, whereas ANA-positive sera from patients with JRA and iridocyclitis were not reactive against human ciliary body. Other studies have noted differences in ANA reactivity using various substrates. Kozin et al.[4] noted absence of reactivty of SLE and scleroderma sera when tested against mouse fibroblasts, compared to high reactivity against rat liver and HEP$_2$ cells. They did not study any differences in reactivity between the two diseases.

Blaszczyk et al.[5] compared the reactivity of ANA sera from patients with SLE and scleroderma against monkey esophagus, guinea pig lip, and rat liver. They found that the lupus sera reacted equally against all three substrates. However, the scleroderma patients' sera reacted preferentially with monkey esophagus and guinea pig lip when compared to rat liver. They suggest that there may be organ-specific ANA in some diseases, such as scleroderma, but not in others, such as lupus.

Our findings that SLE sera reacted well against mouse kidney and ciliary body can be explained by assuming that SLE sera are not organ specific as suggested above. The negative reaction of JRA sera against ciliary body may reflect a more organ-specific type of ANA such as seen in scleroderma. However, the inverse correlation between sera from patients with JRA and eye disease and absence of ciliary body substrate reactivity requires further study.

Acknowledgments

Supported in part by grants from the Marlene Apfelbaum Foundation and the Knights Templar Eye Foundation.

References

1. Schaller JG et al.: *Arthritis and Rheum* **17**(4):409–416, 1974.
2. Kanski JJ: *Trans Ophthalmol Soc UK* **96**:123–130, 1976.
3. Lorincz LL et al.: *Int J Dermatol* **20**(6):401–410, 1981.
4. Kozin F et al.: *Am Soc Clin Pathol* **74**(6):785–790, 1980.
5. Blaszczyk M et al.: *J Invest Dermatol* **68**:191–193, 1977.

chapter 71

Leukocyte Migration Inhibition Test (LMIT) in Behçet's Disease with Retinal Soluble Antigens

Junichi Sakai, Fumiharu Seki, Masatada Mitsuhashi, and Masahiko Usui

Organ-specific retinal antigen can induce experimental autoimmune uveoretinitis (EAU) in various animal species, resulting in humoral and cell-mediated immune responses. Although the observed effects depend on the animal species and the injected doses of antigen, EAU may simulate many different inflammatory disorders of the uvea and retina or degenerative conditions of the pigment epithelium and photoreceptor cells. We previously performed leukocyte migration inhibition tests (LMIT) utilizing soluble extracts of human outer segments in patients with uveitis of various etiologies, and we obtained positive results in a group of Behçet's disease cases.[1]

Therefore, in the present study, we have chosen specifically to study patients with Behçet's disease, and LMIT was performed using porcine-purified retinal S antigen. The results were evaluated by classifying them according to the stage and type of disease and the use or nonuse of colchicine.

Materials and Methods

We performed leukocyte migration inhibition tests (LMIT) using an agarose plate technique with HEPES buffer.[2]

Antigen

S antigen was purified from swine retina according to a method described by Takano, et al.[3] S antigen was added at a dose of 50μl (protein 35μg) per 2.5×10^8/ml of leukocytes. This was estimated to be a subtoxic antigen dose on the basis of a dose-response curve. For controls, 50μl PPD (protein 25μg) and 50 μl buffer were used.

Subjects

The subjects consisted of 20 patients with Behçet's disease (complete type: 9 cases; incomplete type: 11 cases) who visited the Department of Ophthalmology of Tokyo Medical College; 18 normal individuals were utilized as controls. Nine cases were able to be tested at the active stage and were also tested at the remission stage. Therefore, LMIT's were performed on a total of 29 samples from patients with Behçet's disease.

Assessment

All experiments were performed at least in triplicate, and the migration index (MI) was calculated according to the following equation.

$$\text{MI } (\%) = \frac{\text{mean migration area with antigen}}{\text{mean migration area without antigen}} \times 100$$

Department of Ophthalmology, Tokyo Medical College. Tokyo, Japan.

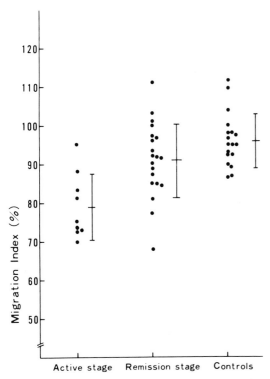

FIG. 1. LMIT with purified S antigen in Behçet's disease.

Cases that showed a decreased MI (by more than 2 SD as compared with the average MI value in the normal individual group) were judged to be positive.

Results

The average MI obtained from 18 normal individuals was 96.3 ± 6.7%. The MI of the patients with Behçet's disease was 69.9–103.3% (average 87.3 ± 10.8%). The differ-

ence between these values was statistically significant ($p<0.005$). Considering cases whose MI values were less than 82.9% (average MI from normal individuals ± 2 SD) to be positive, 10 tests were positive of 29 samples tested (34.5%). It is noteworthy that of nine samples taken at the active stage of the disease, seven samples were judged to be positive (77.8%); the difference from the control value was statistically significant ($p<0.000005$). By contrast, only three samples were positive of a total of 20 samples taken during the remission stage (statistically insignificant). The differences between the values of samples taken at the active stage and remission stage were statistically significant ($p<0.01$) (Fig. 1, Table I). Nine patients tested in the active stage were also tested in the remission stage. The results showed that of seven positive samples at the active stage, five samples were found to be negative in the remission stage. By contrast, negative samples at the active stage never turned positive at the remission stage.

The average MI of the samples from the patients with complete and incomplete type Behçet's disease was 84.9 ± 9.6% and 89.5 ± 11.7%, respectively; these values were not significantly different. In addition, the MI of the samples from the patients with Behçet's disease who were given colchicine at the time of examination was 88.1 ± 8.2%, which was not significantly different from the value of the group not receiving colchicine (86.2 ± 14.0%).

Discussion

Retinal autoimmunity has recently become an important entity in the study of or-

TABLE I
LMIT in Behçet's Disease: Comparison of Active Stage and Remission Stage

	No. of samples	MI (%)	No. positive	t value
Behçet's disease	29	87.3±10.8	10	$p<0.005$
Active stage	9	79.0± 8.4	7	$p<0.000005$
Remission stage	20	91.0± 9.7	3	n.s.
Normal individuals	18	96.3± 6.7		

Positive < average MI of normal individuals − 2SD = 82.9.
n.s.: not significant ($p>0.05$).

TABLE II
Results of LMIT

Diagnostic category	No. of samples	MI (%)	No. positive	t value
S antigen				
Behçet's disease	29	87.3±10.8	10	$p < 0.005$
Pigmentary retinal degeneration	43	86.4±12.8	18	$p < 0.005$
Normal individuals	18	96.3± 6.7		
HSS				
Harada's disease	16	96.9± 7.3	0	n.s.
Retinal detachment	38	89.3±14.4	14	$p < 0.05$
Normal individuals	11			

S antigen: Swine purified S antigen.
HSS: Soluble extract of human outer segment.

gan-specific immunity. With respect to EAU produced by immunization against S antigen, both cell- and antibody-mediated immunity against retinal tissue may probably be interrelated to some extent but their respective roles may be different, depending on the animal species and the injected dose of antigen. However, the role of retinal autoimmunity in human ocular disease is unclear.

In the present study, we performed LMIT as an index of cellular immunity utilizing purified S antigen. Significant inhibition was observed in a group with Behçet's disease, especially at the active stage. Although differences in clinical classification and the use of colchicine did not produce significant changes in the results, there was a statistically significant difference between the samples taken at the active stage and those taken during remissions ($p<0.01$). Colchicine is known to inhibit the chemotactic ability of leukocytes but was not considered to have any special influence upon LMIT. Cellular immunity against S antigen was difficult to observe at the remission stage. These findings are in agreement with Faure's results,[4] showing that there is a good correlation between the therapeutic effect of desensitization with retinal antigen upon EAU and the degree of depression of cellular immunity.

We have previously applied LMIT to Harada's disease, retinal detachment, and uveitis of other causes using soluble extracts of human outer segments as an antigen.[1,5]

We have also performed tests with purified S antigen on patients with pigmentary retinal degeneration (PRD).[6] The results are listed in Table II for a comparison with the present results in Behçet's disease. Many positive cases were found in the group suffering from PRD and retinal detachment, and these were significantly different from the results of the control group. However, it should be noted that there was no positive case in the Harada's disease group. Many positive cases were observed in the group of PRD whose pathological characteristics consisted of degeneration of the outer segments of the photoreceptor cells and hyaline degeneration of the retinal vessels. This result is interesting when it is compared with the results obtained in Behçet's disease in which the principal pathological characteristics consisted of retinal vasculitis and degeneration or disappearance of photoreceptor cells.

There are several reports describing the results of examinations of humoral and cellular immunity against S antigen in Behçet's disease. Faure et al. reported that 65% of patients with Behçet's disease or the Behçetoid syndrome were positive in LMIT.[7] However, Nussenblatt examined the blast transformation of lymphocytes in two cases of Behçet's disease and reported both cases to be negative.[8] Although, in our experience, a significant inhibition rate was noted in Behçet's disease, especially at the active stage of the disease, repeated testing of immune responses during the course of the disease appears to be important.

References

1. Sakai J et al.: *Acta Soc Ophthalmol J* **83**:1179–1190, 1979.
2. Nakagawa J et al.: *Ryumachi Jpn* **14**:270–273, 1974.
3. Takano S et al.: *Folia Ophthalmol Jpn* **32**:491–496, 1981.
4. Faure JP et al.: In *New Directions in Ophthalmic Research,* Sears ML (Ed.). New Haven and London Yale University Press, London, 1981, pp. 31–45.
5. Seki F et al.: *Folia Ophthalmol Jpn* **31**:235–243, 1980.
6. Seki F: *J Clin Ophthalmol* **36**:361–366, 1982.
7. Faure JP and de Kozak Y: In *Immunology of the Eye, Workshop 2, Autoimmune Phenomena and Ocular Disorders,* Helmsen RJ et al. (Eds.). Information Retrieval Inc., New York, 1981, pp. 33–48.
8. Nussenblatt RB et al.: *Am J Ophthalmol* **89**:173–179, 1980.

chapter 72

A Murine Model for Heterotopic Transplantation of Corneas

Proof that Langerhans' Cells Are Essential Ia-Antigen-Presenting Cells

Patrick E. Rubsamen, James McCulley, Paul R. Bergstresser, and J. Wayne Streilein

We have recently described a method by which corneal grafts prepared from eyes of mice can be grafted heterotopically to the thoracic cage of recipients and accurately evaluated for the process of allograft rejection.[1] This approach has the advantage that slit-lamp biomicroscopy can be used to scrutinize the corneal grafts in detail. Several years ago, we demonstrated that corneal grafts from donor mice disparate from recipients at Class I H-2 alloantigens were rejected when placed at heterotopic sites on appropriate recipients.[2] Moreover, Class I *H-2* disparate corneal grafts were able to induce second set alloimmunity in that recipients were able subsequently to reject in accelerated or second-set fashion skin grafts syngeneic with the corneas. By contrast, corneas prepared from donors disparate from potential recipients only at Class II regions (Ia) of *H-2* were not rejected when placed at heterotopic sites and could still be observed as healthy grafts after 45 days. Moreover, these grafts failed to induce anti-Ia immunity in the same recipients; this was revealed when they were challenged subsequently with skin grafts known to bear the same Ia alloantigens. At the time these experiments were conducted, we became aware of the earliest work of Redslob[3] who reported that the cornea is deficient in dendritic cells. By histochemical staining for cell-surface ATPase as well as by the use of fluoresceinated anti-Ia monoclonal antibodies, we demonstrated that the epithelium of nonlimbic portions of the murine cornea were devoid of Langerhans' cells.[4]

Based on these observations as well as on circumstantial evidence collected from studies of Langerhans' cells in body wall skin,[5] the hypothesis was generated that Langerhans' cells are the only important immunogenic source of Ia alloantigens in skin.[6] Since the cornea is a specialized region of the cutaneous surface, the absence of Langerhans' cells in this tissue, as well as its failure to respond immunologically to Ia alloantigens, was felt to be strong circumstantial (albeit negative) evidence in favor of that hypothesis. Subsequently, a corollary to the original hypothesis has been generated: If Langerhans' cells could be forced or enticed to infiltrate the corneal epithelium, corneas should then be able to immunize appropriate recipient mice to Ia alloantigens, which are thought to be expressed predominately and only on these epidermal cells.

To examine this corollary hypothesis, the following experimental approach was

Departments of Cell Biology, Internal Medicine, Ophthalmology, and Dermatology, The University of Texas Health Science Center at Dallas, Dallas, Texas.

adopted. First, the corneal surface was per-turbed with an irritant after which Langer-hans' cells could be observed to have infil-trated the central epithelium. Second, Langerhans' cell-containing corneas were examined for their capacity to immunize Ia-disparate hosts by determining whether the latter would reject, in second-set fashion, body wall skin syngeneic with the cornea. Finally, the susceptibility of Langerhans' cell-containing corneas to allograft rejection directed at Ia alloantigens was examined.

Induction of Langerhans' cells into corneal epithelium

Several irritants were examined for their capacity to induce Langerhans' cell emigra-tion into corneal epithelium. A method de-scribed originally by Rowden was finally adopted.[7] Five microliters of 0.5% dinitro-fluorobenzene (DNFB) was placed on a small gauze pad, cut-to-size so that it fit di-rectly over the surface of a mouse cornea. The DNFB-containing pad was maintained in place for 30 seconds and then removed. Corneas were excised from mice treated in this fashion periodically thereafter and ex-amined histochemically for the presence of ATPase-positive cells in the epithelium. By 7 days after DNFB application, ATPase-positive dendritic cells were easily identi-fiable within the central cornea epithelium. A wide range of surface densities was ob-served: 60–220 per mm^2. This density of Langerhans' cells is approximately ¼ to ¹⁄₁₀ that found in normal body wall skin of mice.[4] Based on these findings, it was decided that corneas treated seven days previously with DNFB would be used for subsequent graft-ing studies.

Ia-immunogenicity of Langerhans' cell-con-taining corneal grafts

The two recombinant, congenic *H-2* inbred mouse strains—A.TL and A.TH—were selected for these studies because these strains differ only at the *I* region of the *H-2* complex. A.TL carries the Ia^k al-loantigens, whereas A.TH carries the Ia^s al-loantigens. Corneas of donor A.TL mice were painted with DNFB, and 7 days later grafts were fashioned therefrom. Care was

TABLE I

Ability of Langerhans' Cell-containing A.TL Corneas to Induce Second-set Anti-Ia Immunity in A.TH Mice

First Graft	N[a]	Median survival time of test A.TL skin grafts[b] in days (95% confidence limits)
Cornea A.TL Normal	10	12.5 (11.2–13.8)
Cornea A.TL DNFB-treated[c]	17	6.6 (6.1–7.2)
Skin A.TL	8	6.0 (5.9–6.1)
None	9	12.2 (11.1–13.0)

[a] Number of graft recipients.
[b] Test graft applied 30 days after initial graft.
[c] Corneas of donor mice treated 7 days earlier with DNFB.

taken to excise widely limbic attachments of the cornea as described previously.[6] Wide excision insures that the corneal graft will not be contaminated with the small numbers of Langerhans' cells that are normally pres-ent at the site of attachment to the limbus. Three individual corneas were placed, en-dothelial side down, on freshly prepared graft beds on the thoracic wall of A.TH re-cipient mice. The grafts were held in place with vaseline-impregnated gauze and plaster of Paris bandages as described previously.[1,6] In control experiments, one panel of A.TH mice received normal corneas from A.TL donors, and a second panel received grafts prepared from body wall of A.TL donors. A final panel (negative control) received no primary grafts. Plaster of Paris bandages were removed at day 10.

Thirty days after initial grafting, mice from these panels received a second graft of normal body wall skin from A.TL donors. Bandages were removed from the second grafts 7 days later, and the grafts were ob-served for evidence of rejection. The data, presented in Table I, indicate that DNFB-treated A.TL corneas immunized recipient A.TH mice to Ia^k alloantigens because these animals rejected subsequent A.TL skin grafts in second-set fashion. The data also confirm that normal corneas from A.TL do-nors fail to immunize their recipients, who then reject test A.TL skin grafts in conven-tional first-set fashion. Thus, corneas that normally are free of Langerhans' cells fail to induce allo-Ia immunity, whereas perturbed corneal surfaces, in which Langerhans' cells have been induced to infiltrate, induce vigorous second-set immunity to the same Ia alloantigens.

Vulnerability of Langerhans' cell-containing corneas to rejection by preexisting anti-Ia immunity

We next examined the possibility that DNFB-perturbed corneas, which are known to contain Langerhans' cells, might even be susceptible to rejection if placed on recipients in which anti-Ia immunity had already been induced. In these experiments, A.TH mice were sensitized to Ia^k alloantigens with orthotopic skin grafts prepared from A.TL donors. Twenty-one days later, DNFB was applied to the corneas of donor A.TL mice. Seven days later, these corneas were excised and grafted heterotopically, as before, to the A.TH mice that were now immune to Ia.^{k} Vaseline-impregnated gauze covered by plaster of Paris bandages were applied. In control experiments, DNFB treated corneas of A.TL donors were grafted to the thoracic wall of syngeneic A.TL recipients.

Inspections of these grafts were made at 7, 10, and 14 days after grafting. Bandages were reapplied after the seventh and tenth day inspections. Methods of observations by gross inspection as well as biomicroscopy have been described previously.[1] Cornea grafts were observed for presence, clarity, edema, as well as for congestion of the graft bed. Based on these observations, which were scored on an arbitrary 0–4+ scale, a rejection index was calculated as described previously.[1] The results are depicted in Figure 1.

A modest inflammatory reaction was observed in and around syngeneic DNFB-treated corneal grafts. However, the intensity of the inflammation was much greater in and about the allografts at the earliest inspection interval, and the disparity between syngeneic and allogeneic grafts in this regard continued to increase. By day 14, the rejection index of the allografts reached 10.5; by contrast, that of control grafts was 2.5. In fact, of 18 individual corneal allografts placed on A.TH recipients, only five were detectable on day 14 as opaque remnants. By contrast, eight of 11 syngeneic DNFB-treated corneal grafts were present on day 14; these grafts were clear and displayed little evidence of edema and/or congestion of the graft bed.

Thus, Ia alloantigenically disparate corneas are susceptible to rejection if they have been altered in such a manner that Langerhans' cells have been induced to infiltrate the epithelial surface. This is a rather surprising finding, since the Ia-bearing Langerhans' cells in these DNFB-treated corneas comprise a very minor component of the total mass of tissue. Nonetheless, these corneas were rejected.

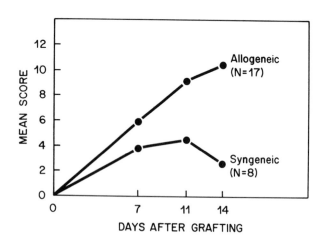

FIG. 1. Rejection index of Langerhans' cell-containing A.TL corneas grafted heterotopically to A.TH mice pre-immune to Ia^k.

Discussion and Conclusion

The circumstantial evidence that exists to implicate Langerhans' cells in the expression of Ia alloantigens in skin and cornea is considerable. As mentioned before, we have previously reported that there are no Langerhans' cells in the epithelium of the central portion of the cornea[4] and that corneas are unable to induce anti-Ia immunity.[2] Other investigators have supplied confirmatory data,[8,9] although the matter still remains controversial. Recently, we have demonstrated independently that Langerhans' cells provide the major immunogenic source of Ia antigens in normal body wall skin. The strategy employed to demonstrate this aspect of Langerhans' cell function was to strip the stratum corneum from skin by repeated applications of cellophane tape. Such treatment, when carried to glistening on the cutaneous surface, produces an epidermis which, during the next 3–4 days, becomes depleted of resident Langerhans' cells. Skin grafts fashioned from tape-stripped areas of mouse skin were found to be inefficient at inducing anti-Ia immunity.[11] Thus, these two independent lines of evidence suggest strongly, in a negative sense, that Langerhans' cells are crucially important for the expression of Ia alloantigens in both skin and cornea. Our current studies tested in a positive sense the corollary to our original hypothesis by demonstrating that a maneuver that induced the emigration of Langerhans' cells into corneal epithelium conferred on that tissue both Ia immunogenicity and vulnerability to anti-Ia rejection immunity.

This formal proof of our hypothesis is important on two accounts. In the first, HLA-D/DR disparity is coming to be realized as the dominant force in the success or failure of renal allografts in man.[12] D/DR-matched kidneys fare better than do D/DR mismatched kidneys, irrespective of whether there is matching for HLA regions A, B, and C. When one considers corneal allografts in man, the incidence of acceptance is rather high even without attempts to match tissue types. However, as tissue typing has been examined in cornea grafting in man over the last few years, evidence suggests that HLA typing matters in predicting the survival of cornea grafts.[13] Our evidence would suggest that HLA-D/DR disparity would be a trivial concern when normal (that is non-Langerhans' cell containing) cornea grafts are utilized. However, if perturbed cornea grafts are employed (grafts containing infiltrating Langerhans' cells) then HLA-D/DR incompatibility may make an important difference in the clinical outcome of the graft.

On the second account, our concern about whether Langerhans' cells alone account for Ia immunogenicity of cutaneous surfaces relates to the correlation between cells that express Ia and the capacity of these cells to present nominal antigen to the immune system. Langerhans' cells, which normally are present in the epidermis of body wall skin, are thought to be the primary antigen-presenting cells of that tissue and perhaps to play a crucial role in rendering immunogenic viral antigens as well as tumor-specific antigens that arise within the epidermal environment. Since the cornea is normally deficient in Langerhans' cells, this structure is relatively incapable of presenting nominal antigens effectively to the immune system. As a consequence, the cornea may be especially vulnerable to certain pathogens who take advantage of its lack of Langerhans' cells to gain entry into the anterior segment of the eye and there to initiate disease.

Acknowledgment

Supported in part by U.S.P.H.S. Grants EY-03119 and AI-17363.

References

1. Streilein JW et al.: *Invest Ophthalmol Vis Sci* **23**: (in press).
2. Streilein JW et al.: *Nature* **282**:326–327, 1979.
3. Redslob E: *Ann Oculist* (Paris) **159**:523–537, 1922.
4. Bergstresser PR et al.: *J Invest Dermatol* **74**:77–80, 1980.
5. Toews GB et al.: *Immunol* **124**:445–453, 1980.
6. Streilein JW et al.: *J Invest Dermatol* **75**:17–21, 1980.
7. Rowden G: *J Invest Dermatol* **75**:22–31, 1980.
8. Lang R et al.: *Invest Ophthalmol Vis Sci* **20** (Suppl):1, 1981.
9. Tagawa Y et al.: *Invest Ophthalmol Vis Sci* **20** (Suppl):2, 1981.
10. Chandler JW et al.: *Invest Ophthalmol Vis Sci* **20** (Suppl):1, 1981.
11. Streilein JW et al.: *J Exp Med* **155**:863–871, 1981.
12. Ayoub G and Terasaki P: *Transplantation* **33**:515–517, 1982.
13. Batchelor JR et al.: *Lancet* **1**:551–554, 1976.

chapter 73

Pathogenesis of Contact Lens Papillary Conjunctivitis

A Hypothesis

Jack V. Greiner,[a,b] Donald R. Korb,[c] and Mathea R. Allansmith[d]

Contact lens papillary conjunctivitis[1] (Fig. 1) purportedly is manifested as the result of a sequence of tissue anatomic alterations that precede the actual onset of the disorder.[2] The initial clinical presentation is an increase in conjunctival mucus-secreting with or without conjunctival hyperemia.[1] Light and electron microscopic examination reveals a corresponding marked increase in the number of mucus-secreting vesicles in nongoblet epithelial cells.[3-6] As the syndrome progresses, the upper tarsal conjunctiva develops lymphocyte-infiltrated papillae.[1] The size of the papillae correlates directly with the magnitude of the lymphocyte infiltration.[1] In conjunction with the increased conjunctival surface area, caused by the ensuing enlargement of the papillae, mucus-secreting goblet and nongoblet cells proliferate proportionately,[5-6] thereby increasing the capacity of the conjunctiva to secrete mucus. The inflammatory cell profile of the affected tissue may become altered further, characterized principally by a marked influx of lymphocytes, basophils, macrophages, eosinophils, and polymorphonuclear leukocytes (PMN) into the stroma and the appearance of basophils and PMN in the epithelium.[1,7,8]

At this point consideration should be given to how the contact lens may initiate or contribute to the development of the immunologic responses described above. Contact lens-induced mechanical trauma to the conjunctival epithelium and altered conjunctival tolerance to accumulated lens deposits are possible mechanisms that may participate in the development of this syndrome. Epithelial cells that cover the upper tarsal conjunctival surface have been shown to be altered in contact lens wearers,[2, 11-14] a manifestation that could be attributed to abrasion by the lens during blinking. Further support for the lens-induced mechanical trauma mechanism is the demonstration that the zonal distribution of papillae[15-17] is related to the size of the contact lens and, therefore, to the area influenced by the lens edges (Fig. 2). Thus, the smaller hard contact lens most profoundly affects zones 2 and 3 (Fig. 3), while the relatively larger soft contact lens affects predominately zones 1 and 2 (Fig. 3).

The accumulation of lens deposits, which are common to both hard[1] and soft contact lenses,[18] may also be an important etiologic factor in papillary conjunctivitis. These deposits appear to consist of mucus,[19] environmental debris, and, in some instances, bacteria.[20] The relationship between conjunctival intolerance to accumulated lens

---

[a]Department of Ophthalmology, University of Illinois Eye and Ear Infirmary, Chicago; [b]Department of Pathology, Chicago College of Osteopathic Medicine, Chicago, Illinois; [c]Private practice, Boston; [d]Harvard Medical School, Boston, Massachusetts.

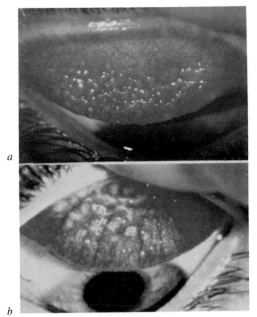

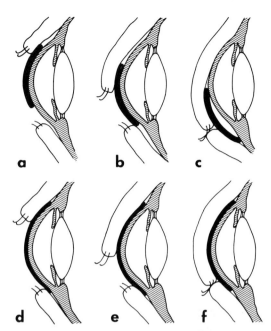

FIG. 1. Clinical photographs of the everted upper eyelid demonstrating the distribution of papillae of the tarsal conjunctiva in a soft contact lens wearer (*a*) and a hard contact lens wearer (*b*) with papillary conjunctivitis.

FIG. 3. Demonstration of movement of hard contact lens (*a–c*) and soft contact lens (*d–f*) during lid closure (blinking).

deposits and papillary conjunctivitis is suggested by the observation that substitution of new lenses or thorough cleansing of presently worn lenses results in a decrease in papillary conjunctivitis symptoms, and, occasionally, a complete disappearance of all signs and symptoms.[1,21] Furthermore, when lenses with unpolished peripheral curves are polished and lens edges are modified, papillary conjunctivitis has been observed to resolve.[22]

Thus, experimental evidence exists demonstrating that both chronic conjunctival mechanical trauma and intolerance to accumulated lens deposits are important factors contributing to the development of papillary conjunctivitis; however, the underlying mechanism responsible for the sudden onset of this syndrome is unresolved. We propose, therefore, that an increased immunosensitivity to mechanical trauma is responsible for initiating the sequence of inflammatory responses that characterize this syndrome. This hypothesis suggests that mechanical trauma alters the superficial layers of the normal conjunctival epithelium so that degranulation of mast cells and disruption of the epithelial surface, which normally acts as a physical barrier to the tearfilm and external environment, results. The mucus-coated contact lens now serves as a vehicle for the presentation of environmental debris, i.e., antigen(s), to the compromised epithelium. Such exposure may result in the release of inflammatory

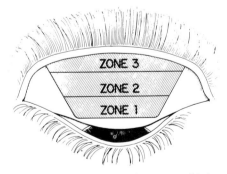

FIG. 2. Diagram of an everted upper eyelid demonstrating arbitrary division of the tarsal conjunctival surface into three zones.

mediators by the immune surveillance cells in the lymphoid conjunctival stroma and thus act to initiate the changes observed in the inflammatory cell populations. Furthermore, it is likely that newly acquired lymphoid cells are deposited in regions of the stroma where there are pre-existing aggregations of lymphoid cells, resulting in the formation of follicle-like structures. Increased aggregation of lymphocytes in these foci would result in a concomitant expansion of the stroma and overlying epithelium into the conjunctival sac, as observed clinically. Further growth of these follicles would result in formation of the surface papillae that enlarge as the lymphocytes increase in number.

In summary, the available data indicate that both immunologic and mechanical components are interrelated in the pathogenesis of papillary conjunctivitis. The contact lens purportedly serves as an abrasive surface (foreign body) for the development of a compromised epithelium that enhances the immunosensitivity of the conjunctiva to external antigen(s).

Acknowledgments

Supported by NIH Grants EY-02099 and EY-02882.

References

1. Allansmith MR et al.: *Am J Ophthalmol* **83**:697–708, 1977.
2. Greiner JV et al.: *Am J Ophthalmol* **86**:403–413, 1978.
3. Greiner JV et al.: *Invest Ophthalmol Vis Sci* **18** (ARVO Suppl):123, 1979.
4. Greiner JV et al.: *Arch Ophthalmol* **98**:1843–1846, 1980.
5. Greiner JV and Allansmith MR: *Ophthalmology* **88**:821–832, 1981.
6. Greiner JV et al.: Histochemical analysis of mucus secretory vesicles in conjunctival epithelial cells. *Acta Ophthalmol* (in press).
7. Allansmith MR et al.: *Am J Ophthalmol* **86**:250–259, 1978.
8. Allansmith MR et al.: *Ophthalmology* **85**:766–778, 1978.
9. Allansmith MR et al.: *Am J Ophthalmol* **87**:171–174, 1979.
10. Allansmith MR et al.: *Arch Ophthalmol* **99**:884–885, 1981.
11. Greiner JV et al.: *Am J Ophthalmol* **83**:892–905, 1977.
12. Greiner JV et al.: *Am J Ophthalmol* **85**:242–252, 1978.
13. Greiner JV et al.: *Ann Ophthalmol* **14**:288–290, 1982.
14. Greiner JV et al.: *Arch Ophthalmol* **98**:1253–1255, 1980.
15. Korb DR et al.: *Ophthalmology* **88**:1132–1136, 1981.
16. Korb DR et al.: *Br J Ophthalmol* **67**:733–736, 1983.
17. Korb DR et al.: *Am J Ophthalmol* **90**:336–341, 1980.
18. Fowler SA et al.: *Am J Ophthalmol* **88**:1056–1061, 1979.
19. Fowler SA et al.: *Invest Ophthalmol Vis Sci* **18** (ARVO Suppl):144, 1979.
20. Fowler SA et al.: *Arch Ophthalmol* **97**:659–660, 1979.
21. Korb DR et al.: *Arch Opththalmol* **101**:48–50, 1983.
22. Korb DR: Unpublished observations.

Index